endorsed by
edexcel

EDEXCEL GCSE SCIENCE

CHRIS CONOLEY, MARY JONES
and DAVID SANG

Hodder Murray
THE HODDER HEADLINE GROUP

This high quality material is endorsed by Edexcel and has been through a rigorous quality assurance programme to ensure that it is a suitable companion to the specification for both learners and teachers. This does not mean that its contents will be used verbatim when setting examinations nor is it to be read as being the official specification – a copy of which is available at www.edexcel.org.uk

Orders: please contact Bookpoint Ltd, 130 Milton Park, Abingdon, Oxon OX14 4SB. Telephone: (44) 01235 827720. Fax: (44) 01235 400454. Lines are open 9.00 – 5.00, Monday to Saturday, with a 24-hour message answering service. Visit our website at www.hoddereducation.co.uk

Cover photo shows a Hubble Space Telescope image of the planet Saturn, showing the varying cloud levels in its atmosphere. Two moons are also visible: Dione (yellow, lower left), and Tethys (green, upper right).

First published in 2006 by
Hodder Murray, an imprint of Hodder Education,
a member of the Hodder Headline Group
an Hachette Livre UK Company
338 Euston Road
London NW1 3BH

Impression number 5 4 3 2
Year 2010 2009 2008 2007

Cover photo NASA/ESA/STScI/Science Photo Library
Illustrations by Oxford Designers and Illustrators Ltd
Typeset in 11.5/14 pt Goudy by Stephen Rowling/Springworks
Printed and bound in Italy

A catalogue record for this title is available from the British Library

ISBN: 978 0340 907 290

Contents

Biology

1 Environment

A Ecosystems and energy flow

In 1977, the remotely controlled, deep-sea submersible Alvin passed over a vast rift in the floor of the Pacific Ocean, about 3500 m beneath the surface. Scientists already knew that huge plumes of hot water seemed to be coming up from beneath the ocean floor through vents. But they were amazed to find that these vents – completely dark, and full of highly poisonous chemicals – were surrounded by thriving communities of living organisms. They saw giant tube worms, clams, mussels and crabs. What could they be feeding on? With no light, there could be no photosynthesis, so no ordinary food chain could exist.

Figure 1.1 These clams are part of the extremely unusual food chain around a black smoker

Since this first discovery, many more so-called 'black smokers' have been found. More than 300 species of organisms have been discovered living in or around them. Their energy source is the sulphur-containing chemicals brought up from beneath the Earth's surface in the hot water. Bacteria use the energy in these chemicals to make their food. Some of these bacteria are eaten by animals such as tube worms, clams, mussels and shrimps. These in turn are eaten by predators such as crabs and fish.

Food chains

A black smoker is a very unusual **ecosystem**. An ecosystem can be defined as a group of organisms and their habitat, which interact with each other.

All organisms – living things – need energy. Table 1.1 lists some of the processes in plants and animals that use energy. We get our energy from the carbohydrates, fats and proteins in the food that we eat.

Table 1.1 Why organisms need energy

Function	Description
active transport	Every living cell uses energy to move substances into itself and out of itself, against their concentration gradients. For example, plants take up nitrate ions into their root hair cells. Our nerve cells constantly pump out sodium ions and pump in potassium ions
movement	Our muscles, including those in the heart, use energy when they contract. Movement also takes place inside cells; for example, chloroplasts inside palisade cells move around within the cytoplasm
making new substances	Energy is needed to build large molecules from small ones. For example, amino acids are linked together inside cells to make proteins
keeping body temperature constant	Mammals and birds are homeothermic – that is, they keep their body temperature constant. To do this, they generate heat inside the body by respiration of carbohydrates

You probably remember that a **food chain** is a way of showing how energy is passed from one organism to another. In most ecosystems, energy first enters a food chain as **sunlight**. Plants use this light energy to photosynthesise. Carbon dioxide and water react and produce carbohydrates. The energy is locked up inside the carbohydrates. When we eat plants, or eat animals that have eaten plants, we take this energy into our bodies.

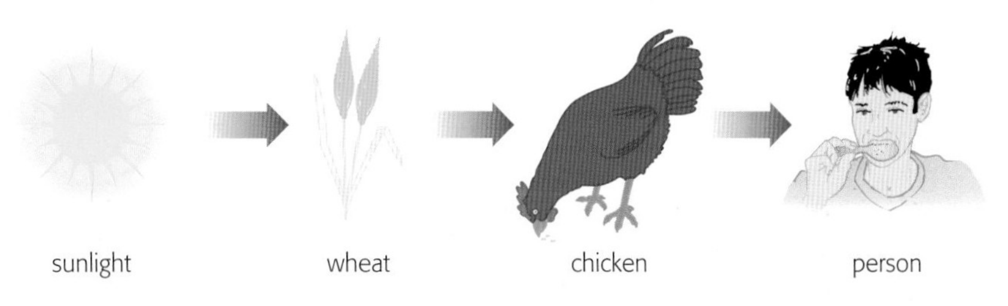

Figure 1.2 A food chain

All the energy you have in your body originally came from the Sun.

In the food chain in Figure 1.2, the wheat is the **producer**. It produces food by photosynthesis. The chicken and the person are **consumers**, because they consume food made by the producer. The chicken is a **primary consumer** and the person is a **secondary consumer**. Primary consumers are **herbivores** and secondary consumers are **carnivores**. The arrows in the food chain show how energy flows from one organism to the next.

Pyramids of biomass

As energy flows along a food chain, a lot of it is lost along the way. For example, the chicken uses energy to move around. In its muscles, glucose is broken down in **respiration** to release energy for movement. This energy is no longer stored in the chicken's body, so it cannot be passed on to the person who eventually eats it.

Because energy is used up like this by each organism, there is less and less energy available to pass on at each step in the chain. The steps are called **trophic levels**. The energy losses mean that less living tissue, called **biomass**, can be supplied with energy as you go further along the chain. We can show this by drawing a **pyramid of biomass**, which shows a food chain **quantitatively**. This means that it not only shows which organisms are in the food chain, but also *how much* of each kind of each organism there is in the chain.

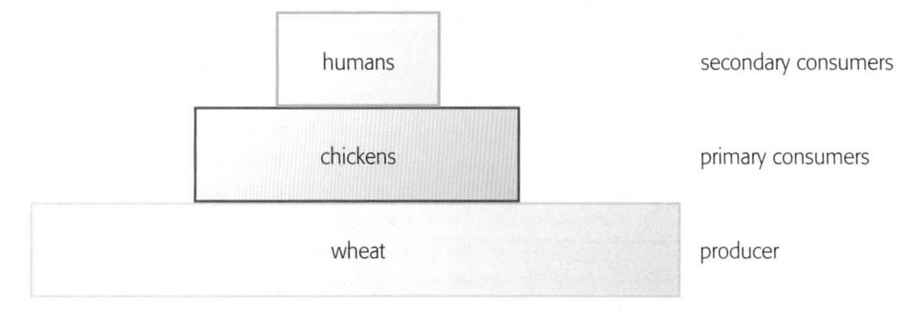

Figure 1.3 A pyramid of biomass for the food chain in Figure 1.2

In a pyramid of biomass, each box represents a trophic level. The area of each box represents the amount of biomass of all the organisms at that level. The pyramid in Figure 1.3 shows that you need a lot of wheat biomass to support less chicken biomass, and that even less human biomass can be supported by this quantity of wheat.

A pyramid of biomass is like a sideways-on bar chart.

Growing food

We can use knowledge of energy transfer along food chains to think about how we produce food.

Figure 1.4 Is it better to use a field to grow wheat or to rear cattle?

A farmer could grow a crop of wheat in a field. The wheat plants trap some of the energy in the sunlight and use it in photosynthesis. They make starch and protein, and store it in the grain (seeds). The grain is harvested, ground into flour and made into bread.

Another possibility would be to keep cattle in the field. Grass grows, photosynthesises and traps some of the sunlight energy in its leaves and roots. The cattle eat the grass and get some of the energy from it. But they don't get very much, because:

- they don't eat all the grass – some of it is inedible because it has been trampled or cow pats have landed on it, and of course the roots are underground and don't get eaten
- the grass uses some of the energy for itself, so only a part of the energy is available for the cattle
- the cattle's digestive systems can't digest the grass completely, so some of it passes straight through and is lost in their faeces.

The cattle could be killed and eaten as beef. Lots more energy is lost here, because we don't eat *all* of the cattle – not their skins, hooves, intestines and so on.

Figure 1.5 The more steps in a food chain, the more energy is lost

So, if we grow wheat in the field and eat the grain, we get much more energy for ourselves than if we keep cattle in it and eat beef.

Summary

- An ecosystem is made up of all the living organisms in a particular habitat, plus their non-living environment. The organisms and their environment interact with each other.
- Energy is passed between organisms in the form of chemical energy in food. Food chains or food webs use arrows to show the direction of energy flow from one organism to the next.
- In most ecosystems, energy enters the ecosystem as sunlight. Green plants are the producers; they transfer this light energy to stored chemical energy in carbohydrates and other organic substances. Animals are consumers; they get their energy by eating plants or other organisms.
- Energy is lost as it passes along a food chain.
- Each step in a food chain is known as a trophic level.
- A pyramid of biomass shows the quantity of living material (biomass) present at each trophic level. This usually decreases as you go up the levels, because of the energy losses.
- It is more cost effective, in terms of energy, to produce a field of wheat compared with a field of beef cattle.

Questions

1.1 Construct a food chain for a 'black smoker' (see page 2). Which organisms are the producers?

1.2 Sketch pyramids of biomass for these food chains:

wheat ➔ humans

grass ➔ cattle ➔ humans

Label each trophic level with its correct biological term.

1.3 Present an argument for humans becoming vegetarians. Can you think of any arguments *against* vegetarianism?

1.4 Suggest explanations for each of these facts.

- **a** Food chains rarely have more than five links in them.
- **b** Mammals and birds need to eat more food than other animals of the same body mass.
- **c** Even though a lot of energy is lost when cattle eat grass, we can get much more energy from eating beef than from eating grass.

B Populations

The red squirrel used to be a common sight in English woodlands. It feeds mostly on tree seeds such as hazelnuts and acorns, but will also eat many other foods, such as fungi and berries.

Now you would be very lucky if you saw a red squirrel. In the late 19th century another species of squirrel – the grey squirrel – was introduced into Britain. Within a short period of time, the grey squirrel population was growing fast, and the red squirrel population was decreasing. Today, it is estimated that there are 30 000 red squirrels in England and 2 000 000 grey squirrels.

Why has the grey squirrel population increased like this, while the red squirrel has declined?

Figure 1.6 Competing species – the red squirrel and the grey squirrel

Both red and grey squirrels are well adapted to live in woodland. They are fast climbers, with good eyesight to spot predators such as hawks and pine martens. They have an excellent sense of smell to help them to find food. Their thick fur helps them to keep warm in winter, when food is relatively scarce.

Red squirrels are smaller than grey squirrels. A grey has a mass of about 550 g, compared with the red at about 300 g. There doesn't seem to be any aggression between them, so earlier ideas that greys might attack reds are now discounted. But greys seem to be able to exploit a wider range of food sources than reds. They also put on more body fat before the winter, which probably gives them an advantage over the reds, as they are more likely to survive the cold weather. It is now thought that the main reason for the decline of the red squirrel is that it fares less well than the grey as they compete for limited food resources. The grey squirrel is simply better adapted for living in English woodland.

Adaptation

Every living organism is adapted for living in a particular habitat. An adaptation is a feature of the organism that helps it to survive in the place where it lives. Figure 1.7 shows an example.

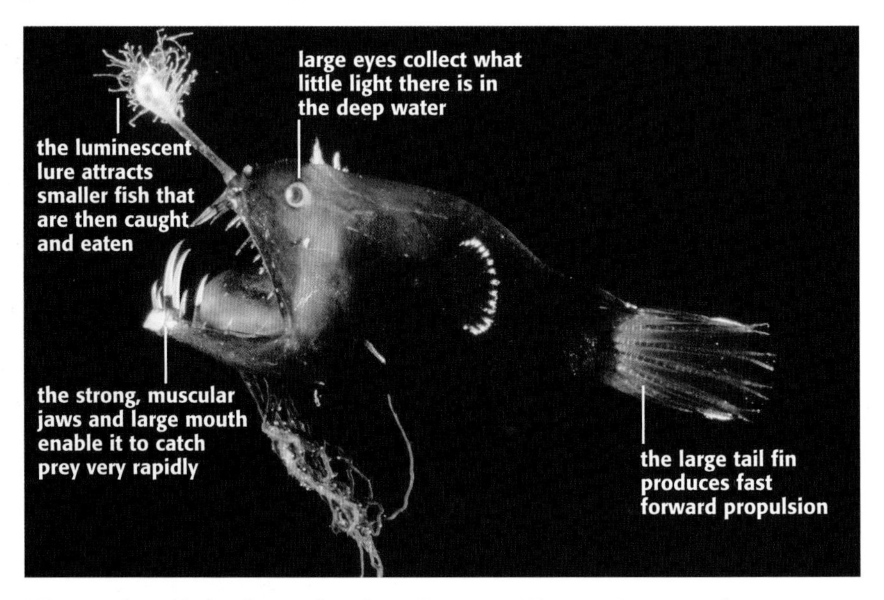

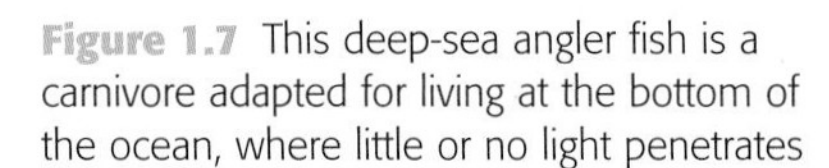

Figure 1.7 This deep-sea angler fish is a carnivore adapted for living at the bottom of the ocean, where little or no light penetrates

These adaptations have developed over millions of years. The characteristics that help the organism to be successful have been inherited from its parents, who themselves inherited those characteristics from their parents. They are handed down in the genes.

Competition

When organisms compete, the ones with the best adaptations are most likely to win.

Many of the things that a living organism needs to stay alive may be in short supply. When two or more organisms need the same resource, and there is not enough of it to go round, they **compete** for it.

For example, bumble bees nest in holes in the ground. If there is not enough undisturbed ground, or only a few holes, then not all the bumble bees will be able to nest. The bees that find the holes first and begin to nest in them will have a better chance of surviving and reproducing than the others. Even though they may not actually fight over them, the bees are competing for the nesting sites.

Figure 1.8 Putting a bumble bee nest box in your garden will reduce competition, so more bees can nest and breed

Competition between the bumble bees stops the number of bees in the area from increasing. The limited number of nesting places limits the size of the bumble bee population. This is a very common situation. For most species, competition between individuals limits the number that can live and breed in a habitat. Table 1.2 shows a few more examples.

Table 1.2 Competition between members of the same species

Organism	Some examples of resources that individuals compete for
frog tadpoles	food, which may be in limited supply in the pond
blackbirds	breeding territories – each pair needs an area of a certain size in which they can collect food for themselves and their young
grass plants	light for photosynthesis

Competition between the bumble bees is an example of **intra-species competition** – that is, competition between members of the same species. Competition can also take place between individuals of *different* species. We have already seen how competition between red squirrels and grey squirrels – for limited supplies of food and other resources – has reduced the size of the red squirrel population.

Predator–prey relationships

Another factor that can affect the size of a population is the interaction between a **predator** and its **prey**. A classic example of this is the relationship between the populations of snowshoe hares and lynxes.

Figure 1.9 A lynx chases a snowshoe hare. What adaptations do they each have to help them to survive in their habitat?

Snowshoe hares and lynxes live in northern Canada. The lynxes feed on snowshoe hares. Snowshoe hares feed on grass and other vegetation.

Figure 1.10 shows how the population sizes of these two animals fluctuated between 1845 and 1935. You can see that, in the late 1840s, the hare population rose. This meant that there was more food for lynxes, so in the following years their population rose as well. As the number of lynxes increased, more hares were killed and eaten, so their population fell. So there was less food for lynxes and therefore their population fell too. With fewer lynxes to eat them, the hare population began to rise, and so on. You can pick out this pattern all through the graph.

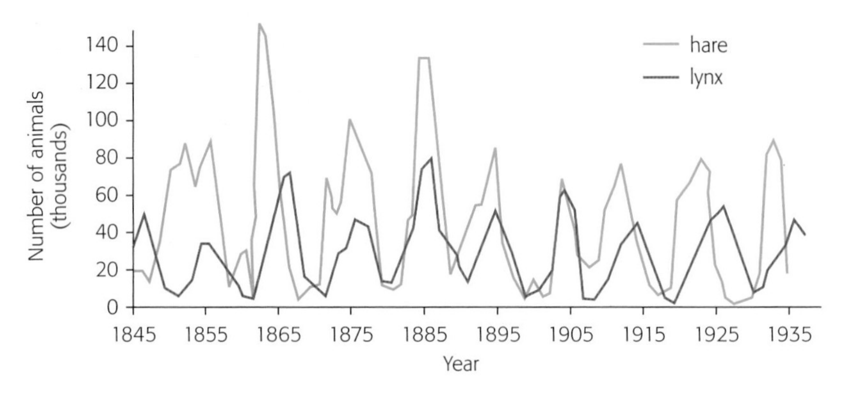

Figure 1.10 The interdependent populations of snowshoe hares and lynxes

The data about the lynx and hare numbers came from records kept by a company that bought skins from fur trappers.

The lynx and the hare populations are **interdependent**. This means that the size of each population is affected by the size of the other.

Predator–prey relationships don't always work like this. This pattern tends only to occur if the predator relies almost entirely on one prey species, and if the prey is killed by only one predator species. Most predators actually eat many different species, so if one becomes less common it can eat more of another.

Computer models

Finding out what is affecting the size of a population isn't easy. Sometimes we can use computers to help us to work this out.

The red grouse lives on heather moorlands. It eats the tips of young heather plants.

Figure 1.11 A red grouse

Landowners can make money by letting people shoot the grouse. But the grouse population fluctuates (goes up and down) wildly – some years there are a lot and other years very few. The owners of the moorland wanted to know why this was happening, so that they could predict what the population might be next year, and maybe do something to stop the population falling.

Various possible reasons for the fluctuations were suggested. These included the amount of food the grouse had, the number of predators that were eating them, the number of grouse that were shot, and the numbers of parasites that were growing in them.

Biologists used the information they already had about the grouse population to try to work out which of these factors might be causing the fluctuations. They wrote a computer program, putting in data about food supply, predators, shooting and parasites. Then they used the program to try to match up each of these factors with the changes in the grouse population. They found that the factor giving the closest correlation with the grouse numbers was the number of parasites. When the grouse were treated with drugs to kill the parasites, the population stayed steadier (Figure 1.12).

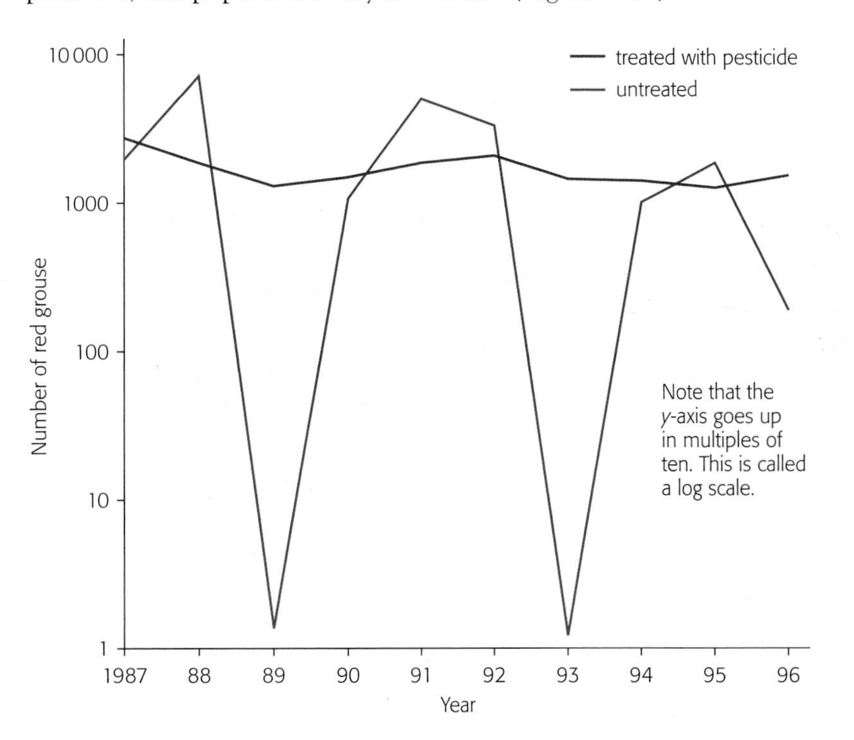

Figure 1.12 Red grouse population change, with and without pesticide treatment

Computer models are very helpful in predicting what might happen to the size of a population in the future. For example, we can use them to predict what will happen to the cod population in the North Sea if we continue fishing as we are now, or if we reduce the amount of fishing. This can help us to make sure that we don't completely wipe out the population of cod.

But computer models are never perfect. There will always be factors affecting the population that we haven't thought of. The way to *really* know what is happening to the size of a population is to go out and count the animals. The best thing to do is to use computer models and real data together, because each of them gives us information that we can't get any other way.

A computer model is only as good as the information that we put into it.

Summary

- When two organisms in an ecosystem need the same resource which is in short supply, they are said to compete for it. Competition occurs between members of the same species, and also between different species.
- Competition between members of the same species – intra-species competition – can affect the size of the population.
- When a predator depends on a single species of prey for most of its food, the size of the predator population is affected by the size of the prey population. The size of the prey population may also be affected by the size of the predator population.
- Computer models can help us to work out what is affecting changes in the size of a population, and to predict what might happen to the population in the future.

Questions

1.5 Write definitions of each of the terms in this list:

competition   predator   prey

Use these words in your definitions:

resource   kills   in short supply   eats   is eaten by

1.6 Ladybirds eat greenfly. Sketch a graph, similar to Figure 1.10, showing how you predict the populations of ladybirds and greenfly might change during one summer. Explain the shape of your graph.

1.7 Explain why the population of a predator is almost always smaller than the population of its prey.

1.8 Give two uses of computer models in the study of populations.

1.9 Explain why computer models can't completely take the place of real data.

C Classifying living things

Marc van Roosmalen guessed there might be something interesting in the bag, as soon as he saw the old fisherman carrying it towards him. Everyone who lived near the Brazilian research centre where Marc worked knew that he was always on the lookout for interesting monkeys, and two or three turned up most weeks. Most were species he knew well. But this old man was more canny than the rest. He knew that Marc was only interested in something genuinely unusual and that a good reward might appear if he brought Marc a kind of monkey he had never seen before.

Marc peered into the wriggling bag, and recognised straight away that the monkey looked similar to other Brazilian monkeys belonging to the genus *Callicebus*. But this little monkey was clearly not quite like any of them. She had a silvery body with a black, furry forehead and bright red, fluffy sideburns.

Marc spent several days comparing the monkey carefully with already known species of *Callicebus*. Eventually, he decided that she belonged to a new species. He named the species *Callicebus stephennashi*, after an artist called Stephen Nash who worked for the organisation Conservation International.

Figure 1.13 A species of monkey discovered in 2002, *Callicebus stephennashi*

Naming species

Since 1980, 40 new species of monkeys have been discovered. And the number of new species of small animals – such as beetles – is many times greater. It has been estimated that one tree in a tropical rainforest may contain 1200 species of beetles. One hectare of rainforest probably contains up to 12 450 beetle species. If you want your name to be given to a new species, then being a beetle hunter might be the thing to do.

Figure 1.14 This biologist is in hot pursuit of flying insects in a rainforest

So far, scientists have discovered and named more than 1.5 million species of animals. We really have no idea how many species of animals there are on Earth, let alone plants, fungi and bacteria. Estimates of the total number of animal species range from 2 million to 100 million.

To help us to study and understand this vast number of species, scientists **classify** them into different groups. We know that species **evolve** from other species, and we classify them by putting species that we think are closely related to each other into the same group.

Each species is given a two-word name, called a **binomial**. The first word is the **genus** to which that species belongs, while the second word is the name for that particular **species**. The new monkey was put into the same genus – *Callicebus* – as other similar species of monkeys. This was done because she shared several features with them, but was clearly not exactly the same. Her species and all the other *Callicebus* species probably evolved from a common ancestor a long time ago.

The person who invented the idea of giving every species a two-word name was Carl Linnaeus, who lived in the 18th century.

Species that are closely related are put into the same genus (plural: genera). Similar genera are put in the same family, similar families into the same order, and so on up through phyla (singular phylum) to kingdom. Figure 1.15 shows how our species, *Homo sapiens*, is classified.

Kingdom:	Animals
Phylum:	Chordates
Order:	Mammals
Family:	Primates
Genus:	*Homo*
Species:	*sapiens*

Figure 1.15 Classifying *Homo sapiens*

Notice that the names of the genus and species are written in *italic*. The genus has a capital letter and the species has a small letter. This is an international convention that helps people all over the world to know that they are reading the scientific name of a species. The same scientific name is used in every country, no matter what language is spoken.

Problems with classification

Biologists try to classify organisms according to their relationships with one another. But this is difficult to do, because we don't usually have a lot of good evidence for these relationships. We have to do the best we can by comparing their structures or their DNA, or by using fossils.

Even deciding whether two organisms belong to the same species is difficult. Biologists often define a species as:

> a group of organisms that share similar body structure and function, and that can interbreed with each other to produce fertile offspring, but cannot breed successfully with members of another species

This definition of a species isn't much use if you are trying to classify fossils or other museum specimens.

That makes it sound as though it should be easy to decide whether a new animal belongs to a new species. You just have to compare it with other species and see if it looks different, and then see if it can breed successfully with animals from other species. But it is actually far from easy, and biologists often disagree over whether or not two groups of animals belong to the same species or to different ones. We can't always test whether they can interbreed with one another, and it is difficult to decide how different two animals need to be before we say they belong to different species.

Summary

- **Organisms are classified according to how closely they are related to one another. This is difficult to do because we can never be sure of these relationships.**
- **Each known species has been given a two-word name called a binomial. The first part of the binomial is the name of the genus, and the second part is the name of the species.**
- **Similar species are put into the same genus. Similar genera are put into the same family. This continues up through order, phylum and kingdom.**

Questions

1.10 Give at least one feature of yourself that makes you:

- **a** an animal (and not a plant)
- **b** a chordate (and not an invertebrate)
- **c** a mammal (and not any other kind of vertebrate)
- **d** an example of the primate *Homo sapiens* (and not a monkey).

1.11 Give two features of yourself that suggest you are more closely related to monkeys than to other kinds of mammals such as cats.

1.12 There are many breeds of domestic dog, many looking very different from one another. They are all classified as the same species. Suggest why.

1.13 Find out the binomials of four species of cats or four species of dogs (including foxes and wolves). Do any of them belong to the same genus?

1.14 A biologist finds two stuffed monkeys in a cupboard in a museum. She thinks they might belong to different species. Explain why it would be very difficult for her to prove that she is right.

D Evolution

In 2004, news of an amazing discovery was reported in the media. A skull and other bones found on an Indonesian island suggested that, not very long ago in archaeological time, our species had shared the Earth with another human species.

The bones were found by a team of Australian archaeologists, at a depth of nearly 6 m. At first, they thought they were the remains of a child, but it turned out that the bones had belonged to a tiny adult, about one metre tall.

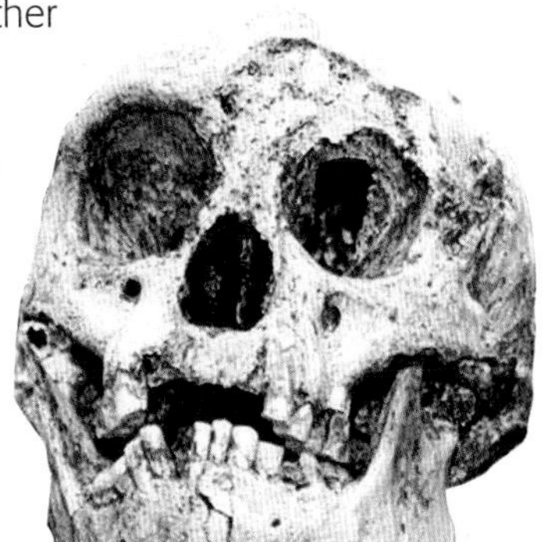

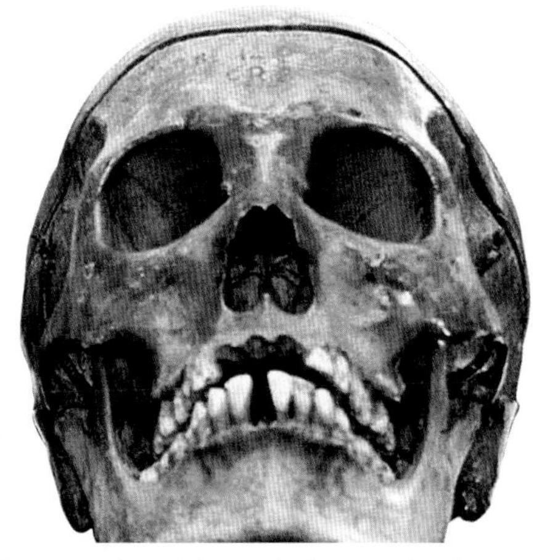

Figure 1.16 The skull on the left belonged to a member of the newly discovered, extinct species *Homo floresiensis*. On the right is a *Homo sapiens* skull

The newly discovered species was named *Homo floresiensis*, because the bones were found on an island called Flores. Later, more bones were found. They were dated, and the results showed that these tiny people lived on Flores as recently as 12 000 years ago. They made tools, and hunted the dwarf elephants and giant tortoises that also lived on the island. At this time, our own species, *Homo sapiens*, was living in the same area. Is this perhaps the origin of fairy tales about leprechauns, dwarves and elves?

Fossils

Fossils are the remains of living organisms preserved in rocks. The parts most likely to be preserved are hard parts, such as bones and teeth. Over thousands of years, as the material around them solidifies and forms rocks, the bones change into other minerals.

Figure 1.17 We only know about dinosaurs because of the fossils that have been found. This is the fossilised skull of *Tyrannosaurus rex*, about 70 million years old

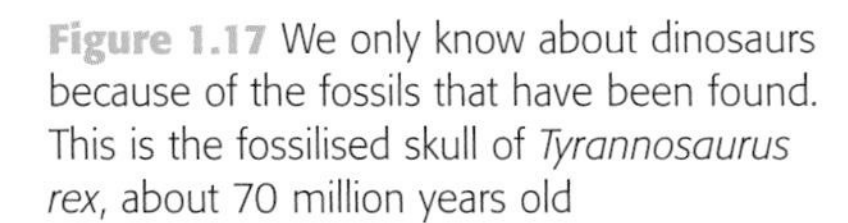

The *Homo floresiensis* bones were not fossils, but very old bones.

Fossils can tell us about organisms that lived millions of years ago. They provide evidence that there are many species that are alive on Earth today that were not alive in the past, and also that there were many species alive then that are not alive now. What is more, they suggest that some species have changed – **evolved** – into others.

A really good example of evolutionary change is shown by fossils of horses, as you can see in Figure 1.18.

a

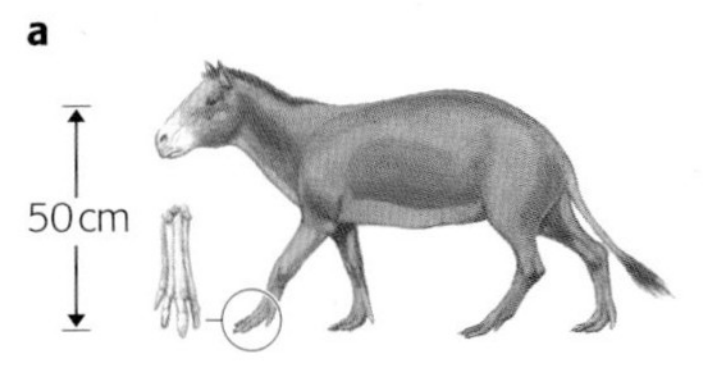

a Fossils in rocks that are around 54 million years old tell us that the horses living then were tiny. They were roughly fox-sized, and had feet with spreading toes. It is thought they were adapted to live on soft, boggy ground.

b

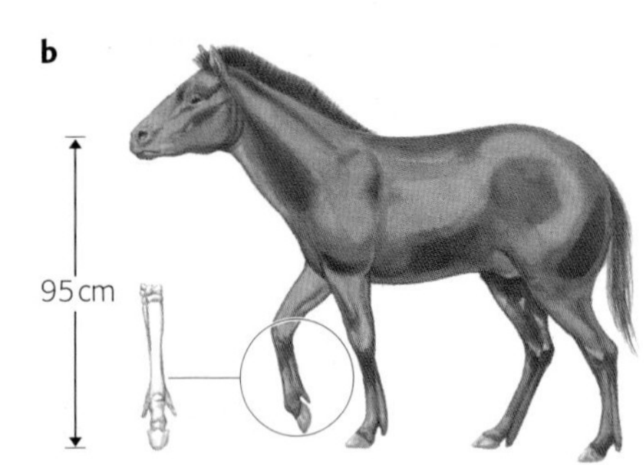

b By 26 million years ago, horses were much bigger, up to 1 metre tall. Now the large central toe had become further enlarged to form a hoof.

c

c The oldest fossils of modern-looking horses are about 1 million years old. The horses were bigger still, about 1.6 m tall. One toe on each foot had become a hoof, with the other toes becoming extremely small. Modern horses are adapted for running fast on firm ground.

Figure 1.18 We know about the evolution of the horse from the fossil record

Natural selection

The sequence of fossils of prehistoric horses strongly suggests that tall, one-toed modern horses have evolved from tiny, many-toed species of horses. How could this have happened?

We will never know for sure, but it could have happened something like this. 54 million years ago, the little many-toed horses were well adapted for living on soft, boggy ground. They could run over it without sinking in, and so had a good chance of escaping predators. But in time the climate changed, or perhaps the horses spread to new areas, so that now they walked on firmer ground.

Within the population of horses, not all of them were exactly the same. There was **variation**. Perhaps one horse had a larger middle toe than the others. This might have allowed it to run faster, so that it had a better chance of escaping predators. It survived and reproduced, passing on the gene for the big middle toe to its offspring. So they, too, had enlarged middle toes. Over time, the number of individuals with large middle toes increased. The ones with many toes eventually died out, because they were less well adapted to the new, drier environment.

This process is called **natural selection**. We can summarise it as:

- Within a population of a species, there is always some variation in characteristics. Some of this variation happens because the individuals have slightly different genes.
- Usually, many more young are born than will survive to adulthood and reproduce. There is competition between them.

- Only those offspring whose characteristics are well adapted to their environment survive. They are the ones that reproduce and pass on their advantageous genes to their offspring.

Natural selection doesn't always change things. If the environment doesn't change, and no new and advantageous characteristics appear in the species, then the species will stay the same.

If this happens time and time again, generation after generation, then there may be a change in the species. The process can even produce new species. This is what seems to have happened to the little horses that lived 54 million years ago. By 26 million years ago, they had become so different that we classify them not only as different species, but also in different genera.

The little horses that lived millions of years ago are now extinct. Changes in their environment, and competition with the longer-legged, hoofed and better adapted horses, meant that they could no longer survive.

Darwin's big idea

In 1859, an amazing and contentious book was published. It was called *The Origin of Species*, and its author was Charles Darwin.

Darwin was born in 1809, into a well educated and well-off family. It took him quite a while to find out what he wanted to do with his life. He tried medicine but couldn't stand the blood of operations. He was always interested in natural history, but it wasn't until 1831, when he was offered a place on an expedition ship, *The Beagle*, which was sailing to South America, that he began to see the direction in which his life might move.

Figure 1.19 Darwin was intrigued by the seaweed-eating marine iguanas he found in the Galapagos Islands

On the five-year voyage, Darwin visited many places on the mainland of South America and also some of the islands around it, including the Galapagos Islands in the South Pacific Ocean. What he saw made him think about why there were different species in different places. He began to realise that species might be able to change into other species. Even more importantly, he thought he could see how this might happen. He had come up with the theory of evolution caused by natural selection.

On his return home, he sat down to write about what he had seen, and about his ideas on evolution. But they worried him. He could

Figure 1.20 One of the cartoons of Darwin that was drawn after the publication of *The Origin of Species*

see that they contradicted the creation story in the Bible, and he knew that this would greatly upset many people. The scientific community mostly believed that Man had been created in the image of God; they thought that humans were unique and completely different from all other animals. Darwin's theory suggested that humans might have evolved from other animals. This challenged not only religious but also current scientific beliefs.

Darwin did not have good health, and the worry about the effect of his book made him more ill. He didn't dare to publish it until 1859, when he was 50 years old. His predictions were right. There was fury and disbelief. Other scientists and members of the Church were horrified, and they gave impassioned talks and published scathing articles about Darwin's theories. Cartoons of Darwin with a monkey's body appeared in the press.

There were a few forward-thinking scientists who could see that Darwin was right. Gradually, his ideas came to be accepted. Much evidence has since been collected which supports the theory of natural selection. With a few updates, we still use this theory today to help us to explain how species may change over time.

Selective breeding

Humans have grown crops and kept animals for food, for skins and for transport for thousands of years. In that time, we have caused some very large changes in the characteristics of the species that we use.

Take maize, for example. It is a kind of grass, and it has grown in America for hundreds of thousands of years. You can still find wild maize growing there. It is called teosinte. Natural selection has ensured that teosinte is very well adapted to its environment, which is often dry with poor soil.

About 7000 years ago, people began to cultivate wild maize. Each year, when they harvested the crop of grain (seeds) for food, they kept back some to sow next year. They kept the seeds from the best plants – the ones that gave the most and the biggest grain. Over many harvests and sowings, the characteristics of the maize plants gradually changed. The cobs (heads of grain) became larger and the grains became bigger and more nutritious.

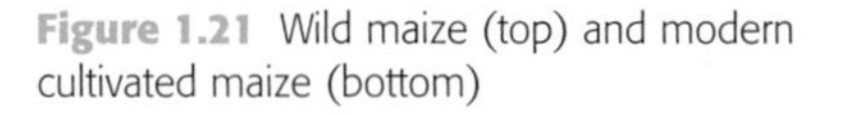

Figure 1.21 Wild maize (top) and modern cultivated maize (bottom)

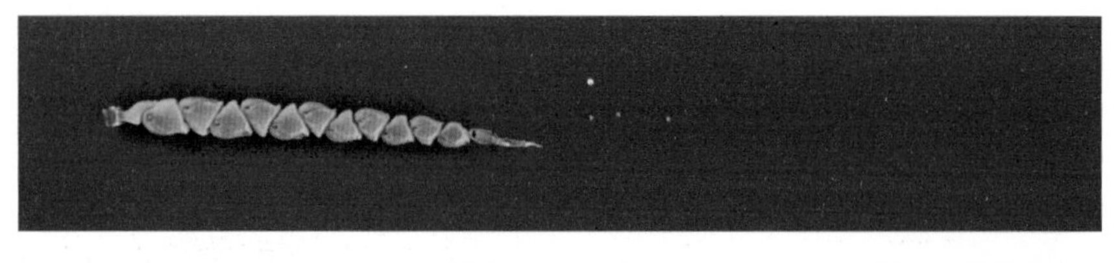

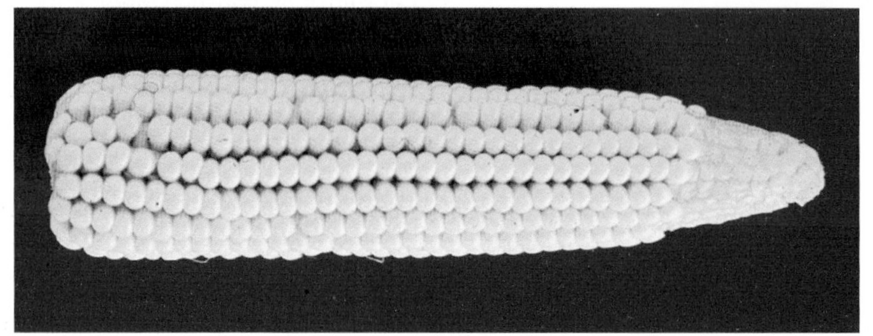

Although they did not know it, these early American farmers were using a technique called **selective breeding**. This is still going on today. In Australia, North America, South America and elsewhere, scientists continue to develop new varieties of maize through selective breeding. Their aim is to produce plants that grow well in the climate in a particular place, that give high yields and are resistant to pests such as weevils (a kind of plant-eating beetle).

As an example of how selective breeding works, consider a sheep breeder who wants to produce a lot of chestnut brown wool. She chooses two types of sheep to breed from – A and B. Sheep A produces a lot of wool, but it is white. Sheep B doesn't grow a thick coat, but its wool is chestnut brown. Can the breeder fulfil her aim by crossing these two types of sheep? Look at Figure 1.22.

A breeder wants to produce a breed of sheep that has long, chestnut-coloured wool.

The breeder crosses a sheep from breed **A** with long wool with a sheep from breed **B** with chestnut-coloured wool.

The breeder chooses the sheep that have the best combination of wool length and colour, and breeds them together.

The breeder keeps doing this for several generations.

Figure 1.22 How selective breeding works

From the first generation offspring, the breeder chooses ewes and rams that best match her requirements, and breeds them together to produce second generation offspring. If she continues to do this for many generations, she should be able to produce a whole flock of sheep with long, chestnut-coloured wool.

Selective breeding often changes the characteristics of a species more quickly than natural selection, and it can select different characteristics. The maize that is grown by farmers has been chosen because it is what we want it to be. It isn't necessarily the kind of maize that would survive best out in the wild.

Figure 1.23 Some people have strong feelings against GM food

GM crops

In the last few decades, another technique for changing the characteristics of a species has become available to us. This is **genetic engineering**. We can sometimes identify a particular gene that produces the characteristics we want. We can again use maize as an example. Maize plants are attacked by many insect pests. Using gene technology, a gene from bacteria that produce a toxin (poison) that kills insects has been inserted into maize plants. The new maize – **genetically modified** or **GM maize** – makes this toxin in its leaves. This can mean that fewer pesticides have to be sprayed onto the fields, and that the farmer will get a larger healthy crop.

Genetic engineering still hasn't taken over from selective breeding. It is very expensive and very difficult. The trickiest part is often identifying the gene that you want to use; and getting it into the new organism is also very hard. What's more, many people don't like the idea of eating GM food, so the technology has not taken off as fast as many thought it would.

Summary

- Fossils give us information about the structures of organisms that lived long ago. By looking at similar fossils from different times in the past, we can sometimes see how particular kinds of organisms seem to have changed over time.
- There is always some variation between organisms within a species. Some of this variation is caused by genes, and so can be passed on to offspring.
- More young are normally produced than will survive to adulthood. The individuals that are best adapted to their environment are more likely to survive and reproduce. This is called natural selection.
- This means that the genes that produce these adaptations are more likely to be passed on to the next generation than other genes.
- Over many generations, natural selection can produce a change in the species. Eventually, a new species may be produced and other species become extinct.
- Humans use selective breeding to change the characteristics of a species so that it becomes more useful to us.
- Genetic engineering involves putting a chosen gene into an organism's cells, which alters its characteristics. This can have faster and more predictable results than selective breeding, but it is much more expensive to do.

Questions

1.15 In each of these three pairs of statements, one is correct and one is incorrect. Decide which is correct, and then explain what is wrong with the incorrect one.

a **i** Only the largest and strongest organisms survive the struggle for existence.

ii Organisms that are better adapted to their environment have a greater chance of reproducing.

b i An organism with a variation that gives it an advantage has a good chance of passing this characteristic on to the next generation.
ii Organisms mutate so that they become better adapted to their environment.

c i An organism that lives in a very cold place and grows thick fur will pass on the characteristic for thick fur to its offspring.
ii Only variations that are caused by genes, not by the environment, can be passed on to offspring.

1.16 How is selective breeding similar to natural selection? How is it different?

1.17 Explain what is meant by a GM organism.

1.18 Why is selective breeding still more widely used than genetic engineering?

1.19 MRSA stands for methicillin resistant *Staphylococcus aureus*. Find out what MRSA is and how natural selection has been responsible for its appearance. Why is it dangerous? What can be done about it?

E Humans and the environment

Tuvalu is an idyllic group of islands in the South Pacific. It is tiny – the total area of land in all of the nine islands is just over 26 square kilometres. Around 12 000 people live there. They grow coconuts and catch fish. Tourism is also a significant contributor to their economy. An ingenious addition to their income has been the sale of the internet suffix .tv to a large, Californian company, which sells it on to television broadcasters.

But Tuvalu is in danger. The islands of Tuvalu are all very low-lying. In 2002, no point on the islands was higher than 5 m above sea level. As global warming causes sea levels to rise, the land area of Tuvalu is steadily decreasing. Storms lash its coast more often than before, increasing the rate of coastal erosion. Fresh water supplies, never abundant, have become contaminated by sea water.

In 2001, the Tuvaluans announced that they would have to evacuate their islands if sea levels continued to rise. Along with other small island communities, they have lobbied hard for a worldwide reduction in emissions of carbon dioxide. So far, this has had little effect. As world population increases, so do emissions of carbon dioxide. It seems inevitable that low-lying countries such as Tuvalu will face complete destruction by the middle of the 21st century.

Figure 1.24 Tuvalu before and after a storm

Global warming

In 1980, the total world human population was 4.5 billion. By 1990 it was 5.3 billion and by 2000 it was 6.1 billion. It is estimated that by 2010 it will be 6.8 billion.

All of these people need food. They also use a great deal of energy – for cooking, heating and lighting, for making things such as metals and plastics, for building roads and houses, and for transport. Much of this energy is generated from fossil fuels. When fossil fuels are burnt, carbon dioxide is released. We know that the concentration of carbon dioxide in the atmosphere is increasing (see Figure 1.25), and it is probable that this is the result of our use of fossil fuels. In 2004, carbon dioxide made up 379 parts per million (ppm) of the Earth's atmosphere. It was 374 ppm in 2002. It is predicted that it will be at least 650 ppm by 2100.

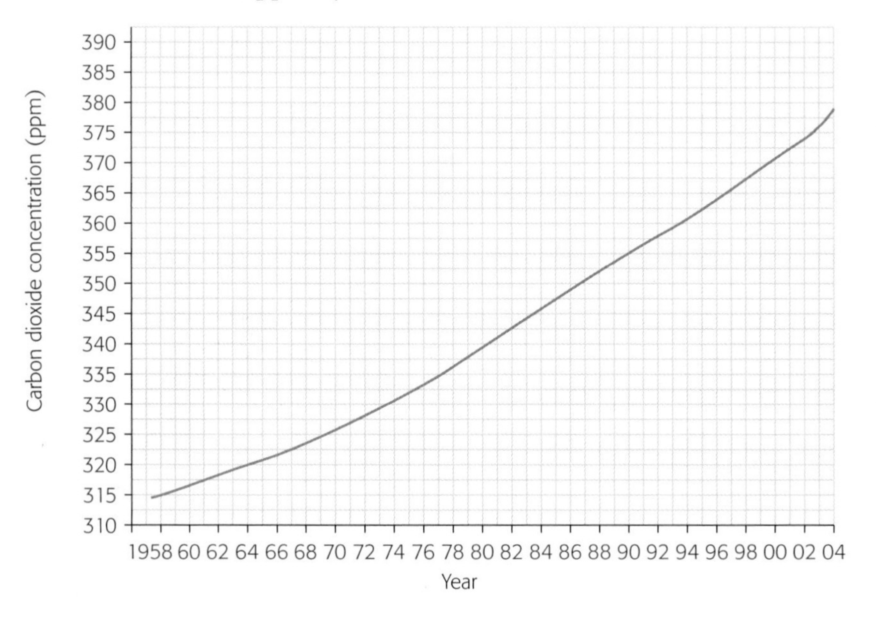

Figure 1.25 Atmospheric carbon dioxide concentrations since 1958

Carbon dioxide is a greenhouse gas. It traps heat energy in the atmosphere, preventing it from escaping out into space. The more carbon dioxide in the atmosphere, the warmer the Earth becomes. If carbon dioxide levels get as high as 650 ppm in 2100, we could expect the Earth's average temperature to increase by at least 2 °C.

This may not sound a very large temperature increase, but it could have devastating effects on Earth. For example, the warmer temperatures are already causing glaciers to melt, increasing the volume of water in the oceans. The water also expands as it gets warmer. Sea levels are predicted to rise by up to 1 metre this century, which would flood not only low-lying countries such as Tuvalu, the Maldives and Bangladesh, but also threaten coastal cities such as Shanghai, Alexandria, London, Amsterdam, Venice and New Orleans.

> There are a few scientists who think that global warming could be happening because of natural causes, and not primarily because of humans burning fossil fuels.

If we know that we are doing this, why can't we stop? In 1997, 84 countries signed an agreement in Kyoto, Japan, pledging that they would limit their carbon dioxide emissions. But not all of them have kept their word. Britain, for example, has failed to meet the

You can read more about global warming in Chapter 7.

targets it set for itself. And the United States of America – the country that emits more carbon dioxide than any other – has backed right away from the Kyoto agreement. Meetings between the heads of government of eight of the most influential countries in the world, held in 2005, still failed to arrive at an agreement that will stop carbon dioxide emissions continuing to rise.

The trouble is that there are more and more of us, and we have come to expect an ever-improving lifestyle. We want to travel more and be able to keep our homes and workplaces at a comfortable temperature. The greatest increases in carbon dioxide emissions are coming from rapidly developing countries, especially China and India. As their industry and economy catch up with those of industrialised countries, they are using more and more fossil fuels and emitting more and more carbon dioxide. They are naturally unwilling to cut their carbon dioxide emissions if other countries are failing to do this.

Figure 1.26 The streets of China's main cities are now full of cars; only a few years ago this street would have been full of bicycles

Organic farming

Ever since humans began to grow food, rather than collect it from the wild, we have been using big areas of land for farming. As the population grows, more and more land is being cultivated.

At the beginning of the 20th century, farms in Britain were mostly small, mixed farms – the farmers grew crops such as wheat and potatoes and also kept animals such as cattle, pigs, sheep and hens. The dung from the animals was used as fertiliser on the fields. There wasn't much a farmer could do if a pest such as fungus attacked the wheat. If there were a lot of weeds, they had to be killed by hoeing.

The Second World War changed attitudes to farming. The government encouraged farmers to grow as much food as they possibly could on their land, so that Britain could be self-sufficient. Throughout the 20th century, farmers concentrated on getting the highest yields they could. This is called intensive farming. Large farms might produce only wheat, for example. Synthetic fertilisers such as ammonium nitrate were added to the soil. Herbicides and pesticides were sprayed onto crops, to kill weeds and pests.

But this way of farming can cause problems. Fertilisers can get into streams and ponds, where they increase the growth of algae and weeds, making the water dark and reducing the amount of oxygen dissolved in the water, so fish can't live there. The herbicides and pesticides may kill plants and animals that we like – meadow flowers, for example, or helpful insects such as ladybirds, hoverflies and bees. Traces of the herbicides and pesticides may even be left in the food that we eat.

Intensive animal rearing can lead to very poor conditions for the animals – cramped stalls for pigs, for example. This in turn can lead to the use of drugs such as antibiotics to prevent diseases spreading.

Registered organic growers are allowed to use some chemicals on their crops – such as copper sulphate to kill an infection of potatoes called blight.

Some farmers have therefore begun to farm **organically**. The idea is to grow crops and raise animals without using any 'artificial' chemicals. The hope is that this is 'better for the environment' – it should cause less pollution and allow other plants, insects, birds and other animals to co-exist on the farm alongside the crops. It also gives better living conditions for reared animals. Reduced use of antibiotics lessens the chance of bacteria developing resistance to them (see page 79).

Figure 1.27 Intensive farming (left) and organic farming (right)

Figure 1.28 Organic food is now easily available, although more expensive

It sounds good, but how does it work out in reality? Organic farming doesn't give such high yields from a particular area of land, often because there is competition from weeds or because pests destroy some of the crop. So farmers have to charge more for their produce in order to make a profit. All the same, a lot of people are prepared to pay extra for organic produce, and all the big supermarkets now stock organic produce of all kinds, from chocolate to beef to runner beans.

There isn't much evidence that organic produce is better for us than 'ordinary' food. There is little if any difference in the nutrients it contains, and most people, despite what they like to think, can't taste any difference between organic and non-organic produce. On the plus side, we know that organic produce won't contain any traces of synthetic pesticides, some of which could possibly pose a long-term health risk.

So is organic farming better for the environment? Certainly it is good not to use synthetic chemicals. But if all the food we ate were to be produced organically, we would need to use much more land because of the lower yield. With the Earth's population still growing fast, it doesn't look as though it is really possible to supply all the food we need by organic farming.

Summary

- The Earth's population is growing. This puts more demands on the environment.
- Carbon dioxide is released when fossil fuels burn. As more humans use more energy, we burn more fossil fuels. This is thought to be the cause of the increasing concentration of carbon dioxide in the atmosphere.
- Carbon dioxide in the atmosphere traps heat around the Earth. Increasing carbon dioxide levels are thought to be causing global warming.
- An increase in the Earth's average temperature is likely to result in rising sea levels.
- Organic farming uses no synthetic fertilisers and few herbicides and pesticides, which can do less harm to living things on and around the farm. But yields are lower, so the cost of the produce is higher.

Questions

1.20 Rising carbon dioxide levels may cause other changes to the environment, not only a rise in sea level. Find out what these are, and then research *one* of them in detail. What is the evidence suggesting that it is happening (or will happen in the future)? How might it affect humans and other species? What can we do about it?

1.21 Find out about how human activities are affecting the ozone layer. Where is the ozone layer, and why is it important? How are we affecting it?

1.22 Some people do not understand the difference between global warming and damage to the ozone layer. Write a short, illustrated newspaper or magazine article that explains this.

1.23 Explain how organic farming can be better for the environment than non-organic farming.

2 Genes

A DNA

Ophelia McKnight went missing on 5 January 1988. Her body was found on 6 February. She had been shot four times.

It wasn't easy to identify her. She had been dead for a month, so her body was decomposing. DNA samples were taken, and compared with samples from her family. They were a close match, so there was no doubt that the body was hers.

Ophelia's killer was not found. But the case wasn't closed, and forensic evidence taken from the scene was stored. Sixteen years later, in 2004, the police matched up DNA samples that had been found on Ophelia's body with DNA from a man who was already in prison for rape and burglary. When confronted with the evidence, he admitted to killing Ophelia.

Figure 2.1 Forensic scientists collecting DNA evidence

Chromosomes and genes

DNA stands for deoxyribonucleic acid. It is perhaps the most amazing chemical in our bodies. It is your DNA that makes you a human being and not a gorilla; it determines what sex you are, your natural hair colour and your eye colour. It influences all sorts of other things about you, such as how good you are at maths, whether you can run fast and how good your memory is.

Your DNA is kept safely inside the nuclei of your cells. You have 46 DNA molecules in each cell. Each DNA molecule is extremely long, made up of a chain of smaller molecules linked together. It is twisted and coiled up tightly to form a **chromosome**. Each chromosome is one, huge DNA molecule.

Figure 2.2 Human chromosomes seen using a scanning electron microscope

The DNA molecules carry a code called the **genetic code**. This code contains information about the proteins that the cells will make. Many of these proteins are enzymes, so they control all the chemical reactions – **metabolic reactions** – that take place in your cells. And it is those reactions that help to build your body in the shape that it is, and affect not only what you look like but also how you behave in different situations.

Each DNA molecule holds instructions for making many different proteins. A stretch of DNA that codes for one protein is called a **gene**. We have about 30 000 genes. Our longest chromosome, chromosome 1, contains 2968 genes. The smallest one, the Y chromosome, has only 231.

You can think of a gene as being a unit of inheritance.

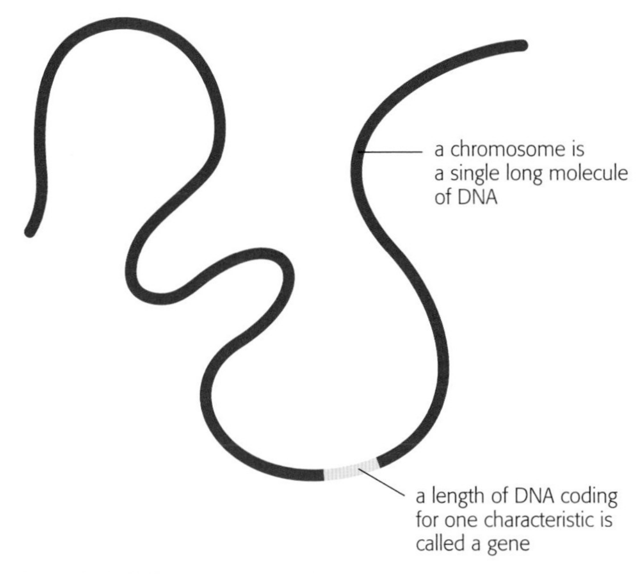

Figure 2.3 A chromosome contains many genes

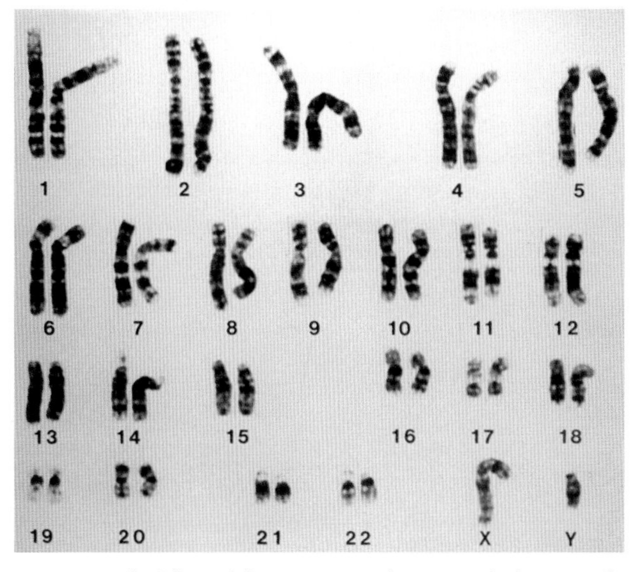

Figure 2.4 The full set of chromosomes, or karyotype, of a human male

The Human Genome Project

The fact that DNA makes up our genes was first discovered in the 1950s. Forty years later, in 1990, scientists around the world joined forces to try to map all of the DNA we have in our cells. The aim of the project was to find the genetic code on all of our chromosomes – the human **genome** – so that we could find out exactly where on each chromosome all our genes are. By 2003, the scientists had succeeded.

Surprisingly, it looks as though only about 2% of our DNA actually makes up working genes. The rest of it is sometimes called 'junk DNA', though it is likely that we will one day discover that at least some of it is doing something useful.

DNA in forensic science

The Human Genome Project has found that there are about 3 million places in our genes where the DNA is slightly different in one person than in another. We can use this to help to solve crimes.

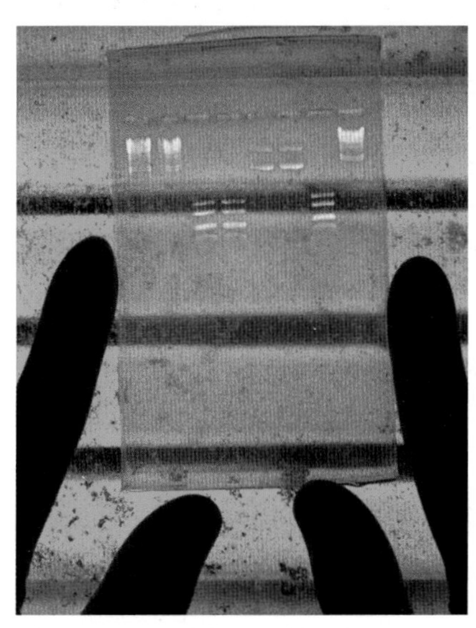

Figure 2.5 Examining a DNA profile

This is sometimes called genetic fingerprinting.

In the Ophelia McKnight case, some cells from her attacker were left on her body. Forensic scientists need only a tiny number of cells to be able to do their work – maybe just a hair that was pulled out, or some saliva, semen or blood. The forensic technicians were able to extract DNA from the cells on Ophelia's body, and match it against DNA taken from all possible suspects. They didn't need to match all of it, just some pieces from areas of the chromosomes where people's DNA usually differs from each other. Unless you have an identical twin, these parts of your DNA are different from everyone else's, so if the forensic team find a match, it is very strong evidence that the person was at the scene of the crime.

DNA is equally useful in proving that someone was *not* involved in a crime. If a suspect's DNA does not match the DNA found at the scene, then he or she probably was not there and is innocent. And if someone else's DNA *was* found there, then even if the police cannot find that person at least they know that their previous suspect probably did not commit the crime.

Cystic fibrosis

One of the great hopes of the Human Genome Project is that we might be able to use our knowledge to cure diseases.

Many of our genes come in more than one variety. The different varieties are called **alleles**. So, for example, the gene for eye colour has an allele that codes for brown eyes and an allele that codes for blue eyes. It doesn't matter too much what your eye colour is, but there are other genes whose effect is much more important. For example, a disease called **cystic fibrosis** is caused by an allele of a gene that is important for making mucus. Mucus is the sticky semi-liquid that you make in the lining of your nose and all the way down the passages to the lungs, where it helps to trap bacteria and stop your lungs getting infected. In someone who has cystic fibrosis, the mucus is too thick and sticky. It gets stuck in the lungs, and they often get infected. Someone with cystic fibrosis has to have therapy every day, during which someone thumps them on their back to loosen the mucus so that they can cough it up. Medical treatment for the disease is now much better than it used to be, but still someone with cystic fibrosis is likely to die at a relatively young age.

In Britain, about one baby in 3300 is born with cystic fibrosis.

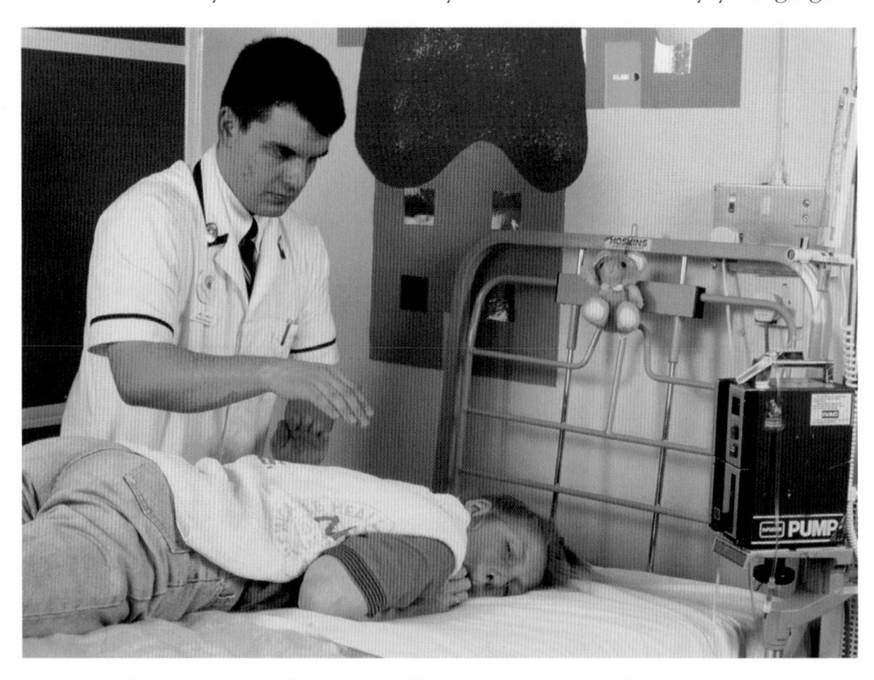

Figure 2.6 Therapy for cystic fibrosis

Now we know where the cystic fibrosis gene is, and we know how the DNA differs in the normal and faulty alleles of the gene. Scientists can make copies of the normal gene and try to put it into the cells of people with cystic fibrosis, but this is very difficult. The cells the DNA needs to get into are down inside the passages leading to the lungs, so it is not easy to reach them. What's more, these cells don't live very long – they die and are replaced every few weeks by new cells – so even if the DNA could be put into the cells, this would have to be done over and over again, every three or four weeks.

This kind of treatment for a genetic disease is called **gene therapy**. It is still in its early days, and even though we know a lot about the genes that cause diseases, it is going to be quite some time before gene therapy can be used to cure them.

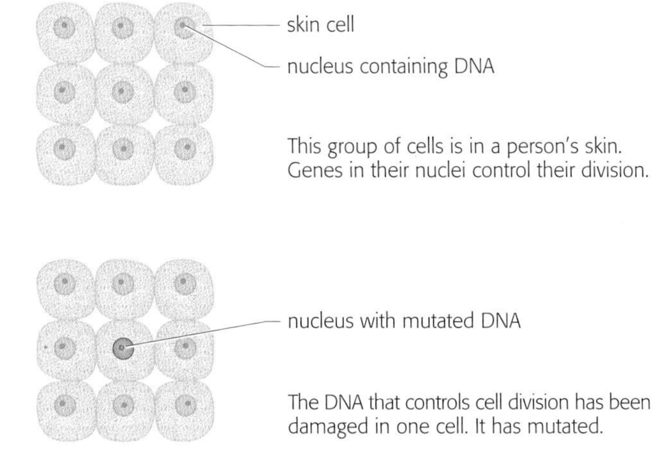

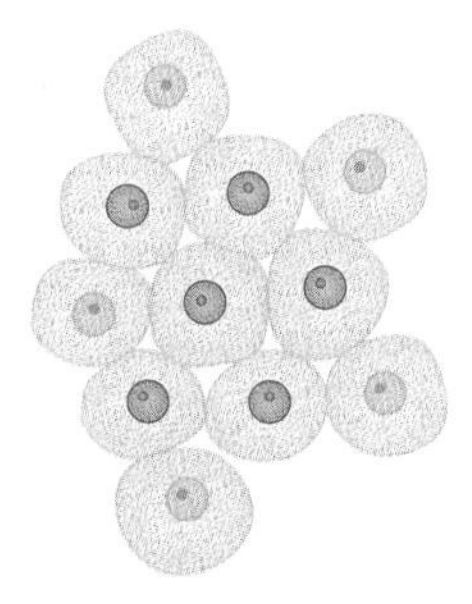

The cell with the mutation divides uncontrollably. It forms many more cells like itself. Eventually they form a lump called a tumour.

Some of the mutated cells may break away and travel to other parts of the body, where they form secondary tumours.

Figure 2.7 How a mutation can cause cancer

Cancer

The DNA in our cells controls practically everything that they do. Some of our genes control when cells should divide.

Sometimes, the genes in a cell get altered. This is called **mutation**. Mutations can happen for no obvious reason, or they can be caused by ionising radiation or by chemicals. These mutations may affect the genes controlling cell division. If this happens, the cell might begin to divide over and over again, forming a big lump of cells called a **tumour**. And sometimes this tumour becomes **malignant**, which means that the cells can break away and start up new tumours in other parts of the body. We call this condition cancer.

A common cancer in women is breast cancer. If breast cancer is caught early, it can usually be treated successfully. But the treatment isn't very pleasant. The woman may have an operation to remove part or all of the affected breast. She may have chemotherapy – drugs that kill the cancer cells – or radiotherapy – bombardment of the cancer cells with ionising radiation that kills them. Both of these treatments affect other cells as well, so she is likely to feel quite ill while the therapy takes place.

What if we could replace the faulty genes in cancerous cells? Cancer treatment would become easy. Just now, there is no likelihood of this happening soon. But scientists are working on it, and it is certainly a possibility for the future.

> Treatment for breast cancer is now very successful. 98% of women live for at least 5 years after being diagnosed, 72% for at least 10 years, and 64% for at least 20 years.

Summary

- The nucleus of a human cell contains 46 chromosomes, each made of a long molecule of DNA. The DNA controls the cell's activity.
- A section of DNA that codes for one protein, or for one characteristic, is called a gene.
- The Human Genome Project has mapped out all the DNA in our chromosomes, and identified many of the genes.
- Some parts of our DNA vary a lot between different people. Forensic scientists can often find DNA at the scene of a crime, and can try to match this up against DNA samples taken from suspects.
- An allele is a particular variety of a gene. Some alleles cause genetic diseases, such as cystic fibrosis.
- Damage to DNA can alter the genes that control cell division, which may result in cancer.
- Gene therapy aims to replace faulty DNA with normal DNA, which could potentially cure genetic diseases and cancer.

Questions

2.1 Copy and complete this sentence.

Chromosomes are found in the .... of a cell. Each chromosome is one .... of DNA. A .... is a length of DNA that codes for one ....

2.2 Explain what is meant by:
- **a** the genetic code
- **b** the human genome.

2.3 How can DNA be used to link a suspect to a crime scene?

2.4 Try to put yourself in the place of someone with cystic fibrosis or breast cancer. Write a paragraph describing how your life changes when a new, genetic treatment is discovered for your disease.

B Inheriting genes

Sal and Tony were amazed when their second child, Josie, turned out to have red hair. They both had brown hair, as did their parents and grandparents. Their first child had also had brown hair. So where did this redhead come from?

Hair colour is determined by a pigment called melanin. There are two versions of it – eumelanin and phomelanin. The eumelanin in your hair can give a range of colour from blonde right through to black; the more eumelanin you have, the darker your hair. Phomelanin gives hair a red colour. If you have a lot of phomelanin and hardly any eumelanin, then your hair is red.

The chemical reactions in the body that make these pigments are controlled by enzymes. The enzymes that are made are determined by your genes. In most of us, there is a gene that codes for an enzyme that converts phomelanin to eumelanin. Josie had somehow ended up with an allele of the gene that didn't do this. So she had lots of phomelanin and hardly any eumelanin.

Figure 2.8 Variation within a family

This red-hair allele is recessive. That means you have to inherit one from *both* of your parents before you get red hair. Sal and Tony must each have had one copy of the red-hair allele, which they inherited from their parents who had inherited it from their parents. It wasn't until Sal and Tony's second child was conceived that two of these alleles came together.

Figure 2.9 A spider plant with 'plantlets'

Asexual reproduction

Sal and Tony are humans, so they reproduced sexually. But many plants, and even a few animals, have a different method of reproduction, called **asexual reproduction**.

In asexual reproduction, there is just one parent. Some of the cells in the spider plant in Figure 2.9 have divided over and over again until they have produced a new plantlet. Each time a cell divided, the two new cells produced were given two complete sets of chromosomes, just like the parent cell. So the new spider plants have exactly the same chromosomes, and therefore genes, as their parent. They are **genetically identical**. They are **clones**.

Sexual reproduction

Like most animals and plants, humans reproduce using **sexual reproduction**. This involves two parents making **gametes** – in our case, eggs in the woman and sperm in the man – that fuse together to form a new cell called a **zygote**. The zygote grows into an embryo, then a baby, a child and eventually an adult.

We've seen that, in each of our cells, we have 46 chromosomes. If you look back at Figure 2.4, you can see that these chromosomes can be arranged in pairs. We actually have two *sets* of chromosomes, each set containing 23 chromosomes.

Usually, when a cell divides, each new cell ends up with two sets of chromosomes, just like the original cell. But when cells in an ovary or a testis divide to form gametes, something different happens. The new cells each get only one set of chromosomes. They have 23 chromosomes in them, not 46. Figure 2.10 explains why this is important. It makes sure that, when two gametes from two parents fuse, the zygote gets only two sets of chromosomes, not four.

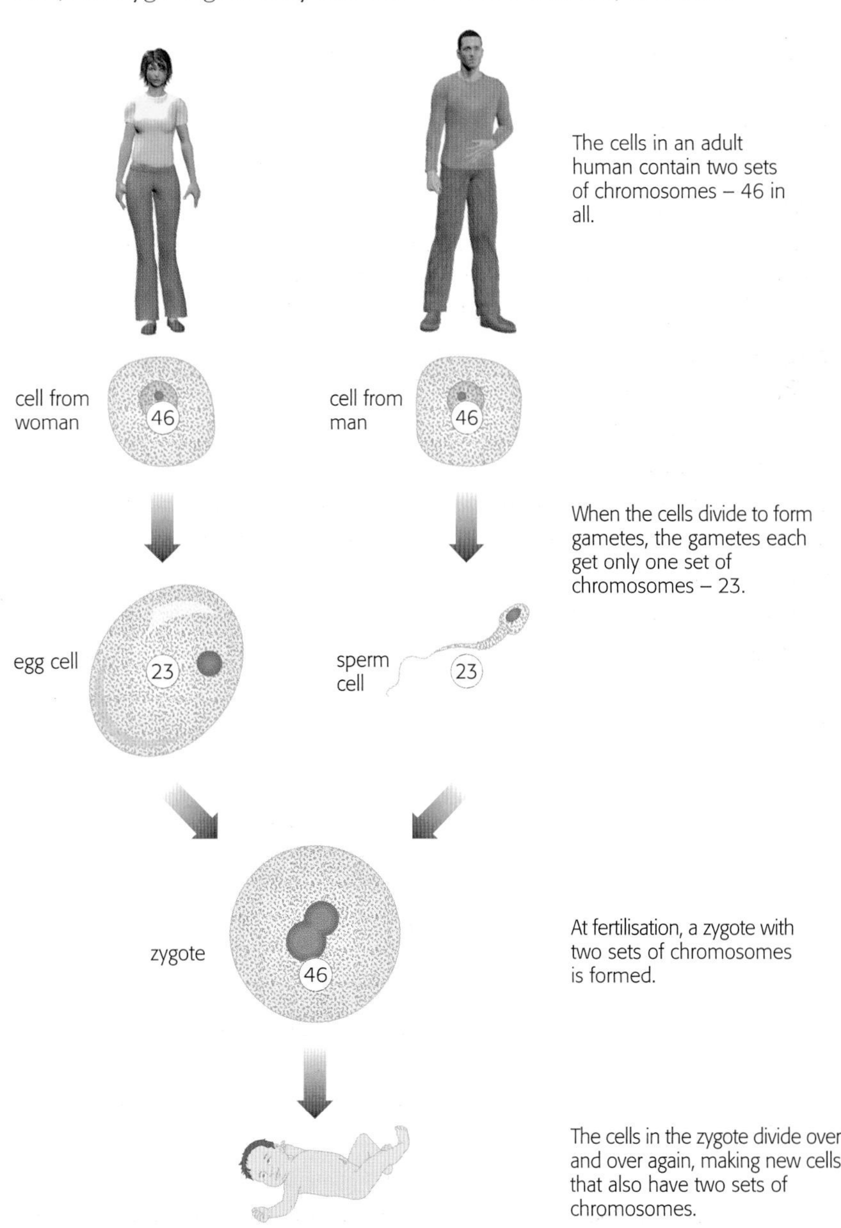

Figure 2.10 What happens to the chromosomes in sexual reproduction

Genetic variation

One set of chromosomes in a person's cells came from their father, and one from their mother. When gametes are made, these sets get all mixed up. A sperm cell might, for example, get a chromosome 1 that originally came from the man's father, a chromosome 2 from his mother, a chromosome 3 also from his mother, and so on. It is just chance how these are shared out. See Figure 2.11.

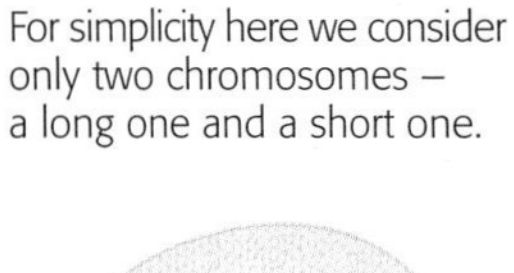

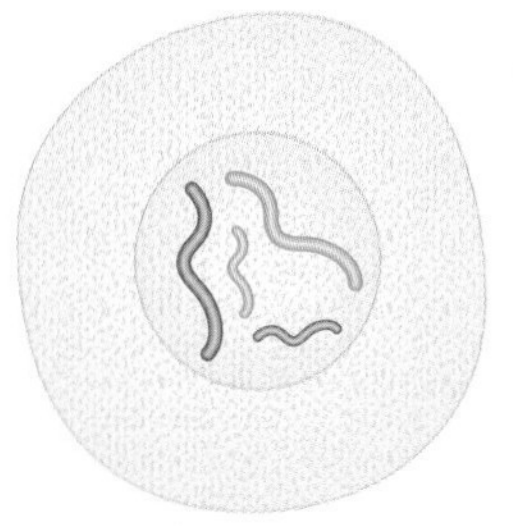

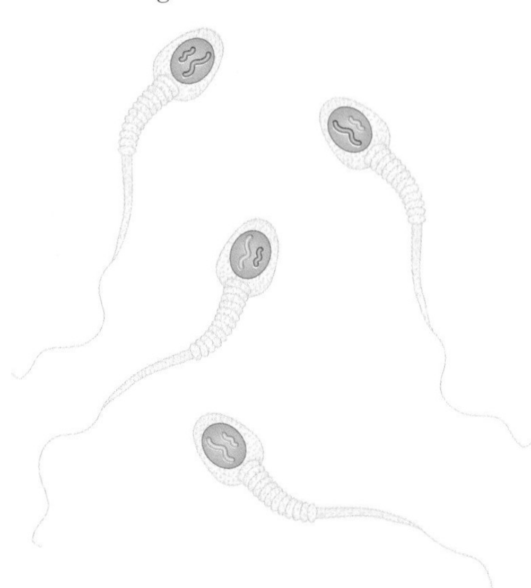

Figure 2.11 How chromosomes are shared out in gametes

So each sperm has a different mix of chromosomes. How many different combinations would there be if the body cell in Figure 2.11 had two sets of *three* chromosomes? Imagine the possibilities with 23 chromosomes. And the same number of combinations occur with the eggs. Then another possible difference comes into play. Any sperm could fertilise any egg. This means that each zygote can have almost any combination of chromosomes – and so genes – inherited from its parents. Every zygote is therefore different from every other zygote – unless they are identical twins who have developed from the same egg and sperm.

This variation in our genes is called **genetic variation**. Sexual reproduction produces genetic variation.

The genetics of cystic fibrosis

We've already seen (page 26) that there is a faulty allele of one gene that causes the disease cystic fibrosis. Because this disease is caused by genes, it can be inherited.

H

It's easier to think about the different alleles of a gene if we use letters to represent them. We can use f for the faulty allele and F for the normal allele. Because we have two sets of chromosomes, we have two copies of each gene in our cells. So there are three possible combinations of these alleles:

FF	which doesn't cause cystic fibrosis
Ff	which also doesn't cause cystic fibrosis
ff	which causes cystic fibrosis

You can see from this that you have to have two copies of the faulty allele, f, in order to get cystic fibrosis. This allele is therefore said to be **recessive**. The other allele, F, is **dominant**. Even if only one of your copies of this gene is the F allele, you won't have cystic fibrosis.

A person with genotype Ff is said to be a 'carrier' of cystic fibrosis.

Imagine two parents who each have the alleles Ff in their cells. When sperm and eggs are made, each will get only one copy of this gene. Half of the sperm will get an F allele and the other half will get an f allele, and the same for the eggs. So, when fertilisation happens, it is possible for a sperm with an f allele to fertilise an egg that also has an f allele. The zygote gets two f alleles. The person who develops from this zygote has cystic fibrosis, even though neither of the parents had it. See Figure 2.12.

Many other genetic diseases and disorders are inherited in the same way.

Figure 2.12 Possible allele combinations at fertilisation

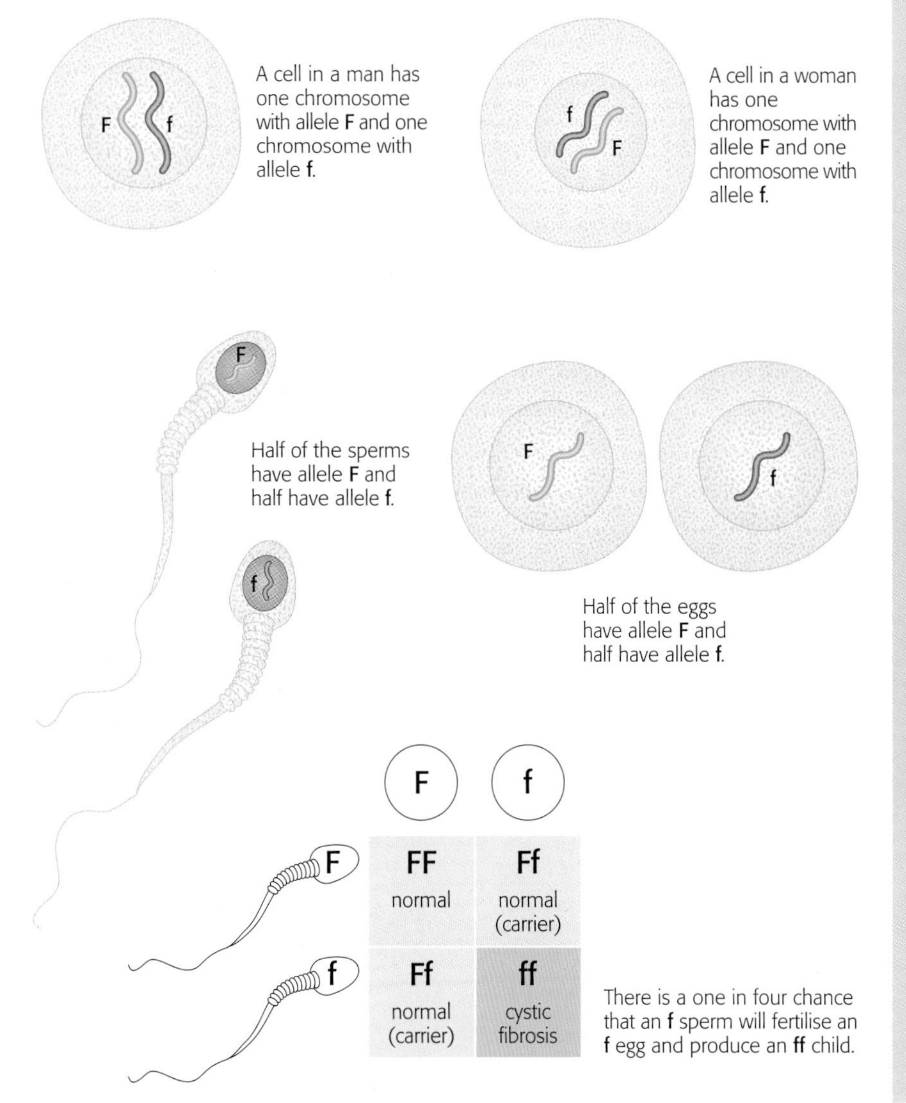

Environmental variation

Cystic fibrosis is a condition that is caused purely by genes. If you have alleles ff, then you will have cystic fibrosis. It makes no difference what you eat or how fit you are. But many of our inherited characteristics can be affected by our **environment**.

Imagine, for example, a child who is born to two tall parents. She inherits the genes for tallness from her parents. She could grow tall. But her parents are very poor, and the place where they are living is war-torn and drought-stricken. She doesn't get much to eat, and her diet is extremely low in protein. So, despite her genes, she doesn't grow tall. Her environment has had a big influence on her height.

The same is true for plants. No matter what genes a plant inherits, it won't grow tall, sturdy and green unless it has all the **minerals** that it needs in the soil where it is growing. To test the effect of the lack of a mineral, plants can be grown in culture solutions that contain all the minerals needed except one. See Figure 2.13.

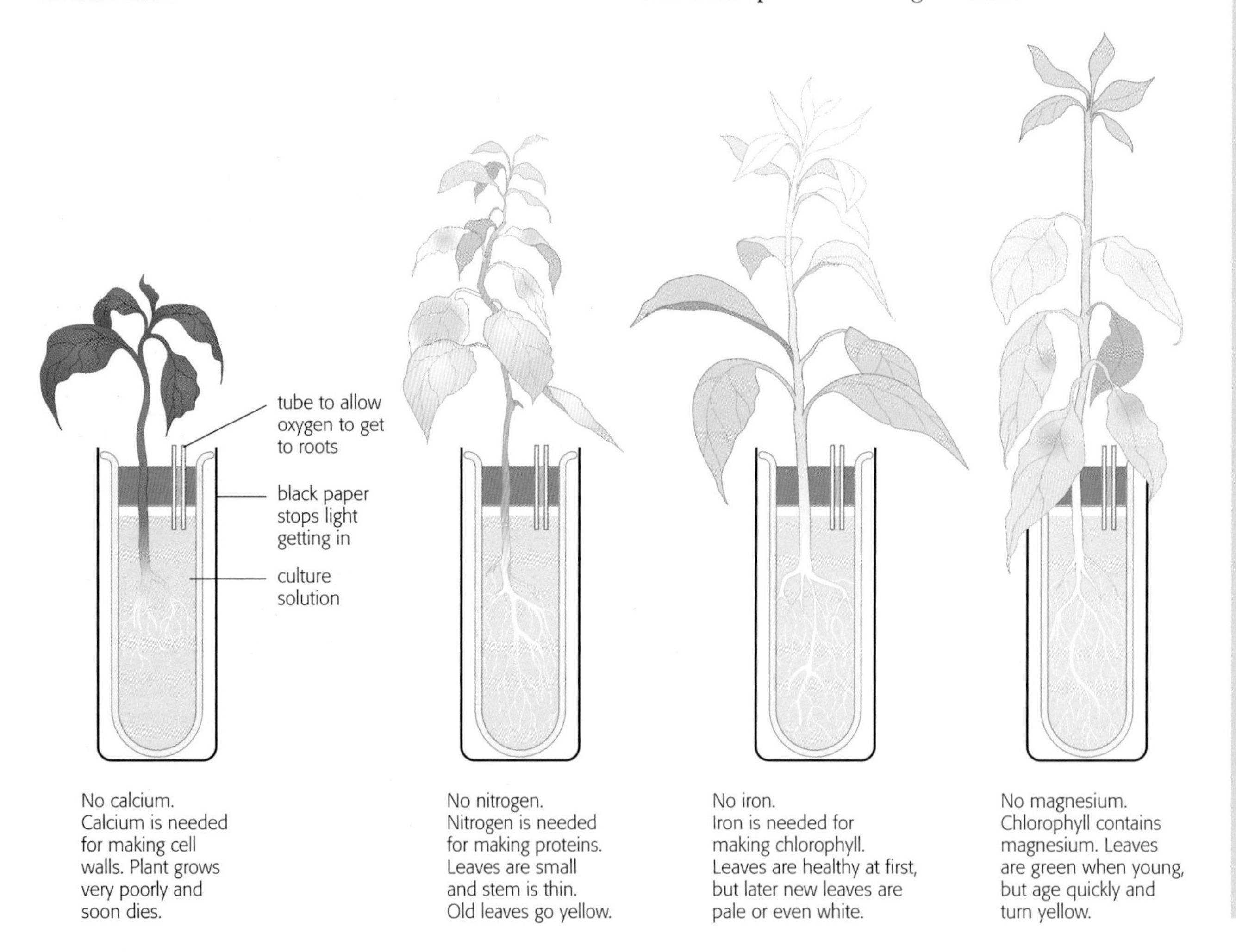

Figure 2.13 Mineral deficiencies in plants will cause variation

Summary

- Sexual reproduction involves gametes and fertilisation.
- Sexual reproduction produces genetic variation. This is because gametes contain different sets of genes, and any two sperm and egg can fuse, so each zygote has a different combination of genes from any other zygote.
- Genes can have different forms, called alleles. Some alleles cause diseases that can be inherited. Cystic fibrosis is an example of an inherited disease.

H

- A dominant allele always shows up, but a recessive allele only shows if there isn't a dominant allele present.
- Some inherited characteristics can be modified by environmental conditions. For example, human height is affected by genes and by diet. The growth of a plant is affected by its genes and by the mineral resources in the soil where it is growing.

Questions

2.5 Copy and complete this table.

	Asexual reproduction	**Sexual reproduction**
How many parents?		
Does it involve gametes?		
Does it involve fertilisation?		
Is a zygote formed?		
Are the offspring genetically identical?		
Are the offspring clones?		

2.6 Why does a person have a mixture of the characteristics of both parents?

2.7 Explain why two brothers are not exactly alike.

2.8 Sometimes, a mistake happens when a sperm is made, and it ends up with two sets of chromosomes instead of one. How many chromosomes would the zygote have, if this sperm fertilised a normal egg?

H

2.9 The colour of a plant's flowers is determined by a gene. This gene has two alleles, R and r. If the plant has alleles Rr, it has red flowers, but if it has rr then it has white flowers. Use this example to explain the meaning of the terms *dominant* and *recessive*.

2.10 Huntington's disease is caused by a dominant allele of a gene. Is it possible for two parents who do not have this disease to have a child who does have it? Explain your answer.

2.11 Two sunflower seeds were sown, one in a pot and one in the soil in the garden. One grew much taller than the other. Explain how both genes and the environment could cause this variation. How could you find out which factor had the greatest influence?

C Gene technology

In April 2005, Professor 'Twink' Allen, the father-in-law of champion jockey Frankie Dettori, became the first person in Britain to be given official permission to try to clone horses.

The first successful cloning of a horse was done in Italy in 2003. A nucleus from a skin cell from a mare was fused with an egg cell from which the nucleus had been removed. The resulting cell therefore contained two sets of the mare's chromosomes, and it behaved like a fertilised egg. It was put into her uterus, where it grew into a foal. The foal, called Prometea, had identical genes to its mother.

Figure 2.14 Prometea, the first cloned horse, with her mother

Professor Allen will do experiments with horse cloning, to investigate questions such as how much a horse's genes, and its environment while it is in the uterus, affect its characteristics. But many people would like to be able to clone horses for other reasons.

For example, most male horses are castrated when they are young – that is, their testes are removed – so that they are easier to handle. If one of them turns out to be a brilliant racehorse, the owner can't breed from it. But if cloning were allowed, then it would be possible to use some of the horse's cells to produce identical offspring. At the moment, the cloning of horses used for sport is banned.

Cloning

Cloning means producing new organisms that are genetically identical to another organism. Spider plants do it naturally. But it does not happen naturally with any mammals.

The first mammal to be cloned was Dolly the sheep. She was born in 1996, and died in 2003. She was made by taking a body cell from one female, taking the nucleus out of an egg cell from another, and then fusing the two together. This 'fertilised egg' was then put into a sheep's uterus, where it grew into Dolly.

Figure 2.15 How Dolly was made

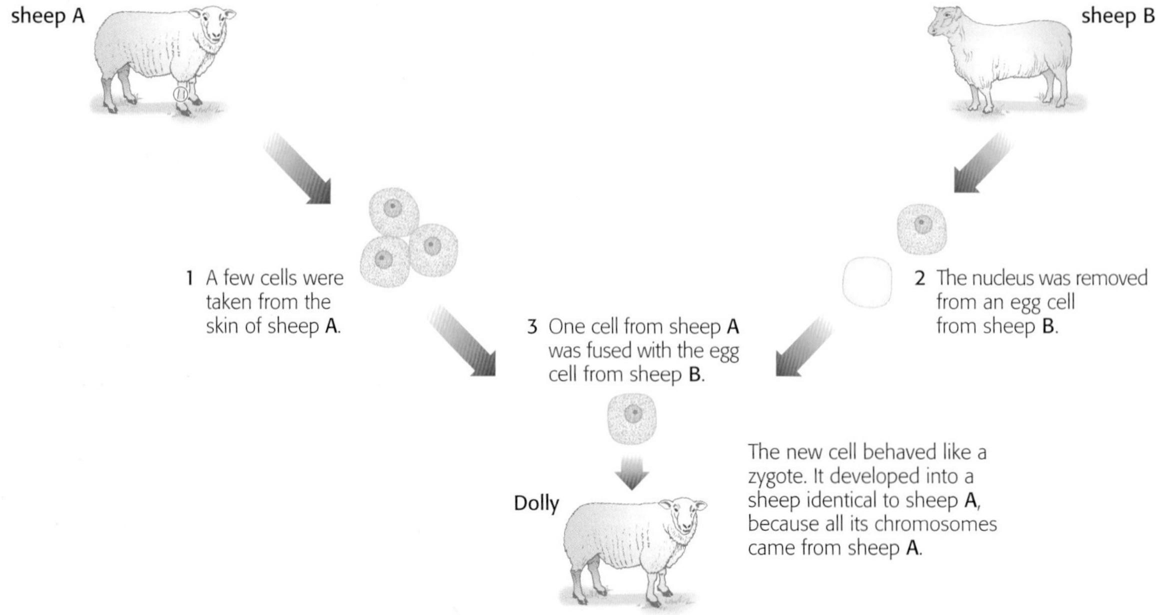

sheep A

sheep B

1 A few cells were taken from the skin of sheep **A**.

2 The nucleus was removed from an egg cell from sheep **B**.

3 One cell from sheep **A** was fused with the egg cell from sheep **B**.

Dolly

The new cell behaved like a zygote. It developed into a sheep identical to sheep **A**, because all its chromosomes came from sheep **A**.

This is nowhere near as easy as it sounds. The scientists fused 277 body cells and egg cells together, but only 29 developed into embryos. Then 28 of these died before they were born. Scientists are getting better at the techniques so their success rate is improving, but it is still an expensive and laborious process.

There is also some concern about how healthy cloned mammals are. Dolly died young – most sheep live almost twice as long. This seems to be happening with many of the other cloned mammals that have been produced since then. Some researchers think these are 'teething problems' and they will eventually be able to produce completely healthy clones. Others are not so sure.

Why do people want to clone mammals? We've already seen one reason – if they have a successful racehorse they might want to make another just like it. A farmer might want to reproduce from a fine meat-producing bullock (castrated bull). Someone might want to clone a dog that they love. You can probably think of some other reasons.

Human cloning

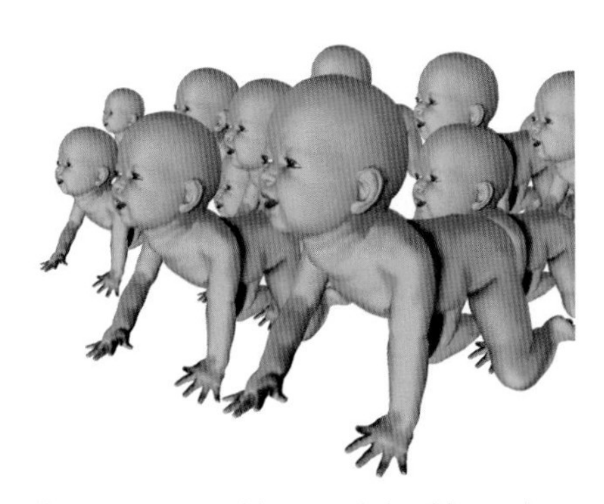

Figure 2.16 Could we, and should we, clone humans?

So far, no one has cloned a human being (although at least one religious sect has claimed to have done it). There is a ban on doing this. Most people feel that it is simply a wrong thing to do. The idea of producing genetically identical copies of a person is just unacceptable to many of us.

But cloning *parts* of humans might be more acceptable to many people. This is allowed and has medical benefits. For example, transplanting an organ such as a liver into a person is difficult, because the recipient's immune system recognises that the liver is foreign, and attacks it. The transplanted liver is often rejected. But if we could take cells from someone with a damaged liver and get the cells to clone themselves to produce a new liver, we would completely get round this problem. The cells in the new liver would have identical genes to the cells in the rest of the body, so the body will readily accept it.

This technology hasn't got very far yet. In 2002, scientists did manage to grow an organ that was a bit like a kidney, using cells taken from a cow, and transplanted it back into the same cow. But we can't yet do it for humans.

H

Another use of human cloning, which is very controversial, is the cloning of human embryo cells for medical research. This is currently allowed in the UK, as long as the embryos used are not allowed to live beyond 14 days. The idea is that cells, called stem cells, from an early-stage embryo can be made to develop into various types of body tissue that could provide effective treatment for conditions such as diabetes, Parkinson's disease and Alzheimer's disease. As with the cloned organs, the ultimate aim would be to produce tissues that are an exact match for the patient. This technology is at an even earlier stage – the first embryonic cells were cloned in the UK in May 2005 – and there is much opposition to this use of embryos.

Designer babies

People who have difficulty in conceiving a child may be given fertility treatment (page 57). Sometimes, this involves collecting eggs from the woman and sperm from the man, and putting them together in a Petri dish so that fertilisation can take place. The zygotes are allowed to divide for three days, until a tiny embryo is formed. One of these is placed in the woman's uterus where, with a little luck, it may grow into a fetus in the normal way. This technique is called 'in vitro fertilisation' or IVF.

Now that we know more about our genome, some new possibilities have opened up, ones that need a lot of careful thought before we can decide how to use them.

For example, say a couple both have the alleles Ff for cystic fibrosis. Each time they have a child, there is a chance that it will inherit an f allele from both its father and its mother. If so, it will have cystic fibrosis. If IVF is used, then it would be possible to screen the embryos to find out which ones have the f allele. These could be discarded, and only an embryo with two F alleles would be put into the mother's uterus. Is that morally acceptable? People's views differ on this.

There are even more difficult questions to answer. In 2004, a story about 'designer babies' made the headlines. A couple had a child, Charlie, with a rare inherited blood disorder. His only chance of a normal life was if he had a sibling with cells very like his own, so that he could be given a transplant of cells from the sibling to replace his own faulty blood cells. The couple wanted to use IVF, and to choose an embryo with cells similar to Charlie's. They were denied the right to do this in the UK. They therefore went to the USA, and had it done there. The new 'designer baby' was born, and Charlie had his transplant.

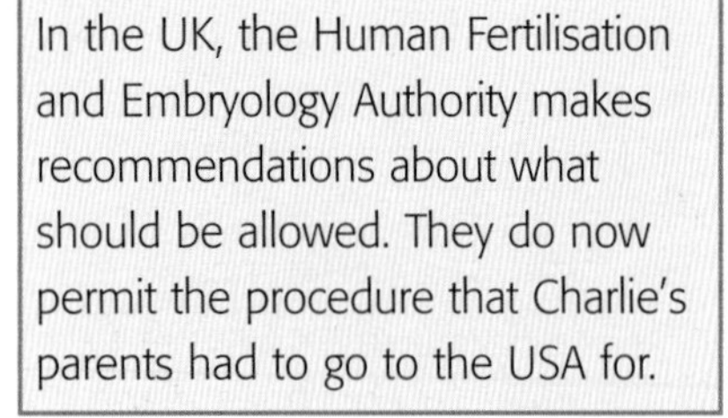
In the UK, the Human Fertilisation and Embryology Authority makes recommendations about what should be allowed. They do now permit the procedure that Charlie's parents had to go to the USA for.

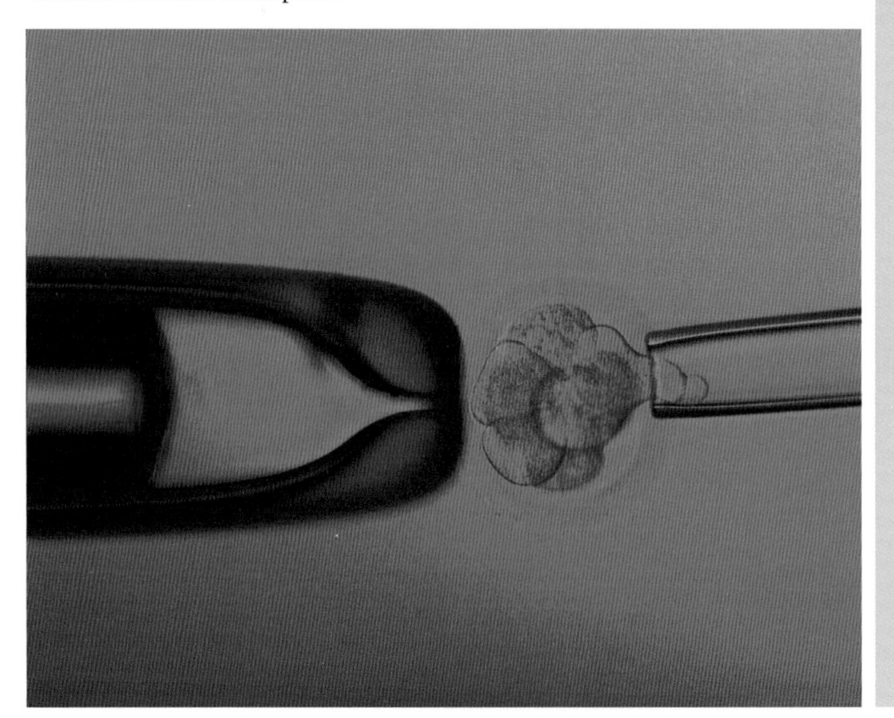

Figure 2.17 This embyro, produced by IVF, is having one of its cells removed for genetic screening

Transgenic animals

Another way in which we are using our knowledge of the human genome is in putting human genes into animals, so that these animals can then make proteins that are useful to us. An animal that has genes from another animal (such as human genes) in its cells is called a **transgenic animal**.

For example, in Iowa (USA) in 2003, a gene for making antibodies was taken from human cells and placed into cow embryos. The embryos were placed into the uteruses of cows, where they developed into calves and eventually grew into cows in the normal way.

Figure 2.18 Transgenic cows don't look different or behave differently from 'normal' cows

When these cows are injected with weakened strains of disease-causing organisms such as anthrax bacteria, their cells respond by making antibodies against them. Because these antibodies are effectively human antibodies, they could be used to help a person's immune system to resist the disease. The scientists who have done this work think that it could be useful in the event of bioterrorism – for example if large numbers of people were exposed to anthrax. Large quantities of antibodies could be produced in this way, which could be injected into people at risk, combating the anthrax bacteria before the human immune system could make its own antibodies.

Another possibility is to introduce genes to cows that would alter their milk. We need to keep an eye on the amount of fat in our diet, as high-fat diets can increase the risk of developing heart disease. It could be possible to introduce genes into cows that cause them to produce low-fat milk.

Although several projects like this have been carried out, so far none of the products have been licensed for human consumption.

People hold widely differing views on the ethics – rights and wrongs – of producing transgenic animals. Some see this kind of technology as having huge potential to help to keep people healthy and cure diseases that are currently incurable. They can't see a downside – the cows are healthy and no different from any other cows. Others are horrified. They instinctively dislike the idea of having cows with human genes in them. Some people are worried because they don't understand the science and imagine the cows developing into Frankenstein-like monsters.

Summary

- Cloning means producing genetically identical organisms.
- Cloning mammals has proved to be difficult, but it is now happening on a small scale.
- No-one has cloned a human yet. Scientists are working on trying to clone human body organs and tissues, so that they can be used in transplants.
- Cloning mammals raises social and ethical concerns. For example, cloned mammals are often not healthy and die young. There are arguments that it is wrong for us to produce them.

H
- When in vitro fertilisation (IVF) is used, it is possible to choose an embryo which does not have harmful alleles, or which does have alleles that produce characteristics that the parents want their baby to have. There are strict rules on the circumstances in which this can be done.
- Genes can be transferred from one animal to another. The animal with the 'foreign' genes is said to be transgenic. For example, human genes for making antibodies can be transferred into cows, which then make human antibodies.

Questions

2.12 Suggest how cloning might help to conserve rare species of animals. Research an example of how this has been done, and write a brief description of it.

2.13 Write a paragraph arguing the case for allowing scientists to clone mammals. Then write another paragraph arguing that this should not be allowed.

H

2.14 Explain how in vitro fertilisation can be used to produce a 'designer baby'.

2.15 What is a 'transgenic animal'? Research an example of the production of a transgenic animal other than a cow, and discuss whether this should or should not have been allowed.

3 Electrical and chemical signals

A The central nervous system

Jane Lapotaire is a highly successful and highly regarded actress, on both stage and screen. On a cold January day in 2000, she was about to do what she had often done before – conduct an all-day Shakespeare master class with 16- to 18-year olds in a school.

She'd been feeling tired for days. She knew she had been overdoing things, but that wasn't unusual. Acting is an all-or-nothing job; if you are out on stage or in front of a camera, then nothing less than your best will do. Jane had been getting headaches for years, and was accustomed to taking painkillers with her everywhere she went. Today was no different. She'd woken feeling awful, with aches and pains seemingly all over. She did what she always did – took some painkillers and just 'got on with it'.

So, at the beginning of the school day, she found herself standing in the school gym, surrounded by excited young faces, listening as the headmaster introduced her to the class. She writes:

Figure 3.1 Jane Lapotaire

> I stop listening to his voice as I become aware slowly that the world has gone awry. Nothing is solid. I think I am going to faint. I think, My God, I am, I'm going to faint. I don't faint. I have never fainted in my life. Get the water, I think. I move, I hope, towards the glass of water. Waves tip the floor from under my feet, ripple the walls and faces in front of my eyes, undulate along my spine and engulf me. Time goes awry.

from 'Time out of Mind', Jane Lapotaire, Virago

Jane had had a subarachnoid haemorrhage. The brain is surrounded by three protective membranes, including one called the arachnoid ('spidery') layer. In Jane's head, a blood vessel had burst and filled the space surrounding the brain with blood.

Emergency surgery saved her life. She remained in hospital for five weeks, and even then was unable to look after herself for months. Like most people who survive such a crisis, she found her personality altered, and had to battle against fatigue and depression for several years.

The spinal cord

The brain and the **spinal cord** make up the **central nervous system** (see Figure 3.2 on the next page). The spinal cord begins at the base of the brain, and extends all the way down through the backbone. The vertebrae have holes in them, forming a protective channel inside which the spinal cord lies.

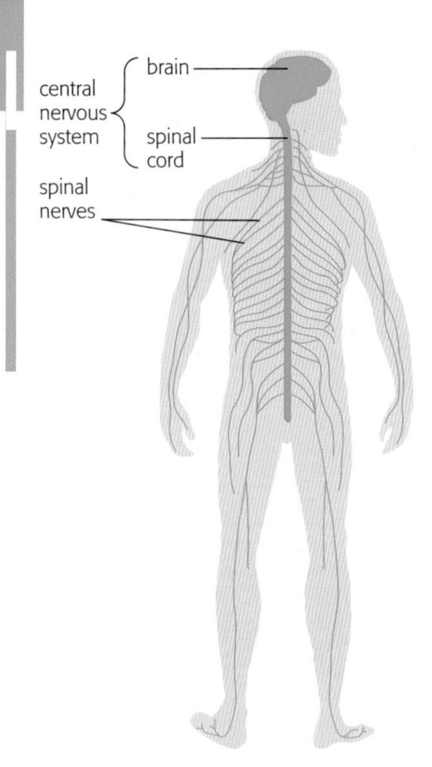

Figure 3.2 The central nervous system and main nerves

Nerves lead out from the brain and spinal cord to the rest of the body. Each nerve contains hundreds or thousands of **nerve fibres**. Some are carrying electrical impulses inwards from the sense organs, while others carry impulses outwards from the central nervous system to muscles and glands.

Figure 3.3 shows a slice across the spinal cord. The cord is about 11 to 12 mm in diameter. In the centre is a canal filled with fluid, called **cerebro-spinal fluid**. This is surrounded by an H-shaped area called **grey matter**, and the rest of it is made up of **white matter**. All of this is surrounded by protective membranes called **meninges**, which enclose a layer of cerebro-spinal fluid. The fluid helps to cushion the spinal cord and to keep its chemical environment correct.

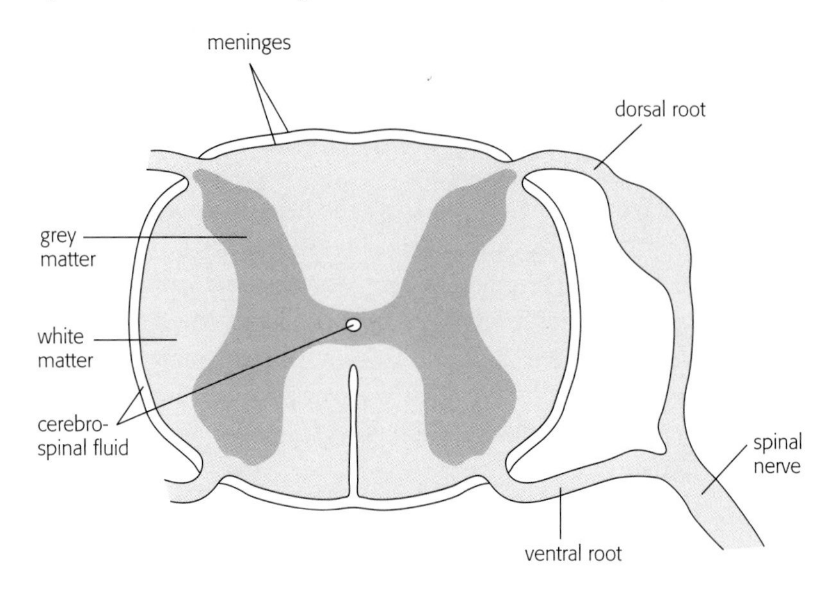

Figure 3.3 Transverse section through the spinal cord

Leading out from the spinal cord are pairs of 'horns', called **dorsal roots** and **ventral roots**, which join together to form **spinal nerves**. They emerge from between the vertebrae all down the length of the back.

If the vertebrae shift, or get squashed closer together, the spinal nerves can get pinched and cause severe pain.

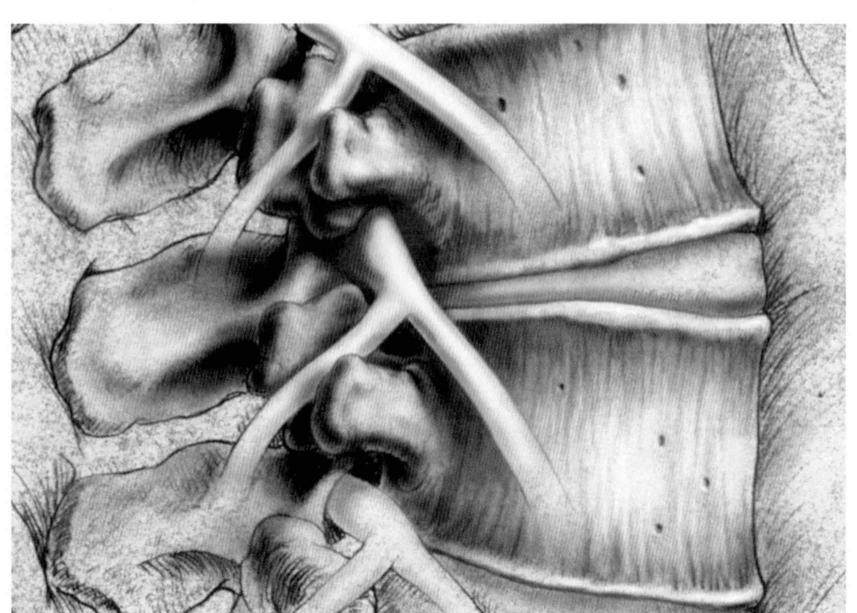

Figure 3.4 Part of a vertebral column, showing pinched nerves due to injury

The brain

The brain is the most complex and least well understood organ in our bodies. It contains around 100 000 000 000 (100 billion) nerve cells, each one connecting up with hundreds of others. It controls the functions that go on in our bodies all the time, which we are largely unaware of, including the beating of the heart, breathing, and the activity of the digestive system. It coordinates movements of muscles during walking, catching a ball, dancing or singing. It holds our memories. It gives us our personalities.

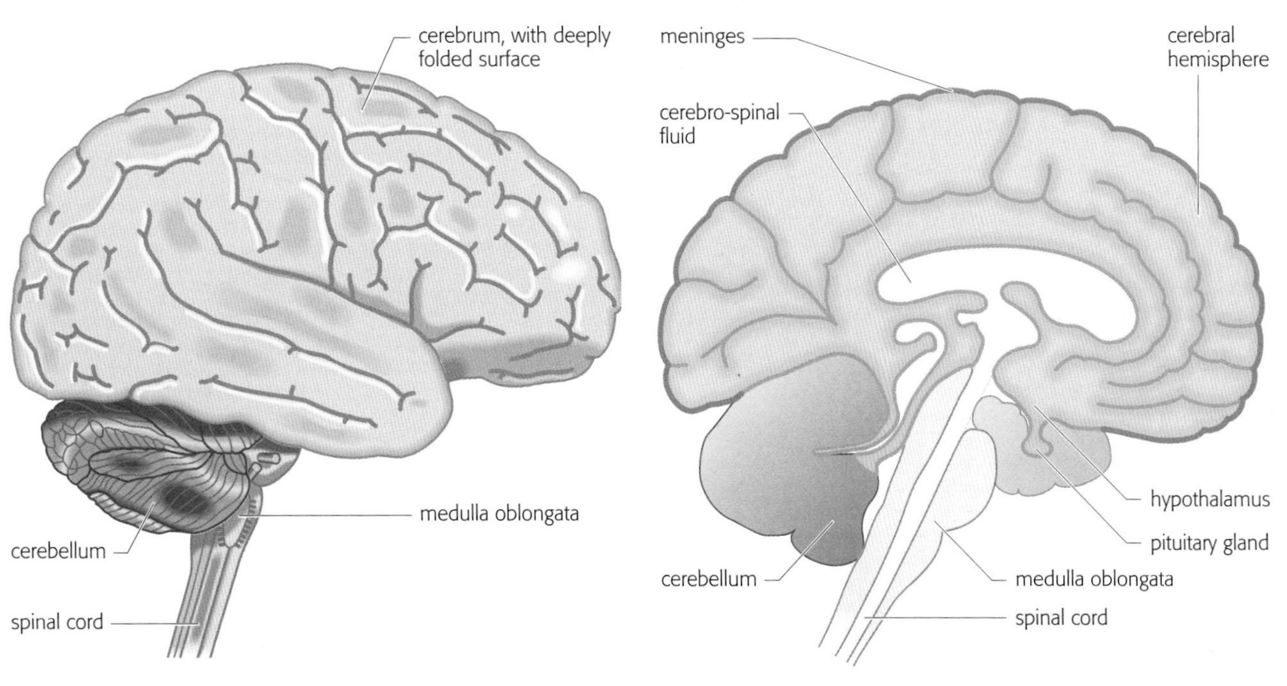

Figure 3.5 The human brain

Figure 3.6 Vertical section through the brain

Figure 3.5 shows a human brain. Most of it is made up of a creamy-white area called the **cerebrum**. This has deep folds all over its surface. It has two halves, called the left and the right **cerebral hemispheres**. This part of the brain is where our conscious thinking and decision-making takes place. An area near the front is responsible for our personality. Other areas deal with language, spatial ability and emotions. The cerebrum receives inputs from all the sense organs, and it is able to integrate these with one another and with your stored memories to help you to make sense of the world and to respond appropriately to it.

Dwarfed by the cerebrum and almost hidden by it is the **cerebellum**. This area controls movement and posture. It receives inputs from the eyes and ears. It is when these two inputs don't match that you can get travel-sick – your balance organs in your ears are saying you are going up and down and side to side, but your eyes are not sending in the same information.

Lastly, just before the brain merges into the spinal cord, is the **medulla oblongata**. This controls your heartbeat and breathing movements, as well as movements of the alimentary canal.

Figure 3.6 shows a section through the brain. Although it is much more complicated than the spinal cord, you can see that there is a lot in common. Like the spinal cord, the brain has spaces inside it

The meninges sometimes become infected with viruses or bacteria, and this causes meningitis. If not diagnosed and treated rapidly, this can be fatal.

filled with cerebro-spinal fluid. It is also surrounded by meninges, which enclose more cerebro-spinal fluid. Head injuries or haemorrhages (like Jane Lapotaire's) can cause bleeding into the space between the meninges and the brain, and the pressure this exerts on the brain can cause unconsciousness and sometimes permanent brain damage.

Neurones

Figure 3.7 shows the structure of a neurone. This is a **motor neurone** – one that carries electrical impulses outwards from the central nervous system to a muscle or a gland.

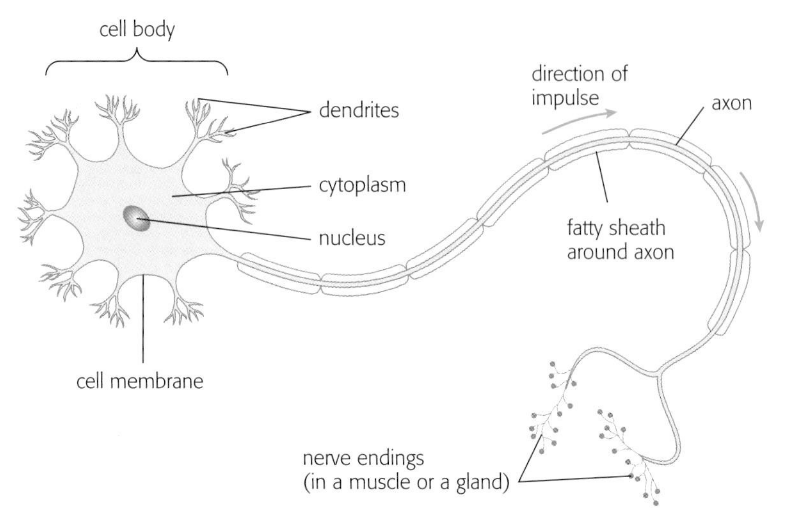

Figure 3.7 A motor neurone

The axon of a motor neurone can be over a metre long, so neurones are by far the longest cells in your body.

A neurone is a single cell. It has a cell body where the nucleus is found. In the motor neurone, little strands of cytoplasm, called **dendrites**, reach out from the cell body and receive electrical impulses from other neurones. These electrical signals flow rapidly along the membrane of the neurone, around the cell body and then sweep along the **axon**.

Figure 3.8 shows the structure of a **sensory neurone**. This carries electrical impulses from a sense organ towards the central nervous system. It has the same basic structure as a motor neurone, but in the sensory neurone one long strand of cytoplasm carries the impulses towards the cell body rather than lots of short ones from the dendrites.

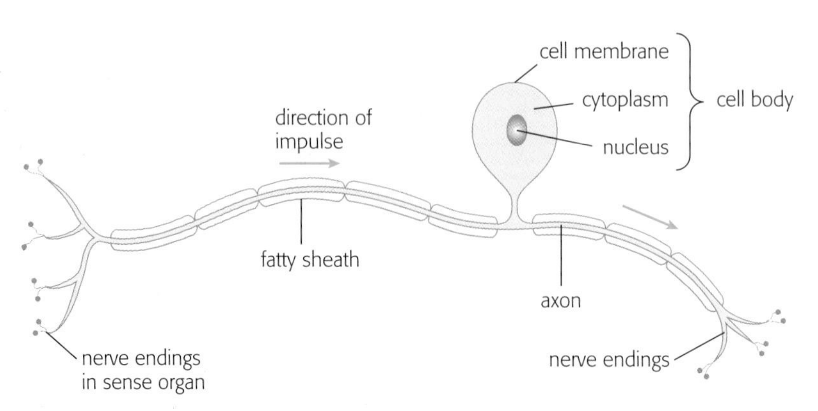

Figure 3.8 A sensory neurone

The path taken by a nerve impulse

When a sense organ is stimulated – let's say you accidentally put your hand down on a drawing pin – an electrical impulse sweeps along the sensory neurone towards the central nervous system. Figure 3.9 shows the pathway this impulse takes. You can see that the cell body of the sensory neurone is inside a little swelling in the dorsal root. From here, an axon runs into the spinal cord.

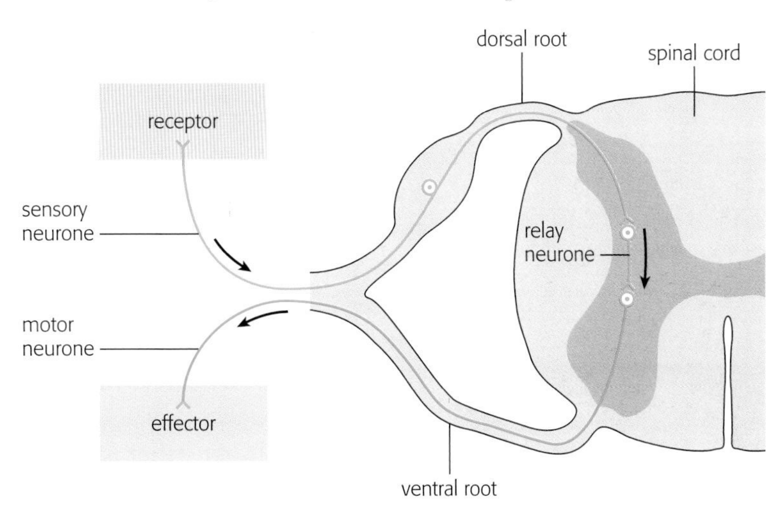

Figure 3.9 A reflex arc

Inside the spinal cord are millions of small neurones called **relay neurones**. They have lots of short dendrites that make connections with many others. The impulse that swept along the sensory neurone is transferred to a relay neurone in the spinal cord, and this in turn transfers it to a motor neurone. You can see that the cell body of the motor neurone is inside the spinal cord. Its axon leads out through the ventral root, and eventually reaches a muscle. Here, the electrical impulse causes the muscle to contract and pull your hand away from the pin. The whole pathway of the nerve impulse is called a **reflex arc**.

In the time it's taken you to read these last two paragraphs, a nerve impulse could have swept along this pathway many, many times. The time it takes between the receptors picking up the information and your muscles reacting to it is your **reaction time**. Most people have reaction times of around 0.25 to 0.30 seconds. Sprinters work hard to improve their reaction times, because a fast start is a crucial part of winning a race that may last less than 10 seconds in total. Their starting blocks contain electronic sensors that measure the time between the gun going off and the first push of their feet. If this is less than 100 milliseconds (0.1 s), then they must have moved before they heard the gun, and are said to have made a false start.

Synapses

If you look carefully at Figure 3.9, you will see that neither the sensory neurone nor the motor neurone quite make contact with the relay neurone. There is a tiny gap between them, called a **synapse**.

The electrical signal can't jump this gap. Instead, when the signal arrives at the end of the sensory neurone's axon, a tiny amount of a chemical called a **transmitter substance** is released. This diffuses across the gap and is picked up by the relay neurone. This starts up an impulse in the relay neurone, which again is carried across the synapse to the motor neurone by a transmitter substance.

Strokes

Jane Lapotaire's brain injury was caused by a blood vessel bursting close to her brain. This is a kind of **stroke**. A stroke is sudden damage to the brain caused by a problem with its blood supply. Jane Lapotaire's kind of stroke is a relatively uncommon one. Most strokes are caused by blockages in a blood vessel supplying the brain. The brain cells don't get oxygen or nutrients, and they die.

What causes such a blockage? It normally happens because blood vessel walls have become damaged over a long period of time. The inside of an artery should be smooth so that blood can flow easily through it. Its walls should be stretchy and strong, so that they can be pushed outwards as the blood is forced through them every time the heart contracts, and then recoil to their normal size when the heart relaxes. You can feel your artery walls doing this when you feel your pulse.

Living with a partner who smokes, even if you don't smoke yourself, increases your risk of having a stroke by six times.

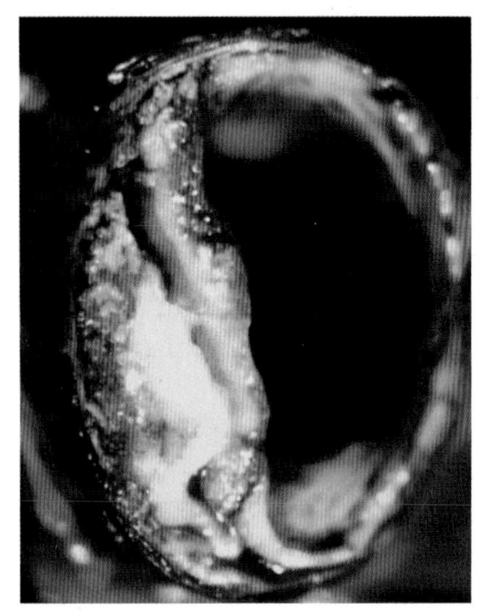

Figure 3.10 Plaques (yellow) in an artery wall

Artery walls get stiffer as you get older – it is a normal part of ageing. But in some people this process starts early and progresses rapidly. This is often caused by a build-up of **plaques** in the artery wall. These partly block the artery, so there is less space for blood to flow through. They also slow down the blood and increase the likelihood of a blood clot forming.

The blood clot can break away and be carried to other parts of the body in the blood. This can cause a problem if it gets stuck in a small blood vessel. If this is a vessel carrying blood to the brain, then the person suffers a stroke.

The effect of a stroke depends on whereabouts it is in the brain. The left side of the brain controls the right side of the body and vice versa. So if someone has a stroke that affects the left side of their brain, they may lose movement in their right leg, arm or face muscles. The ability to speak, or to understand language may be damaged. Memory is almost always affected.

Figure 3.11 This stroke patient is having help from a speech therapist

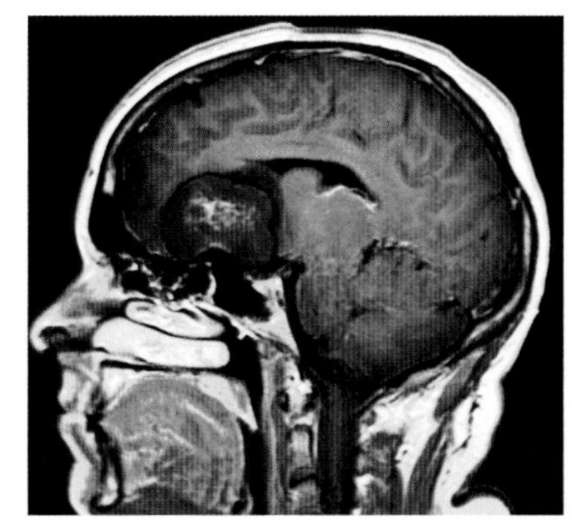

Figure 3.12 Scan of a brain showing a tumour (orange)

Brain tumours

Most of the cells in your body don't divide, once you have finished growing. But sometimes cells begin to divide when they shouldn't. They can form a mass of cells called a **tumour** (see Figure 2.7, page 27).

Tumours can form in the brain. The effect that they have depends on where the tumour is. Many brain tumours can be removed by an operation, and the patient can make a full recovery. But others may be in a part of the brain that a surgeon can't reach. If the tumour keeps growing, it is likely that the brain will eventually be so badly damaged that the person will die.

Parkinson's disease

Figure 3.13 Despite having Parkinson's disease, diagnosed in 1991, Michael J. Fox was able to take part in a celebrity ice-hockey match in 2000 thanks to drugs that control the condition

About one in 100 people in Britain have Parkinson's disease. It is most common in people over the age of 50. A person with Parkinson's disease loses control of their muscles, to a lesser or greater extent – the illness develops gradually over several years. At first the person may be able to move slowly, but eventually their legs simply don't respond when they try to walk. This happens because a little group of brain cells die. These cells should be producing a transmitter substance called **dopamine**. Without dopamine, nerve impulses don't get sent to the muscles.

The disease is treated by giving people a substance called **levodopa**. Cells in the brain convert it to dopamine. The dose has to be carefully controlled. Too little and the muscles don't work. Too much and they can twitch and move uncontrollably.

Epilepsy

We've seen that the brain contains around 100 billion neurones. All the time, even when you are sleeping, electrical impulses are sweeping around in the brain, passing from neurone to neurone.

In some people, these impulses can begin to behave wildly. It is as though all the neurones in one part of the brain become active at once, sending impulses off to every other neurone, all at the same time. The effect this has depends on the part of the brain where it is happening. In the condition called **grand mal epilepsy**, there are uncoordinated muscle contractions that cause convulsions. At first, the body becomes rigid and the person falls to the ground. They may bite their tongue as the jaw muscles contract, and stop breathing as the breathing muscles contract. After a few seconds or minutes, the contractions stop. Gradually the person regains consciousness. They may feel disorientated and confused, and often feel battered because of the strong movements their body has been making. The whole event is called a **seizure**.

H

We don't really know why some people have epilepsy. It can run in families, which means that genes probably have something to do with it. It can follow a head injury.

There are quite a lot of different drugs that people can take to stop them having a seizure, and for most people these work well. Sometimes, if a person has frequent seizures and drugs don't work for them, they may be offered a brain operation. This is most often done if the seizure starts off in a part of the cerebrum called the temporal lobes. If only one temporal lobe seems to be causing the trouble, then this can be surgically removed. It is a serious and difficult operation, but it can have enormously beneficial effects on a person's life if it goes well.

Summary

- The central nervous system is made up of the brain and the spinal cord.
- Nerve cells (neurones) carry electrical impulses. Sensory, relay and motor neurones carry impulses from sense organs, through the central nervous system and then to a muscle or gland.
- Tiny gaps between neurones are called synapses. The impulse is carried across a synapse by a transmitter substance.
- Both the brain and the spinal cord contain cerebro-spinal fluid and are surrounded by membranes called meninges.
- The largest part of the human brain is the cerebrum, which controls conscious actions as well as emotions, language and personality.
- The cerebellum controls coordination of movement and posture.
- The medulla oblongata controls heart beat and breathing movements.
- H Strokes, brain tumours, Parkinson's disease and grand mal epilepsy disrupt the functioning of the brain.

Questions

3.1 Make a simple drawing of the human brain. (You could use Figure 3.5 to give you the outline.) Label the parts that have each of the following functions. Some parts will have more than one label.
- understanding language
- controlling the rate of heart beat
- controlling movement
- conscious thought
- personality

3.2 Explain how the structure of neurones is adapted for their functions.

3.3 Find out more about the different areas of the cerebral hemispheres. The two sides are not the same – how do their functions differ?

H

3.4 It may become possible to treat Parkinson's disease by using stem cells (cells that can divide and turn into different kinds of cells), which could be transplanted into the brain. Find out more about this treatment. Where might the stem cells come from? (See page 35.) Why are some people opposed to their use?

B Responding to stimuli

Even if you've never played cricket, you might be able to imagine what it is like to have a hard, round object coming straight at your head at a speed of more than 90 miles an hour! Practically everyone would respond by ducking to get out of the way.

But if you are a batsman playing cricket then you might not do that. You are there to get runs, so you might have set your brain to 'override' mode – you are going to hit the ball, not escape from it.

That's what happened to one of South Africa's most experienced batsmen during a test match against Pakistan. Gary Kirsten was facing bowling from Shoiab Akthar, who had recently bowled a ball at a world record speed of more than 100 miles per hour. Kirsten misjudged things, and didn't let his reflexes get him out of the way. The ball hit him beneath his left eye and the blow knocked him to the ground. He received a gash just below his left eye and, even though he was wearing a protective helmet, two fractures to the bone in the bridge of his nose.

Figure 3.14 Response to a ball bowled at you must be fast

Receptors

We have receptors in practically every part of the body. A **receptor** is a cell that picks up stimuli from around it and sets up a nerve impulse. The stimuli might come from outside the body (for example light, sound, heat, pressure) or inside it (too much carbon dioxide in the blood, too little water, internal temperature getting too high).

For example, if light enters your eye it will fall onto the retina – the sensitive layer at the back of the eye. The retina contains special receptor cells called rods and cones. When light falls onto them, they send electrical impulses along the optic nerve to the brain.

Rods are able to respond to dim light, but don't allow us to see in colour. Cones give us colour vision, but need brighter light than rods in order to respond.

Some receptors are part of a **sense organ**. Rods and cones are part of the eye. The ear is another sense organ; it contains receptor cells that are sensitive to sound and others that help us with balance. The skin could be classified as a sense organ. It contains receptors that are sensitive to temperature, pain, and pressure.

Figure 3.15 A TENS machine in use

Endorphins are transmitter substances found in the brain.

TENS

TENS stands for transcutaneous electrical nerve stimulation. 'Transcutaneous' means 'across the skin'. A TENS machine produces low-intensity electrical impulses that stimulate receptors and sensory neurones in and beneath the skin.

TENS can sometimes produce relief from pain. Some women use it during labour, and find that it reduces pain significantly.

We don't really know why TENS works. One theory is that it stimulates sensory neurones that don't carry 'pain' signals, so the brain gets only 'non-pain' impulses instead. Another theory is that it stimulates the brain to produce natural painkillers called endorphins.

Reflex actions and voluntary actions

Some of our actions are automatic. They happen quickly and we don't think about them. Ducking to get out of the way of an object moving quickly towards your head is a reflex action. So is pulling your hand away from a pin. The part of the body that responds to the stimulus is called an **effector**.

Figure 3.16 shows the pathway taken by a nerve impulse during a reflex action. It doesn't involve the brain, so you are acting without thinking. But you can see that the relay neurone makes contact with other neurones that do carry the impulse up to your brain. By the time the impulse gets there, your muscle has probably already contracted, so you've already pulled your hand away from the pin. But at least you get to know about it eventually, as the information does belatedly get to your brain. You might check next time for upturned drawing pins on the table!

> Reflex actions are very good at getting us out of trouble. They allow us to respond much more quickly to danger than we would do if we stopped and thought about what to do.

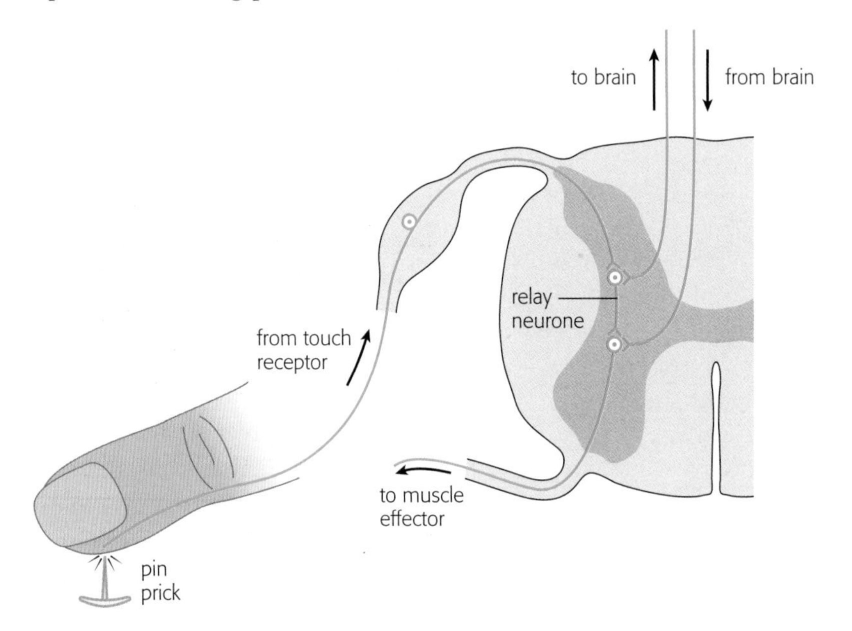

Figure 3.16 Information about a reflex action is passed to the brain

Other actions that we make are not automatic. They are actions we decide on. Talking, walking, reading, cooking, eating – these are all **voluntary** actions. For these, the impulses that make your muscles move originate in the cerebral hemispheres of your brain. Voluntary actions can be much more complex than reflex actions and much more varied. You don't always behave in the same way just because you get the same stimulus. You make up your mind what to do.

Voluntary actions can sometimes override reflex ones. We've seen, for example, that impulses from the cerebral hemispheres can inhibit the impulses that would normally make you duck away from a fast-moving ball.

The iris reflex

The iris reflex is a good example of a reflex action over which we have no voluntary control at all.

The iris is the coloured part of the eye. The dark area in the centre of the iris is the **pupil**. The pupil is a hole through which light passes on its way towards the retina.

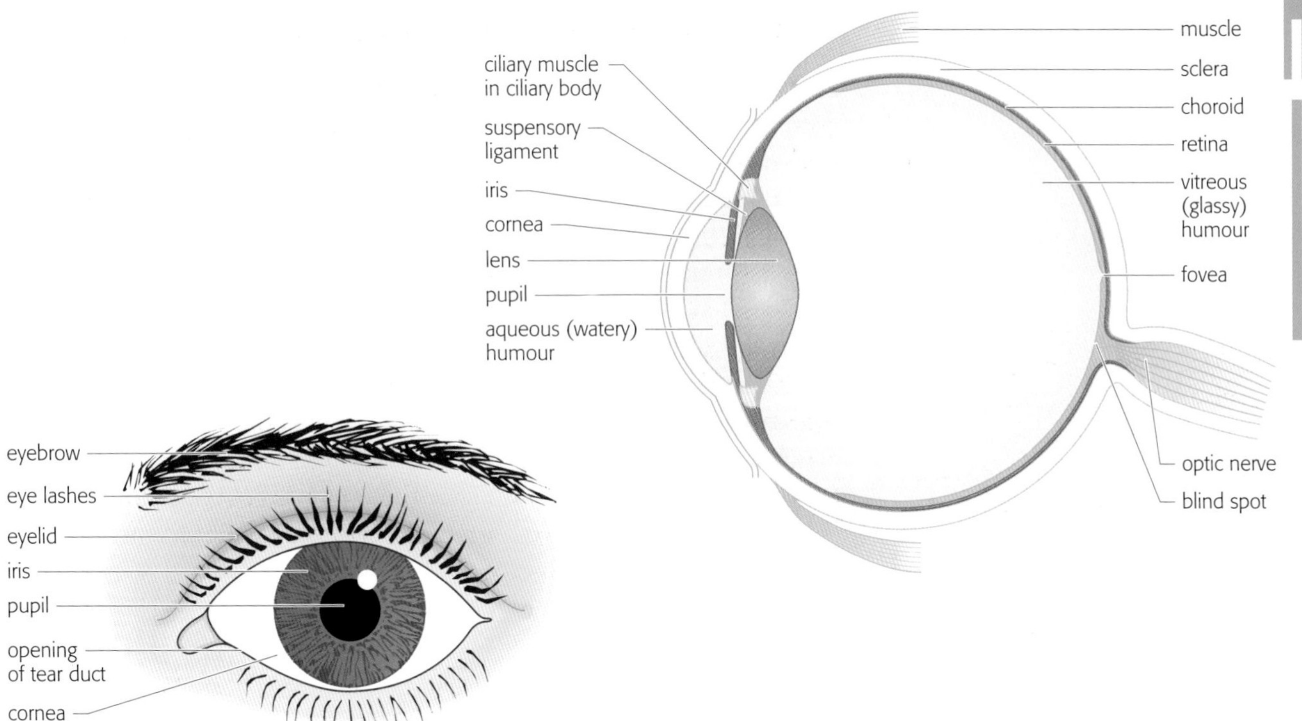

Figure 3.17 The structure of the eye: front view and section

The rods and cones in the retina respond to light by starting off impulses in neurones. The impulses pass along the optic nerve to the brain, where we interpret them as images. Too much light entering the eye can damage these sensitive cells. The function of the iris is to stop this happening.

You can see the iris reflex quite easily. Next time you are somewhere with a mirror, put all the lights on until it is as bright as possible. Then put something dark over your head. Keep your eyes open. After a minute or so, quickly uncover your eyes and look at them in the mirror. You should see that the pupil is wide open just after you uncover your eyes, and then gets much smaller in the bright light.

What is happening? The iris contains muscles – two sets of them, one set arranged in concentric rings and the other set radiating outwards. When you move from dim light to bright light, the receptors in the retina send more impulses to the brain than usual. The brain responds – automatically, not because you have thought about it – by sending impulses along a motor neurone to the circular muscles in the iris. These contract (get shorter), pulling the iris inwards to cover more of the pupil. Less light can get through, and the chance of damage to the retina is reduced.

Figure 3.18 The iris reflex

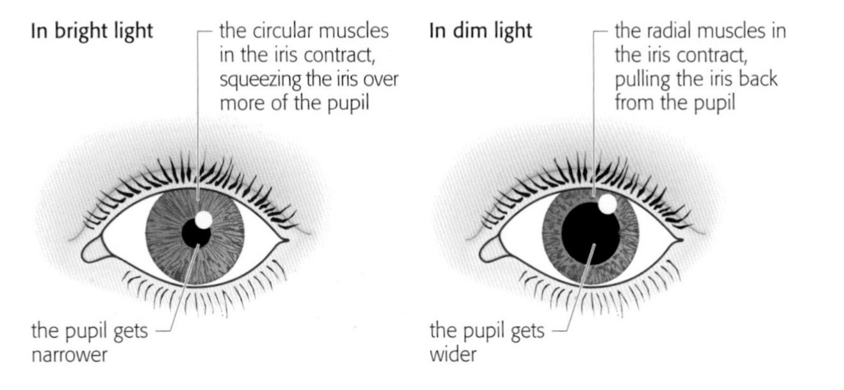

Accommodation

As light enters the eye, it is **refracted** (bent) first by the **cornea** and then by the **lens**. These bend the rays exactly the right amount to bring them to a sharp focus on the retina. If they don't manage to do this, then your brain 'sees' a blurred image instead of a clear one.

Not all light rays need refracting by the same amount. Light rays from a distant object are almost parallel as they reach your eye, and they don't need bending very much. But light rays from a nearby object are spread outwards. They need bending much more.

The cornea stays the same shape all the time, but the shape of the lens can be altered. It needs to be more rounded to focus the diverging rays from a nearby object. It needs to be flatter to focus the parallel rays from a distant object. The way the eye changes the shape of the lens to focus on objects at different distances is called **accommodation**.

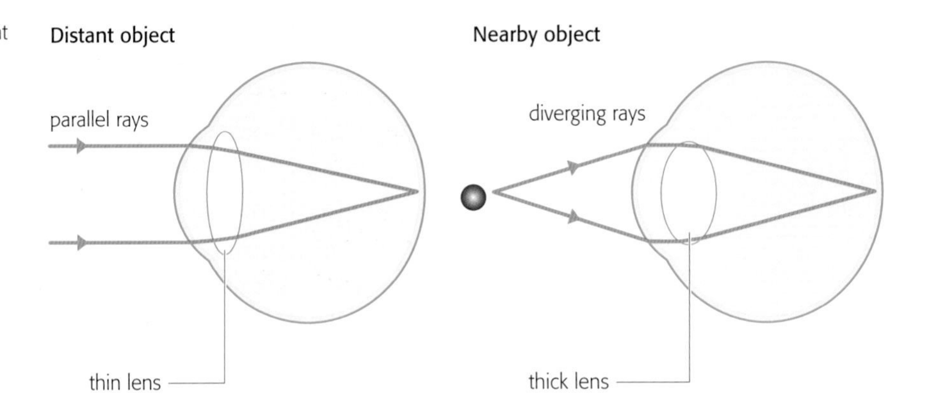

Figure 3.19 Focusing on objects at different distances from the eye

The muscle that brings about the change in shape of the lens is called the **ciliary muscle**. This is a ring of muscle that completely encircles the lens. The lens is attached to the muscle by a ring of ligaments (Figure 3.20).

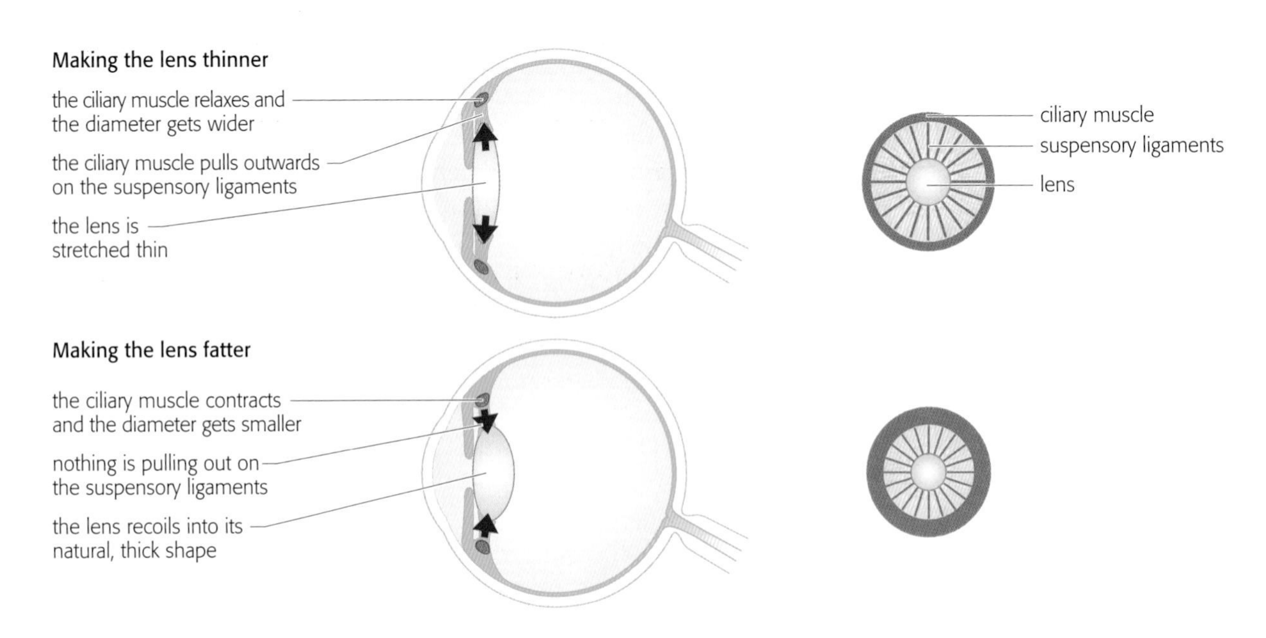

Figure 3.20 How the shape of the eye lens is changed

When the brain receives a fuzzy image of a distant object, a nerve impulse is automatically generated and sent along a motor neurone to the ciliary muscle. This muscle relaxes, making the circle of muscle larger. The muscle pulls on the ligaments, and the ligaments pull on the lens. The lens is pulled thinner.

When you need to focus on a nearby object, the opposite happens. The muscle contracts, so its circle gets smaller. This loosens the tension on the ligaments, which in turn loosens the tension on the lens. The lens springs back into its natural, rounded shape.

Summary

- **A receptor is a cell that picks up a stimulus and generates an electrical impulse in a neurone. Some receptors are part of a sense organ. The stimulus can be either external or internal.**
- **A reflex action is a fast, automatic response to a stimulus. It does not involve conscious thought. Reflex actions are useful for responding rapidly in a dangerous situation.**
- **Examples of reflex actions include the iris reflex, accommodation of the eye, and reacting to get out of the way of a fast-moving object.**

Questions

3.5 Draw an outline of a human body. Then draw label lines to show where you would find receptors for each of these stimuli:

touch light sound movement of the head taste
smell pressure temperature

3.6 Think of one example of a reflex action, other than those mentioned in this section. What is the receptor in this action? What is the effector? Does the reflex action help to safeguard the body?

3.7 Copy and complete this table, to show what happens during accommodation.

Part of eye	Focused on a close object	Focused on a distant object
ciliary muscle	contracted	
suspensory ligaments		
lens		

3.8 People often get mixed up between the iris reflex and accommodation. Why do you think they get confused? Think of a way to help people not to make this mistake.

C Chemical messengers

Debbie was very nervous on the day of the exam. So nervous that she forgot all the usual safety precautions she normally took without thinking. She forgot to test her blood sugar level that morning. She forgot to take a chocolate bar or any glucose tablets in her bag. She felt shaky on the way to school and waiting to go into the exam, but just put it down to nerves.

Once she got started on the exam paper, she felt a bit better. But after a while she realised she was feeling much more tired than she should, a bit sweaty too, and ratty – she felt cross with the examiners for asking so many hard questions. She couldn't concentrate. Eventually it dawned on her – she was hypoglycaemic.

She couldn't believe she had been so stupid. She'd had diabetes all her life, and knew exactly how to keep her blood sugar level right. She should have eaten something sugary ages ago. Now she was getting a bit desperate. If she didn't get her blood glucose level up soon, she could be into a crisis. She had nothing with her, and had to ask the invigilator to find her some glucose or a chocolate bar. A Mars bar did the trick. It took a little while for the sugar to get into her blood, but once it did she felt better almost immediately.

Figure 3.21 The stress of an exam can increase your metabolic rate, so cells need more glucose

Hormones

Nerves are not the only communication system we have in our bodies. We also use chemicals called **hormones**. Hormones are substances made in **endocrine glands**. They are released (secreted) from these glands directly into the blood. They are carried in the **blood plasma** all over the body.

Figure 3.22 The structure and function of the blood

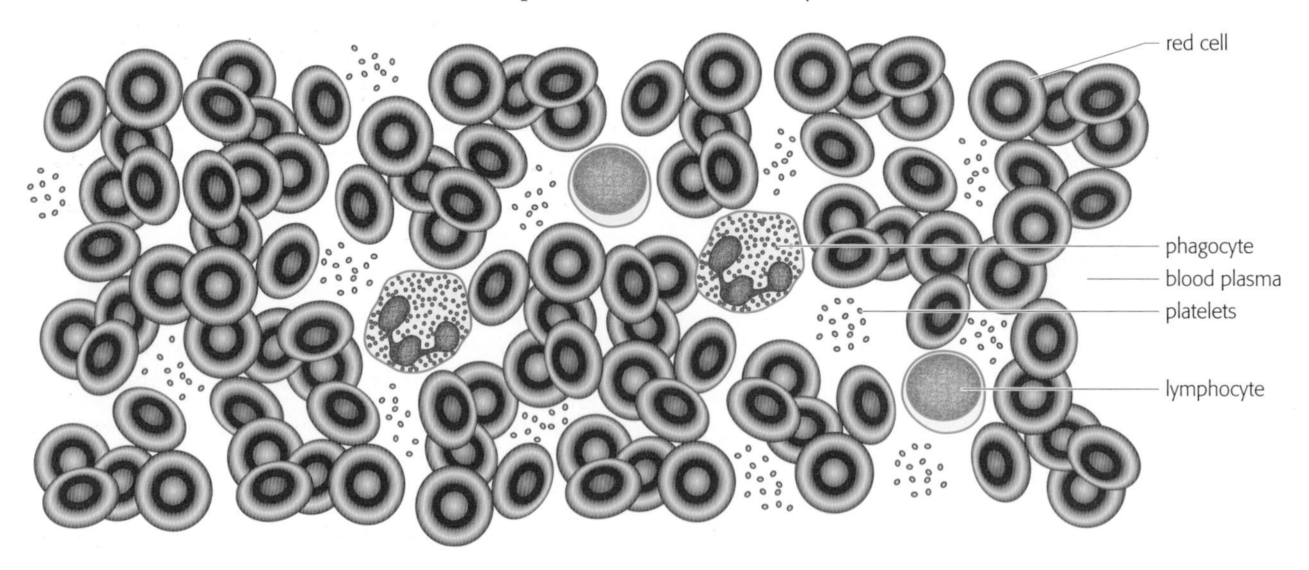

Red blood cells contain the pigment haemoglobin. This transports oxygen from the lungs to all respiring tissue.

Blood plasma is a watery liquid. It transports many different substances in solution. These include glucose, amino acids, hormones, urea and carbon dioxide.

Platelets are tiny fragments of cells. They help in blood clotting, releasing chemicals that cause a soluble protein, fibrinogen, to change to an insoluble one, fibrin. Fibrin forms a web of threads in which red blood cells become entangled.

Phagocytes are white blood cells that engulf and digest pathogens (disease-causing organisms).

Lymphocytes are white blood cells that secrete antibodies to destroy specific pathogens. There are thousands of kinds of lymphocyte, each one responding to one particular pathogen.

Although hormones are carried all over the body, they only affect some organs or cells. These are called their **target organs**.

For example, if you get a fright, two endocrine glands just above each of your kidneys secrete a hormone called adrenaline. Within seconds, it is transported all over your body in your blood. It makes your heart beat faster, makes your pupils open wider, and might even make your hairs stand on end. Its effect on the body helps you to cope with the danger – either to run away from it or stand up to it.

Adrenaline is sometimes called the 'fight or flight' hormone.

Even though the blood transports hormones quickly, it can't do it as fast as an electrical impulse can speed along the axon of a neurone. So hormones take a little bit longer to have their effects than nerve impulses do. Another difference is that hormones last longer – the hormones stay around for a while, whereas a nerve impulse only exists for a fraction of a second. Hormones are eventually broken down by the liver, and their breakdown products are lost from the body in urine.

Insulin

Brain cells are the first to suffer when blood glucose levels drop, because glucose is the only fuel they are able to use.

Your blood contains the sugar **glucose**, dissolved in the blood plasma. Glucose is an essential fuel for many of your body cells. It's important that they get a constant supply of it.

On the other hand, there shouldn't be too much glucose in the blood. If there is, then the glucose concentration in the blood can become greater than that in the cytoplasm inside your cells. This can make water move out of the cells by osmosis. The cells can become dehydrated, and metabolic reactions won't take place as they should.

So it is important that blood glucose concentration should be kept roughly constant. One of the hormones that helps with this is **insulin**.

The **pancreas** contains cells that constantly monitor the concentration of glucose in the blood. If it rises too high, then they release more insulin. The insulin dissolves in the blood plasma. Its target organs are the **liver** and **muscles**.

The arrival of insulin at the liver makes the liver cells remove glucose from the blood. The liver cells string thousands of glucose molecules together to form a long-chain carbohydrate called **glycogen**. Glycogen can be tucked away inside the liver cells as a glucose store.

If the blood glucose level drops too low, then the pancreas secretes less insulin. The liver cells are no longer affected by it, so they stop taking glucose out of the blood. If blood glucose gets very low, then they may have to break down some of their glycogen stores into glucose again, and release it back into the blood.

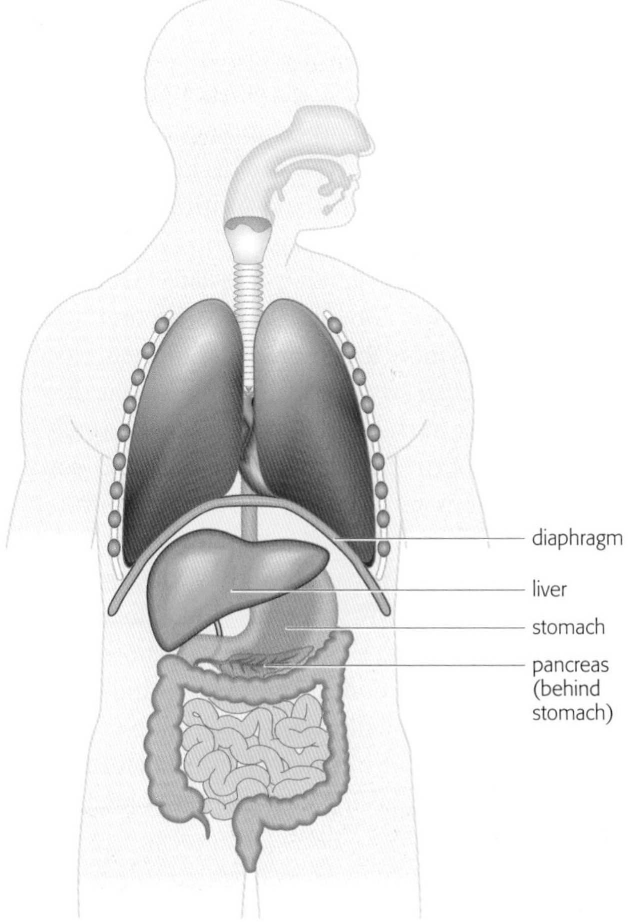

Figure 3.23 The pancreas and the liver are important organs in stabilising blood glucose level

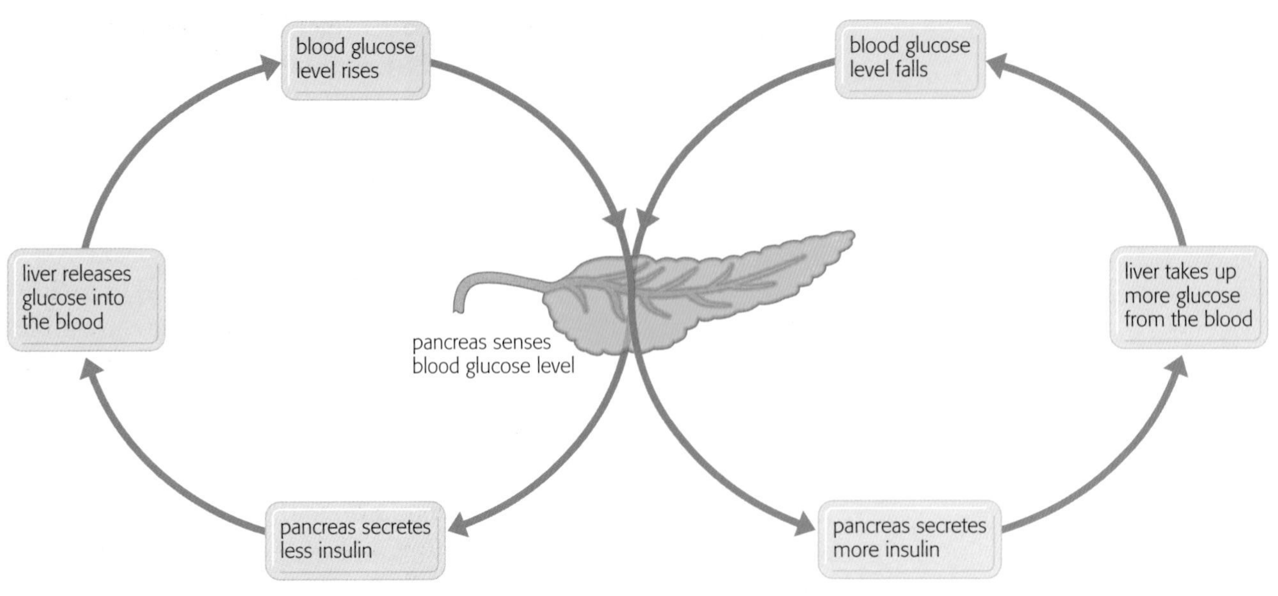

Figure 3.24 How blood glucose level is controlled

Diabetes

Blood glucose concentration always wobbles up and down a little, because neither the pancreas nor the liver can react absolutely instantaneously to every small fluctuation in glucose concentration. In most people, though, it is kept fairly constant. But for some this control system doesn't work.

Debbie (page 52) has had diabetes since she was small. She has Type 1 diabetes. The insulin-making cells in her pancreas don't work. It's thought that Type 1 diabetes is caused when a person's own immune system attacks the insulin-making cells and destroys them.

There's another kind of diabetes, called Type 2. This kind develops later in life. The insulin-making cells are fine, and continue to secrete insulin when they should. But the liver and muscles don't respond to it. The reason for this happening isn't completely understood, but it is known to be very much associated with obesity. It's possible that, if a person regularly overeats and adds more and more sugar to their blood, then the liver and muscle cells get so swamped with insulin that they just stop reacting to it. In the past in the UK, Type 2 diabetes usually developed in a small proportion of the population in middle age, but in recent years it is being seen in more and more young people, and the overall number with the condition is growing rapidly.

No matter which kind of diabetes a person has, they run the risk of damage to their body because of wildly swinging blood glucose concentration. If a lot of sugar gets into the blood over a short period of time, then the blood glucose concentration goes high and stays high. This is **hyperglycaemia**. Usually, the kidneys don't let any glucose go out of the body in urine, but in these circumstances they can't cope. So one of the ways of diagnosing diabetes is to check for glucose in urine.

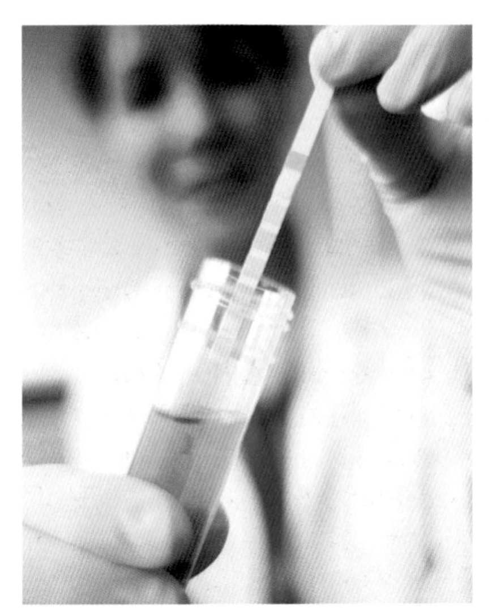

Figure 3.25 Testing for glucose in urine

None of this extra glucose gets stored away as it would in a non-diabetic person. So if, later on, the blood glucose concentration starts to drop, there are no stores to draw on. It just keeps going down. A low blood glucose concentration is called **hypoglycaemia**.

Eventually, the brain cells become so starved of glucose that they stop working. Debbie spotted that this was happening in time. If she'd not done anything, she could have gone into a diabetic coma. This is really dangerous if it isn't dealt with quickly. The treatment is simple – get some glucose into the blood.

Treating diabetes

One day, it may be possible to transplant insulin-making cells into the body of a person with Type 1 diabetes.

At the moment, there is no cure for either type of diabetes.

Everyone who has diabetes needs to take great care over their diet. They need to ensure that not too much glucose gets into the blood, because their body has no control mechanism for bringing the blood glucose concentration down. This means they must not eat too many sugary foods, or starchy foods that are broken down to glucose. On the other hand, they must also take care that blood glucose concentration doesn't get too low, so it is a good idea to have a sugary snack close at hand, to ward off hypoglycaemia.

People with Type 1 diabetes usually need to give themselves regular injections of insulin. They will probably test their blood glucose level at around the same time each day. It is best to try to get into some kind of pattern, so that food intake and level of activity remain around the same each day. That way, the blood glucose concentration is likely to vary in the same way each day, and a routine can be established of having a particular dose of insulin at a particular time. If someone is extra active, they may need less insulin because the body is using up glucose for energy, whereas if they are less active than usual then the body will use up less glucose and they'll need more insulin.

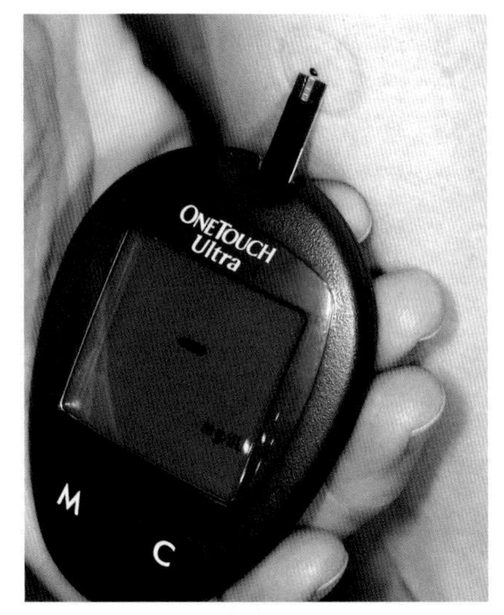

Figure 3.26 Using a biosensor to check blood glucose concentration

GM insulin

More than 200 000 people in the UK have Type 1 diabetes, so a great deal of insulin is needed each year. Until relatively recently, all of this was extracted from the pancreases of cattle or pigs. Since the early 1980s, however, insulin has also been obtained from genetically modified (GM) bacteria.

The bacteria, a strain of the common gut-living species *Escherichia coli*, have had the human gene for making insulin inserted into them. They then follow the instructions on the gene and produce insulin. These GM bacteria are cultured in huge vats, and the insulin they make is extracted and purified.

Everyone thought this new source of insulin would be better than the insulin obtained from cattle and pigs. Cattle and pigs have a very slightly different form of insulin from humans, and so people thought the genuinely human insulin made by the GM bacteria would work better. But that hasn't really proved to be the case. Although the bacterium-produced insulin works well for most people, others still say they feel better using cattle or pig insulin. Nevertheless, most insulin is now produced by bacteria; cattle and pig insulin for human use is only produced in very small quantities. The big advantages of bacterium-produced insulin are that it can readily be produced on an industrial scale, which means it is cheaper to make, and that it is easy to ensure that it is always the same.

Summary

- The blood contains red cells that transport oxygen and white cells that act against disease-causing organisms (pathogens) in the body. Some white cells are phagocytes that engulf and digest pathogens. Others are lymphocytes that secrete antibodies to destroy specific pathogens. Blood also contains cell fragments called platelets that help with blood clotting.
- Blood plasma contains many different substances in solution, which are transported from one part of the body to another.
- Hormones are chemicals produced by endocrine glands. They are transported in the blood plasma and affect target organs or target cells in different parts of the body.
- Insulin is secreted by the pancreas when blood glucose concentration rises too high. It causes the liver and muscles to take glucose out of the blood and store it as glycogen.
- When blood glucose concentration drops too low, less insulin is secreted and the liver breaks down some of its glycogen stores and releases glucose back into the blood.
- In Type 1 diabetes, the pancreas does not produce insulin. A person with this condition must take care with their diet and inject insulin to keep their blood glucose concentration at a suitable level.
- In Type 2 diabetes, the liver does not respond to insulin. This kind of diabetes usually develops later in life, and is associated with obesity.

Questions

3.9 Name the part of the blood that has each of these functions:

- **a** transporting oxygen
- **b** producing antibodies
- **c** destroying pathogens by phagocytosis
- **d** carrying hormones and other substances in solution
- **e** producing a blood clot to seal a wound.

3.10 Copy this table and complete it to compare the transfer of information by nerves and by hormones. Try to think of at least four comparative points.

Nerves	Hormones
information travels as an electrical impulse	information travels as a chemical

3.11 These are incorrect statements that students have made in examinations. For each one, write a correct sentence and explain what was wrong with the student's sentence.

- **a** Insulin turns glucose into glycogen.
- **b** If you have diabetes, you must never eat sugar.
- **c** If there is too much glucose in the blood, the brain tells the pancreas to secrete more insulin.

3.12 Some athletes have illegally taken hormones called anabolic steroids, to improve their performance. Find out what these hormones are and what effect they have on the body. Why can testing an athlete's urine help to check whether they have taken an illegal hormone? Why has it been made illegal for athletes to take these hormones?

3.13 Use the internet to find out about how a person with diabetes might use an insulin pump. What are the advantages of an insulin pump over insulin injections? Why doesn't every person with diabetes use one?

D Controlling fertility

Frozen Embryo Case Goes To Court Of Appeal

Figure 3.27

Natalie Evans and her partner Howard Johnston were very keen to have a baby. But Natalie had cancer and needed treatment that would make her infertile. So, before the treatment began, some of her eggs were removed from her ovaries. Six of them were fertilised by her partner's sperm in a Petri dish, and then frozen.

Now the cancer treatment is over, Natalie wants to have one of these embryos implanted into her uterus. It's her only chance of ever having children. But she has split up with Howard, and he doesn't want to be the father of her children. He will not let Natalie use the embryos.

The law is on Howard's side. The Human Fertilisation and Embyrology Act says that the man's consent must be given before this treatment can go ahead.

Natalie has tried to fight the decision in the courts. The ruling went against her in the High Court, so she took her case to the Court of Appeal. They, too, turned her down.

Infertility

About one in every seven couples in Britain who would like to have children has difficulty in conceiving. Most of them eventually manage to do so, but some still don't manage to start a baby after two years. These couples may be able to get fertility treatment through the National Health Service.

The first thing to do will be to find out the cause of the problem. In around 29% of cases, it will be something to do with just the woman's reproductive system. 32% of cases will be caused by just the man's reproductive system, and 17% by a combination of problems with both the woman and the man. Most of the remaining 22% of cases are never explained.

There are many different ways in which infertility can be tackled. Some of these raise ethical questions. For example, if the man does not produce sperm that are capable of fertilising an egg, the couple might decide to use another man's sperm instead. Is that a morally acceptable thing to do? If so, should the couple know the identity of the sperm donor? Should the child be given the opportunity of meeting his or her biological father?

Figure 3.28 Counselling at an NHS fertility clinic

Now that there is no longer a right to anonymity, there are far fewer sperm donors and a shortage of donated sperm.

These and many other dilemmas associated with fertility treatment are discussed by the Human Fertilisation and Embryology Authority, the HFEA. This sets guidelines that fertility clinics are expected to follow. For example, in 2004 it was decided that in the future a child would have the right to know who their biological father was – in other words, sperm donors could no longer be anonymous.

These kinds of issues are very difficult to make decisions about. And things keep changing. Society's perceptions of what is acceptable and what is not acceptable can become very different over quite short periods of time. Scientific and technical progress are making things possible now that in the past would have been the stuff of science fiction. Many scientists think it won't be long before it becomes possible to clone human beings. Should that ever be allowed? (See page 35.)

Sex hormones

The production of sperm and eggs is controlled by hormones. In men, the main hormone is **testosterone**, while in women the two hormones **oestrogen** and **progesterone** are involved. Testosterone is produced by the testes, and oestrogen and progesterone are produced by the ovaries.

Testosterone stimulates the production of sperm. Once a boy reaches puberty, the quantity of testosterone produced rises and sperm production begins. The testosterone level then stays fairly constant for the rest of his life.

In women, oestrogen stimulates the production and release of eggs. It also causes the lining of the uterus (womb) to become thick and spongy, and filled with blood vessels, in preparation for the arrival of a fertilised egg. Progesterone keeps the uterus lining in this condition. If an egg is fertilised, then progesterone is secreted all through pregnancy.

The menstrual cycle

Unless a woman is pregnant, the quantities of oestrogen and progesterone secreted change over a cycle, usually of around 28 days, called the **menstrual cycle** (see Figure 3.29 on the next page).

In the first few days of this cycle, menstruation (a period) happens. The levels of both oestrogen and progesterone are low. Then, over the next few days after menstruation, the ovaries secrete more and more oestrogen, reaching a peak at around day 13. This stimulates the ovary to release an egg – **ovulation** – which usually happens around day 14. Oestrogen levels drop, and now the ovary secretes progesterone. This carries on for the next week or so, and then – if the egg hasn't been fertilised – progesterone secretion slows right down. As a result, after 28 days the lining of the uterus breaks down and menstruation occurs again – the uterus lining is gradually lost through the vagina during the next four or five days.

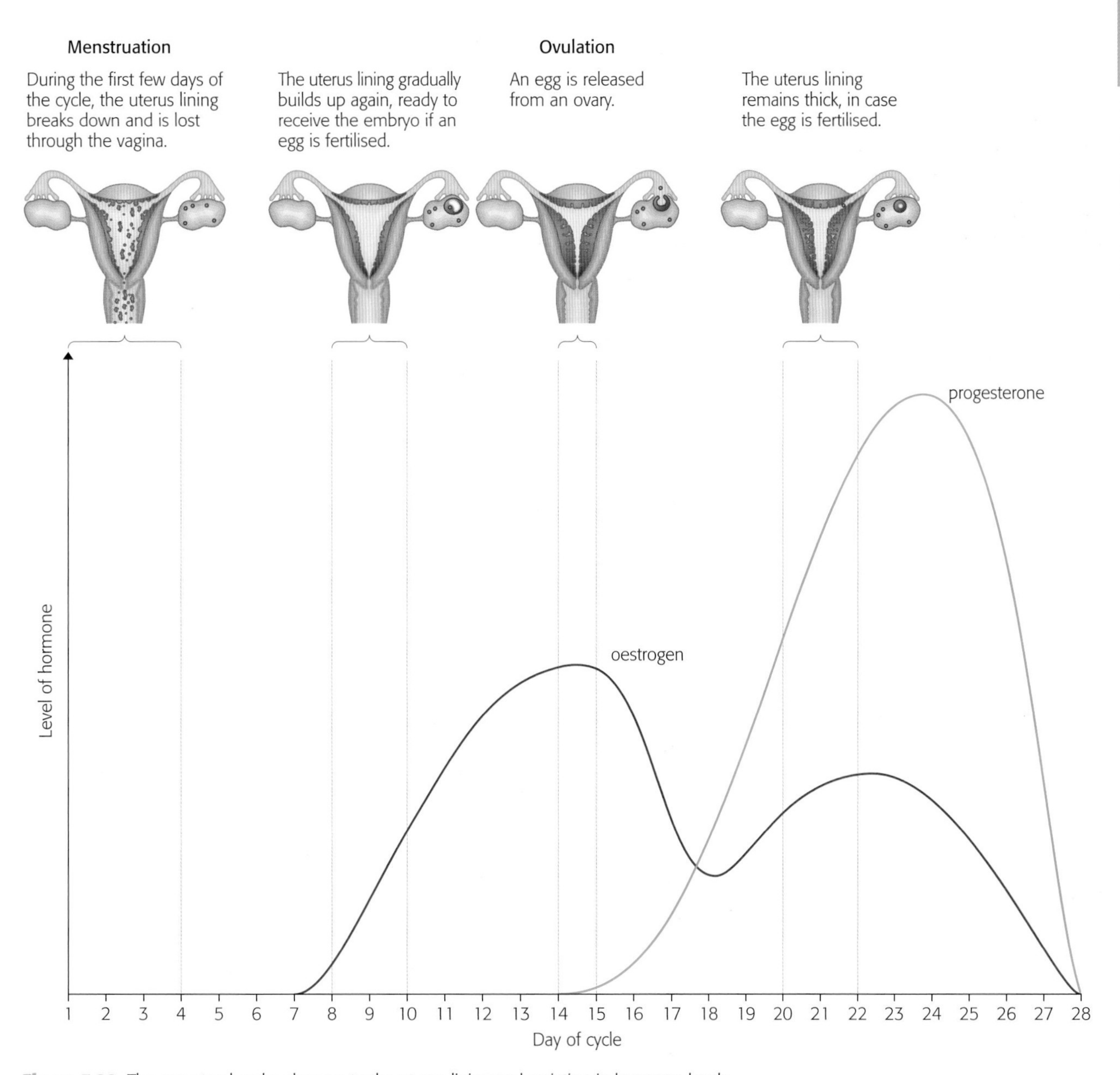

Figure 3.29 The menstrual cycle: changes to the uterus lining and variation in hormone levels

Fertility treatment

There are many different ways in which a couple can be helped to conceive. Most of them involve the use of hormones.

For example, one of the commonest causes of infertility is that eggs are not released from the woman's ovaries. She may be making perfectly good eggs, but ovulation just is not happening. In this case, she may be given hormones to stimulate ovulation. Usually, the hormones are made artificially, which is much easier than getting them any other way. Once she has ovulated, she has a chance of becoming pregnant after sexual intercourse with her partner.

It is quite tricky to get the dose of hormones just right. Too little, and no egg will be released. Too much, and several eggs may be released at once. This can lead to multiple births.

For some couples, this kind of fertility treatment still does not solve the problem. Perhaps the woman has a blockage in her oviducts, so that sperm cannot reach the egg. Perhaps the man's

'In vitro' means 'in glass'. The technique is also sometimes called 'test tube fertilisation', although no test tubes are involved.

Until recently, two or even three embryos were placed in the uterus, to increase the chances that at least one of them would implant and develop. However, it is now recommended that only one is used, to reduce the risk of a multiple birth.

Figure 3.30 Transferring sperm by pipette to a dish of eggs during IVF

Age of woman	Chance of IVF being successful
under 35	28%
35–37	22%
38–39	18%
40–42	10%

Table 3.1 IVF success

sperm are not good swimmers, and cannot get from the cervix (where they are left after sexual intercourse) to the oviducts. In these cases, **in vitro fertilisation**, **IVF**, may be tried.

First, the woman is given hormones to stimulate her ovaries to make several eggs. These eggs are then removed from the ovaries, which involves a small operation. The eggs are placed into a Petri dish, in a warm fluid containing nutrients, and some of her partner's sperm are added. With luck, some of the eggs will be fertilised.

The fertilised eggs are kept in the dish for a few days, until they have divided to form a tiny ball of eight cells – an **embryo**. The embryos can be inspected through a microscope, and ones that look particularly healthy are chosen. One of these is then placed in the woman's uterus. It may implant itself into the uterus lining and then develop in the normal way.

Ethical issues

Fertility treatment is expensive. It costs about £3000 for one cycle of IVF treatment for a couple. Very frequently, several cycles are needed before a successful pregnancy is achieved. The success of IVF drops sharply as a woman gets older (see Table 3.1).

So should this treatment be available from the National Health Service, free for everyone?

Some people say that, for some couples, infertility is devastating. They are desperate to have a child and the NHS should help them. Others think it should not. They argue that being infertile isn't an illness and that there are already too many people in the world. Couples who can't have a family should just accept it, and perhaps decide to adopt a child instead. They say that the money spent on IVF treatment would be better spent on other, life-saving, medical treatments.

If fertility treatment *is* offered free of charge, then who should be offered it? Recommendations on what should be available from the NHS are made by the National Institute for Clinical Excellence, or **NICE**. The recommendation is that IVF should be offered to those who really need it but only if there is more than a 10% chance of success.

It is up to local health authorities to decide which women qualify as 'really needing' IVF treatment, and this can lead to what is termed a 'post code lottery'. For example, in London about 4% of couples who have difficulty in conceiving get IVF treatment. In north-west England, only about 1% get treatment. Many people think this is unfair, and that the guidelines that local health authorities use should all be exactly the same.

Hormones and contraception

While around 1 750 000 couples in the UK are thought to be having problems in starting a family, even more are trying to avoid pregnancy. **Contraception** means stopping conception occurring when sexual intercourse has taken place. Many women choose to use 'the pill'. At the moment, there isn't a contraceptive pill for men, although there probably will be at some time in the future.

The contraceptive pill, also called the 'oral contraceptive' (because the pill is taken by mouth), contains female sex hormones that stop ovulation happening. They are usually artificially made hormones, just slightly different from the ones that the body naturally makes. This is because they are broken down less easily in the body, so they last longer. All contain progesterone. We've seen that this is the hormone that is secreted throughout pregnancy. Progesterone fools the woman's body into behaving as though she was pregnant, stopping eggs from ripening in her ovaries. Most pills also contain some oestrogen.

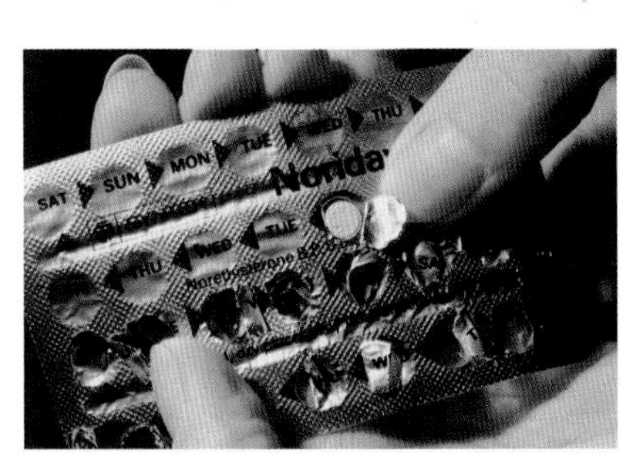

Figure 3.31 Oral contraceptives

With most kinds of oral contraceptives, the woman takes one pill every day for 21 days. She then stops taking them for 7 days, during which she has a period.

Some women may prefer not to have to bother remembering to take a pill every day. They may opt for an implant that releases a progesterone-like hormone steadily over a long time. The implant is a little plastic rod that is inserted just beneath the skin, often on an arm. If the woman decides she wants to become pregnant, the implant can be removed.

Yet another possibility is a long-lasting injection of a progesterone-like hormone. A repeat injection is needed after 12 weeks.

Summary

- **The testes secrete testosterone, and the ovaries secrete oestrogen and progesterone.**
- **During the menstrual cycle, the secretion of oestrogen causes the lining of the uterus to thicken. Progesterone maintains it in this condition. If the egg is fertilised, progesterone is secreted all through pregnancy.**
- **If a woman is not producing ripe eggs, she may be treated with hormones to stimulate her ovaries. She can then conceive in the normal way.**
- **IVF treatment involves putting a woman's eggs and her partner's sperm together in a dish, so that fertilisation can take place. One embryo is then placed in the woman's uterus, to develop normally.**
- **Fertility treatments raise many ethical issues, about which people have differing views.**
- **The contraceptive pill contains progesterone and usually oestrogen. It prevents ovulation.**

Questions

3.14 Write a paragraph explaining why couples should be offered IVF from the National Health Service. Do you think an age limit should be placed on this? Why?
Now write another paragraph explaining why IVF should *not* be available from the National Health Service.

H

3.15 Look at the graphs in Figure 3.29.

a How do the graphs support the idea that oestrogen causes the lining of the uterus to thicken?

b How do the graphs support the idea that progesterone maintains the thickness of the uterus lining?

4 Use, misuse and abuse

A Drugs

Tom woke in the night feeling awful. He had a pain in his middle and he felt sick. He dragged himself to the bathroom and lay on the floor. Every now and then he lifted his head up over the lavatory pan and vomited.

The pain didn't go. It got worse if anything. Tom managed to crawl along to his housemate, Paul's, bedroom and woke him. Paul was instantly worried. 'You look green,' he said.

Paul got Tom into the car and drove him to Accident and Emergency. By the time they were there, the pain was so bad that Tom couldn't stand on his own. He was still retching, but now on an empty stomach. Paul half lifted, half dragged Tom into the reception area.

He was seen after only a few minutes, and taken to lie down in a cubicle on his own. Eventually, a doctor appeared. She told Tom she was pretty sure what it was. 'Looks like acute appendicitis,' she said. 'We'll get rid of the pain for you, then get you X-rayed so we can see what's going on.'

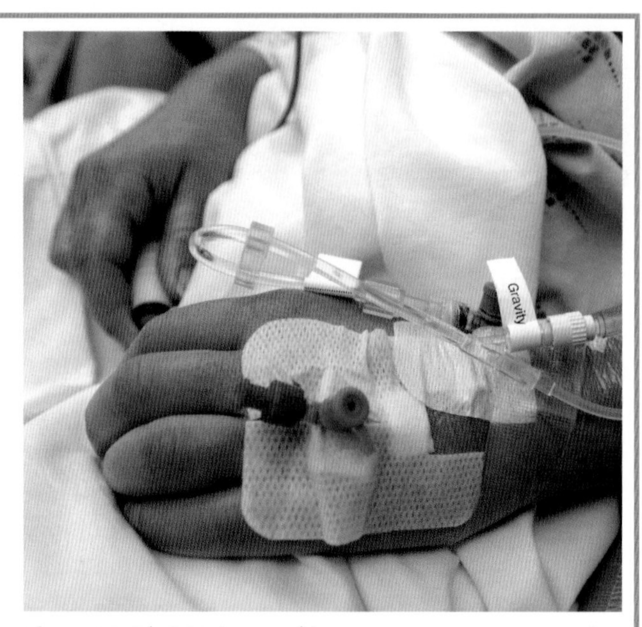

Figure 4.1 Administering morphine

A nurse appeared with a syringe and injected something into Tom's arm. Within a very short time, the pain began to fade. 'Magic,' he mumbled. 'What's that stuff?'

'Morphine,' said the nurse.

Any other time, and Tom might have worried about having morphine squirted into his blood. He knew it was an opiate, and that opiates were classed as hard drugs – dangerous and very addictive. Just now, though, he didn't care. All that mattered was that the pain had gone.

He was glad of it once again, the next morning, after he'd had an operation to have his appendix removed. A tube was inserted into a vein in his arm, and he was given a button to push that released morphine into the tube. He could dose himself with morphine whenever the pain got bad. It was great to be in control himself. By the following morning, he felt he didn't need it any more.

Types of drugs

A drug is a substance that alters the working of the body's metabolism – the chemical reactions that take place inside and around cells.

We hear so much about people abusing drugs that it is sometimes easy to forget that the vast majority of drugs are very helpful to us. We use **painkillers** such as aspirin, ibuprofen or paracetamol to help to relieve headaches and other pain. **Antibiotics** help us to get over infectious diseases caused by bacteria. People with epilepsy use **anti-convulsants** to allow them to lead normal lives with no fear of having fits. Women recovering from breast cancer use **tamoxifen** to reduce the chance of the cancer reappearing. The list is endless. Every year, after many years of rigorous tests, new drugs appear on the market.

Figure 4.2 The range of drugs is enormous

Some drugs have been around so long, and are so accepted by society, that many people take them almost every day. Most of us like drinks containing caffeine, and many people drink alcohol or smoke tobacco. All of these are legal, despite the fact that alcohol and the substances in cigarette smoke are capable of doing a great deal of harm to the body.

Solvents

A solvent is something that is used to dissolve a solute. Water is a solvent, but when people talk about 'abusing solvents' they don't mean water. Many chemicals that we use around the home contain solvents that can be very harmful to health if they are sniffed or swallowed. It is very important to keep these substances away from small children.

Some young people misuse solvents. They sniff glues that contain strong solvents. Sniffing solvents can make you feel different. When solvents are breathed in, they go into the lungs. From here, they quickly enter the blood and are carried all over the body, including the brain. They can damage cells in the lungs and neurones, especially in the brain. Someone who sniffs solvents regularly is likely to find it difficult to concentrate. They may have slurred speech, watery or glassy eyes and dilated pupils. People often have accidents after sniffing solvents, because the solvents affect the brain so they might take risks that they would not normally.

Each year, many young people die after sniffing solvents. Some of them die the very first time they do it.

Alcohol

Alcohol is absorbed into the blood through the walls of the stomach and small intestine. It affects several organs in the body, especially the brain and the liver.

Alcohol slows down the activity of neurones. It is therefore a **depressant**. Because neurones are working more slowly, the person's reaction times are longer. Alcohol inhibits the parts of the brain that we use when we decide how to behave, and so people who have drunk alcohol may lose their inhibitions and do the first thing that comes into their head without thinking about it. So driving after drinking alcohol is extremely dangerous.

Each year, about 150 000 people are admitted to hospital after drinking alcohol. Sometimes this is a result of an accident. Often, it is because they became violent after drinking.

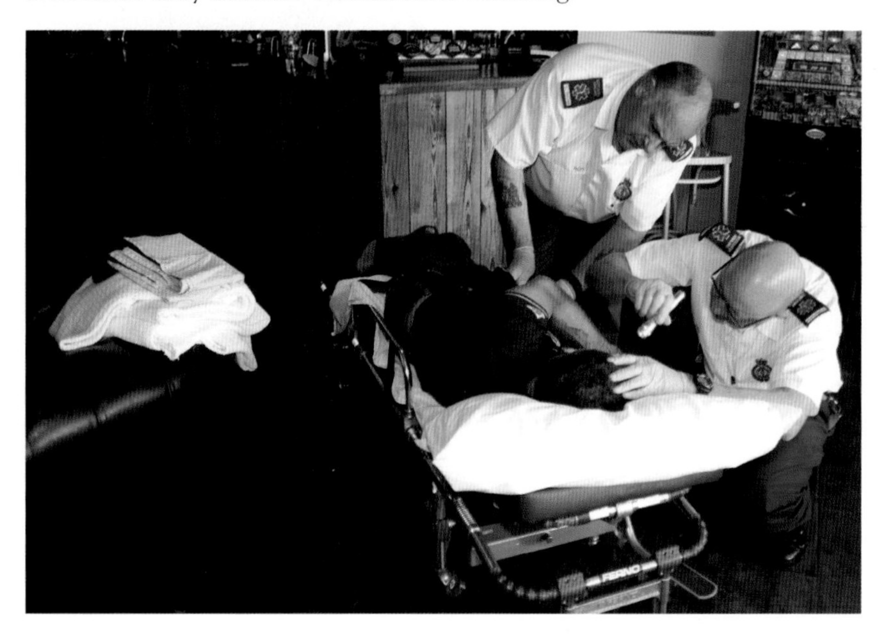

Figure 4.3 About one third of people who arrive at A and E have been drinking alcohol

Alcohol is a poison, and each year people die from alcohol poisoning, after drinking too much alcohol in a short period of time. A person who drinks a lot of alcohol on a regular basis runs a very high risk of suffering permanent brain and liver damage. The liver is the organ that breaks down toxins, including alcohol, and in the process its own cells can be killed.

You can read more facts about alcohol use in Chapter 8.

Tobacco

Tobacco contains the drug **nicotine**. Nicotine is addictive, which is why it is so hard to give up smoking once you've started. Nicotine increases blood pressure, which puts strain on the heart and increases the risk of developing heart disease.

Nicotine is a poison, and it is sometimes used in insecticides.

Tobacco smoke contains many other substances, and several of them do serious damage to the body. **Tar** contains chemicals that can cause cancer. These chemicals can cause cells to begin to divide uncontrollably, forming a tumour. Most cases of lung cancer occur in people who smoke, or people who live or work in places where they have to breathe other people's cigarette smoke. Smoking also greatly increases the risk of developing cancer in other parts of the body.

Chemicals that cause cancer are called carcinogens.

Tar and the tiny particles of ash in cigarette smoke irritate the lungs. This can cause coughing and inflammation, and the walls of the alveoli gradually break down. Eventually, the smoker may end up with many fewer alveoli in their lungs. Gas exchange becomes very poor, and they may feel short of breath. They have **emphysema**. As emphysema progresses, less and less activity is possible, until the person may be confined to their house where they have to breathe oxygen from a cylinder.

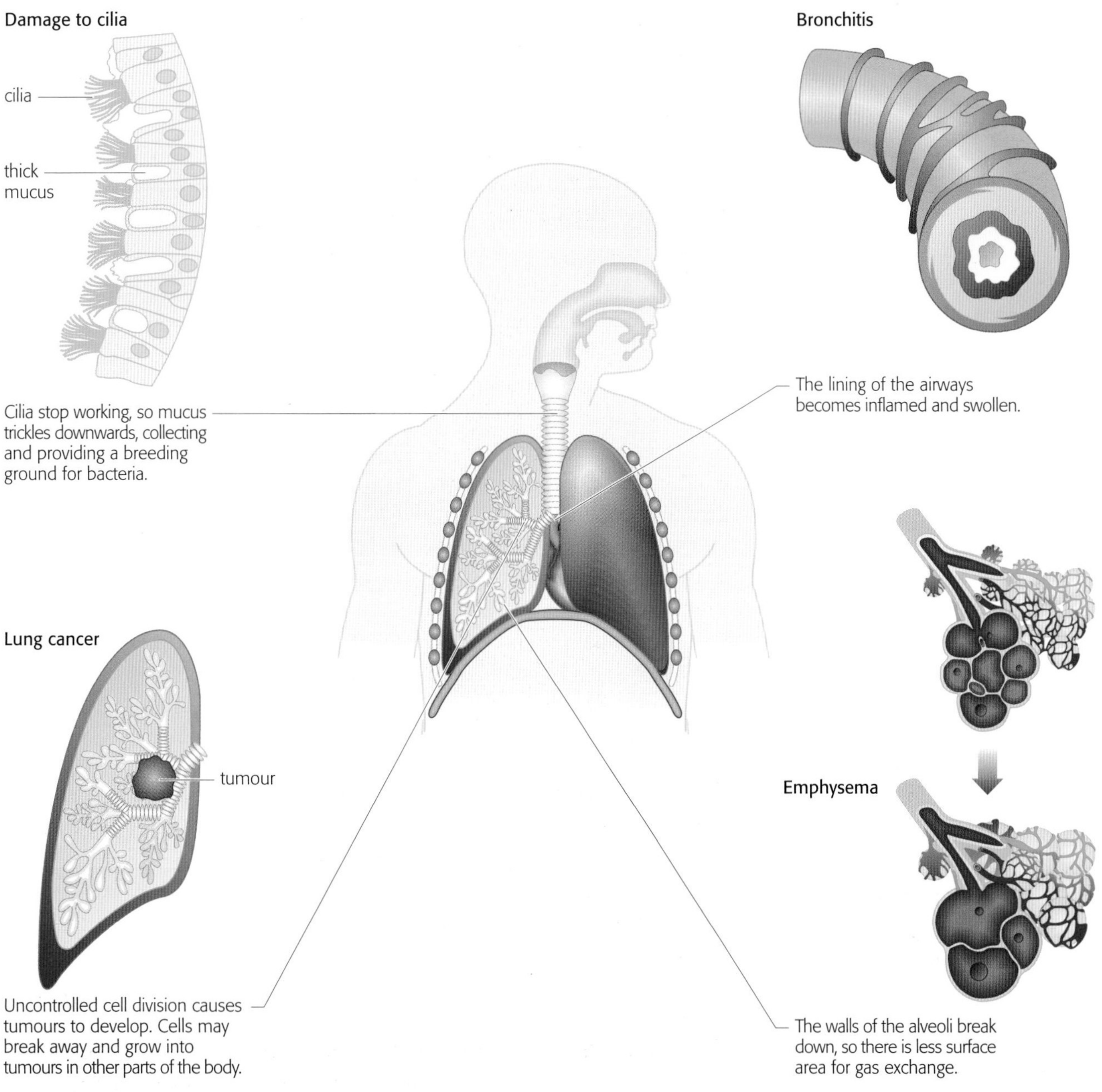

Figure 4.4 How smoking damages the lungs

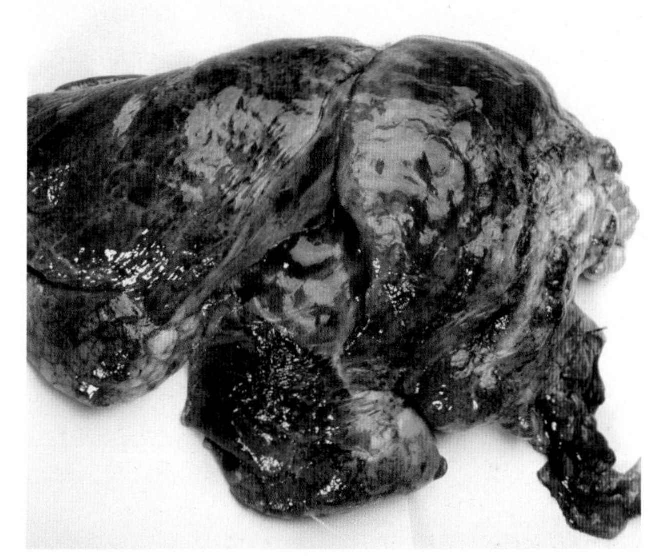

Figure 4.5 A smoker's lungs

Carbon monoxide in cigarette smoke passes through the walls of the alveoli and into the blood. Here it combines with the haemoglobin in the red blood cells, displacing the oxygen that the cells should be carrying. This means body tissues are not supplied with as much oxygen as they need. If a pregnant woman smokes, this happens in her growing baby's blood too, and the baby is likely to have a smaller weight at birth.

Smoking also causes **bronchitis**, which is inflammation of the bronchi caused by a bacterial infection. Chemicals in the smoke stop the cilia which line the airways from working (see Figure 4.20, page 75). At the same time, they speed up the production of mucus. So the mucus just drips down into the lungs, where it collects. It forms an excellent breeding ground for bacteria.

Stimulants

Most of us drink **caffeine** in cola drinks, tea or coffee. Many people find it helps them to stay alert; caffeine is a **stimulant**. Stimulants tend to increase the activity of neurones in the brain. They can make your reactions faster. Most evidence suggests that caffeine does no harm, as long as it is not taken in excessive quantities.

Some stimulants, though, are much more powerful than caffeine and more dangerous. For example, amphetamines make people feel they have huge amounts of energy. Their body works harder than normal, and as the effects of the drug wear off they may feel completely drained and quite unwell. It is illegal to take amphetamines unless you have been prescribed them by a doctor.

Sedatives

A sedative is a drug that slows the brain down and makes you feel relaxed. **Barbiturates** used to be widely prescribed to help people to feel less anxious, to relax and to sleep. They are sometimes called **tranquillisers**. Unfortunately, the amount of the drug that you have to take to get any benefit is not much less than a dose that can kill you. Many people have died after taking too high a dose.

Barbiturates are especially dangerous if alcohol has been drunk.

Barbiturates have been mostly replaced by another group of sedatives called **benzodiazepines**. These sedatives can be habit-forming if taken over a long period of time or in large doses. They are illegal unless you have them on prescription.

Someone who is affected by a sedative should not drive, because their reaction time will be longer than usual.

Figure 4.6 Benzodiazepines may be a factor in at least 10% of road accidents

Paracetamol

Paracetamol is a tried and tested and very safe **painkiller**. It can be bought over the counter in pharmacists and supermarkets.

Paracetamol works by acting on enzymes in the brain to block pain and to reduce fever. It is very effective against mild to moderate pain. Unlike aspirin, it does not harm the lining of the stomach.

Perhaps because it is so easily available, many people who try to commit suicide use paracetamol. For this reason, you are only allowed to buy a limited amount of it at any one time.

Each year, there are up to 200 deaths in England and Wales from deliberate overdoses of paracetamol.

When someone takes an overdose of paracetamol, there probably won't be any symptoms until at least four days later. Then they will begin to feel unwell and their abdomen may feel sore. The paracetamol is damaging their liver. Hospital treatment can usually save them from death, and in many cases the liver recovers completely. In others, it is damaged beyond repair, and they may need a liver transplant.

Opiates

Morphine is an **opiate**. Opium, heroin and codeine are also opiates. Opiates are drugs that we get from the seed pods of opium poppies. They have been used for thousands of years to relieve pain. Apothecaries and herbalists have long used them as a standard ingredient in many of their potions.

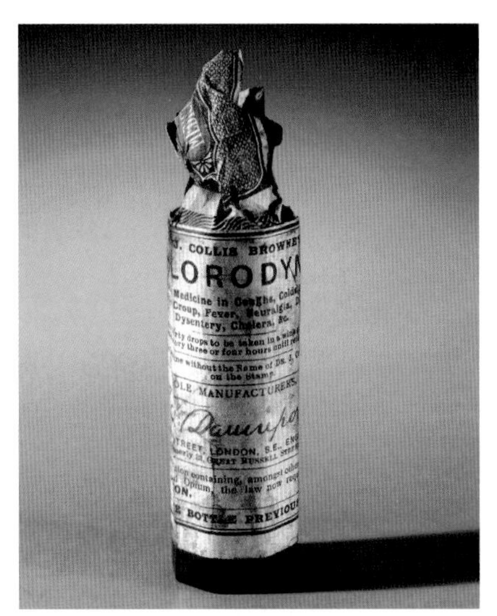

Figure 4.7 These 'chlorodyne drops' which contained opium were a popular remedy in the late 18th century. They were claimed to cure many ailments as diverse as toothache and cholera

Figure 4.8 Opium poppies

Opiates help to dull pain. They work because they behave like the brain's natural painkillers, called **endorphins**. People who've had a really bad injury – for example losing an arm in an accident – often don't feel any pain at all for a while. This is because the body has secreted lots of endorphins. This temporary lack of pain might let you concentrate on getting out of the dangerous situation before even more harm is done.

A person who takes an opiate, for example heroin, will probably have a wonderful feeling – all their problems will have disappeared and they feel completely happy. This is called euphoria. This feeling can draw people into taking opiates such as heroin. For many people, this can rapidly turn into an **addiction**. They feel they cannot manage without the drug. If they stop taking it, they feel completely awful. They may be anxious and restless, with pain all over their body; they may vomit and have diarrhoea. These **withdrawal symptoms** can go on for days. A person who is addicted to heroin will do anything to get their next dose of the drug. So dealers can charge high prices; they have the addicts just where they want them. Many addicts turn to crime to get money to pay for their drugs. They lose their friends and families, and their jobs.

Addicts often find that they need larger and larger doses of heroin to get rid of the unpleasant feelings they suffer between doses. There's always a danger of an overdose. Heroin, like all the opiates, decreases breathing rate. If the dose is too great, then breathing may shut down completely, causing death.

Most heroin addicts end up injecting the drug, because this gets it into their bloodstream faster and deals with their cravings more quickly. This introduces another danger to their health. Often, the same needle is used more than once, so people are injecting themselves with dirty needles that may contain bacteria or viruses. It's even more dangerous if the same needle is used by more than one person. If one of them has a disease such as HIV/AIDS or hepatitis, then the viruses that cause these diseases can be transferred from one person to another.

Being given morphine in hospital to relieve pain won't cause you to become addicted. The doses aren't usually very large, and you probably won't need it for very long. However, someone who is terminally ill and in constant pain may need larger doses. It doesn't really matter if they become addicted. They can be given whatever dose is needed to control their pain.

Unfortunately, an infamous misuse of morphine by a doctor has given it a bad name, and has made doctors think twice before they administer large doses to their patients. In January 2000, Harold Shipman was convicted of killing 15 patients by injecting them with morphine. He was given 15 life sentences, but he somehow managed to kill himself by hanging in his prison cell only three years later. It is thought that he had probably killed at least 236 patients, not only the 15 for whose murders he was tried.

Cannabinoids

A **cannabinoid** is a drug derived from **cannabis**. Cannabis comes from the cannabis plant, sometimes called hemp. Like opiates, cannabis has been used for thousands of years.

People smoke cannabis because it makes them feel relaxed and contented. It isn't addictive, and for this reason it is not classified as a hard drug. Until recently, it seemed that it didn't do too much harm at all. But now a link has been uncovered between smoking cannabis and developing the mental illness **schizophrenia**. Young people are four times as likely to develop schizophrenia if they have ever smoked cannabis. It seems that cannabis can trigger the illness in someone who is susceptible to it.

Figure 4.9 A field of cannabis plants

The use of cannabis is illegal in the UK, but still many people use it. Apart from those who smoke it for pleasure, many people with chronic (long-term) illnesses say that it helps them to feel better. It seems to help especially with illnesses where there is pain caused by damaged nerves, such as multiple sclerosis. Some people have been prepared to defy the law in order to use cannabis for pain relief.

Until recently, there had not been very much research on using cannabis for pain relief, and what research there is has sometimes been contradictory. Some studies have found that cannabis helps to reduce pain, while others found that it doesn't. The trouble is that not many of the studies have been scientifically rigorous. They may have relied just on collecting stories from people who've used cannabis, or asking them to fill in questionnaires.

The active ingredient in cannabis is called **THC**. Researchers are now looking into the possibility of developing drugs derived from THC that might help to relieve pain without having any of the mood-altering effects of cannabis. If they could do that, it would be a big breakthrough. Pain from damaged nerves can't be combated even by the most powerful opiates. If cannabinoids could do this, it would improve the lives of many people with chronic pain.

Finding the facts about drugs

If you type 'cannabis' into Google, you will find an enormous number of websites. How do you decide which will give you good scientific facts?

bbc.co.uk/science/hottopics/cannabis
cannabis.net
lca-uk.org
drugs.gov.uk
talktofrank.com
www.addaction.org.uk/Drugcannabis.htm

Figure 4.10

The best sites will contain **primary data**. These are data that have been collected by researchers. They may have done experiments on the effect of cannabis and recorded the results that they found. They may have compared the health of people who have smoked cannabis and those who haven't. They may have done a well planned survey to see if there is any correlation between smoking cannabis and using hard drugs such as heroin.

Primary data could also come from a person who accurately reports on their own behaviour, feelings or symptoms after taking a drug. This can be useful, but any one person's experience may not be typical. And how sure can you be that they are being entirely honest?

Other sites may use results from primary sources and put their own interpretation on them. They give **secondary data**: the information is secondhand. This can still be excellent information. But take care – sometimes people may have put their own 'spin' on it. Their interpretation may be biased.

For example, someone may disagree strongly with the idea that cannabis does harm. They might just select some data that supports their view, and ignore other data that suggests that cannabis is harmful. Another website might do just the opposite.

On the whole, the best websites for finding good scientific information about the effects of drugs will be ones linked to scientific or medical institutions, such as a university or a hospital. You should be very wary of an individual person's website, or one produced by a group that seems to be campaigning for something.

Summary

- A drug is something that affects the metabolism of the body.
- Paracetamol is a drug used for mild to moderate pain. It is a very safe drug, but deliberate overdoses can kill by harming the liver.
- Opiates are drugs derived from the opium poppy. They are all painkillers. They can be addictive if misused. Morphine is an opiate used in medicine to treat severe pain.
- Cannabinoids are drugs derived from the cannabis plant. They are often used as recreational drugs, but recent research suggests that they are more dangerous than had been thought, because they seem to increase the risk of developing schizophrenia. In some cases, they seem to help to reduce pain.
- A primary source of data is one where data collected by research are given. A secondary source of data is one where someone has interpreted data obtained from a primary source.

Questions

4.1 Which of these statements are correct and which are incorrect? For those that you think are incorrect, write a correct one instead.

- **a** Cannabis is an opiate.
- **b** There is evidence of a link between smoking cannabis and developing schizophrenia.
- **c** Opiates act by mimicking endorphins in the brain.
- **d** Morphine is used as a painkiller in hospitals.
- **e** A paracetamol overdose kills within hours.

4.2 **a** Explain the difference between a primary source of data and a secondary source of data.

- **b** Which of these are primary sources of data?
  - A person's diary of how they feel when they have taken a drug.
  - A magazine article about the addictiveness of opiates.
  - Results of an experiment into the severity of withdrawal symptoms after taking different doses of an opiate.

4.3 When a prisoner is taken into custody, a urine sample may be tested when they enter prison. The sample is tested for evidence of whether the person has recently taken illegal drugs. In 2001–2002, 45% of the prisoners who were tested gave positive results for cannabis, 48% for benzodiazepines, 44% for opiates and 9% for cocaine (a highly addictive stimulant).

- **a** Show these data as a bar chart.
- **b** Explain how this evidence shows that some people had taken more than one drug.
- **c** A newspaper published these data and said that they showed that people who took drugs were more likely to go to prison than those who did not take drugs. Explain why this is not a safe conclusion to make. Suggest how you could carry out a study to test the suggestion put forward by the newspaper.

B Pathogens

'Bird flu' is not uncommon in Asia, and every now and then there are major outbreaks of it. But the bird flu virus is not the same as the human flu virus, and usually bird flu doesn't infect humans. When it does, it is extremely dangerous. In a 1997 outbreak of bird flu in Asia, only 18 people were infected but six of them died.

In 2003–4, there was a further big outbreak of bird flu in poultry in Asia. Many Asian countries acted swiftly. Nearly 100 million chickens and ducks were killed to try to control the outbreak. Once again, some people caught the disease and died. About half of those who were infected died.

At the moment, there haven't been any known cases of this strain of bird flu being passed from one person to another. But many scientists think it is only a matter of time before someone gets infected with both bird flu and human flu at the same time. When that happens, there is a chance that the bird flu virus might gain the ability to spread from person to person, perhaps by picking up DNA from the human flu virus. Then we could face a pandemic of a new strain of flu.

Figure 4.11 Chickens in poor conditions are susceptible to bird flu

A pandemic is a worldwide outbreak of a disease. There were three flu pandemics in the last century. The 1918–19 pandemic was called Spanish flu, and is thought to have killed 50 million people. Almost half of these were young, healthy adults. In 1957–58, Asian flu killed hundreds of thousands, and in 1968–69 Hong Kong flu did the same.

Each of these pandemics happened because a new strain of flu virus appeared, which our immune systems had never met before. No one had immunity to it. Both Asian flu and Hong Kong flu were caused by viruses that had a mixture of DNA from human and bird flu viruses.

If we can keep an eye on the bird flu cases and try to develop vaccines, then we may be able to prevent a new flu pandemic. But it needs a lot of cooperation and information-sharing between different countries, and unfortunately that does not always happen.

In early 2006, bird flu came much closer to Britain, killing several people in Turkey. By the time you read this, there will have been further developments. Has the spread of bird flu got out of control, or are we managing to stay one step ahead?

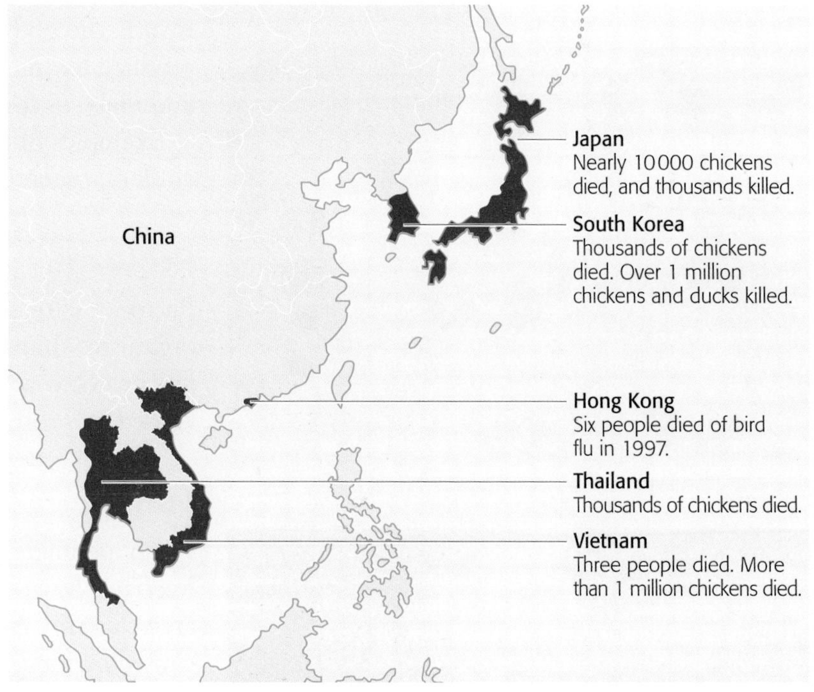

Figure 4.12 The spread of bird flu in Asia by 2003

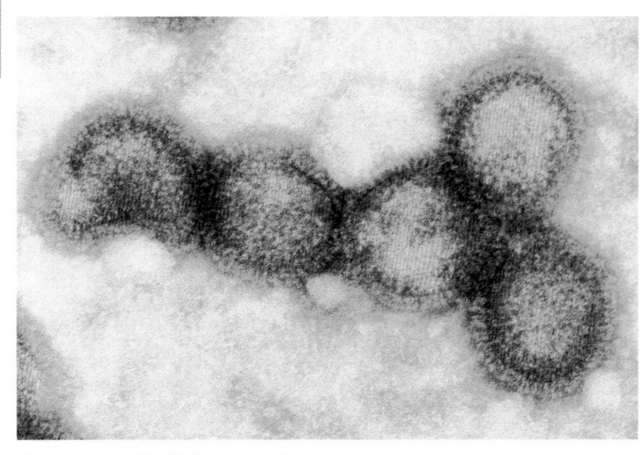

Figure 4.13 Flu (influenza) viruses

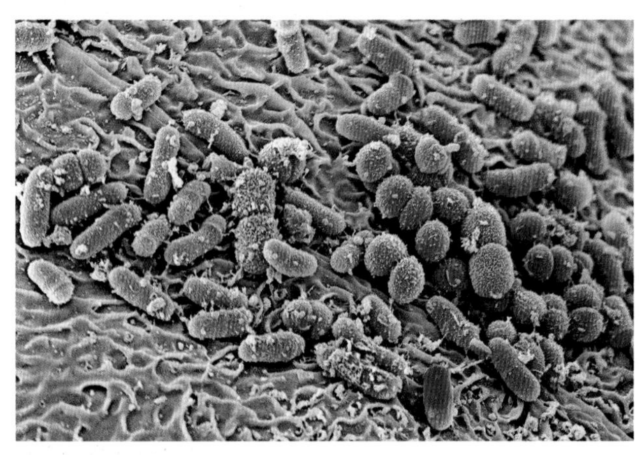

Figure 4.14 Bacteria in the nose

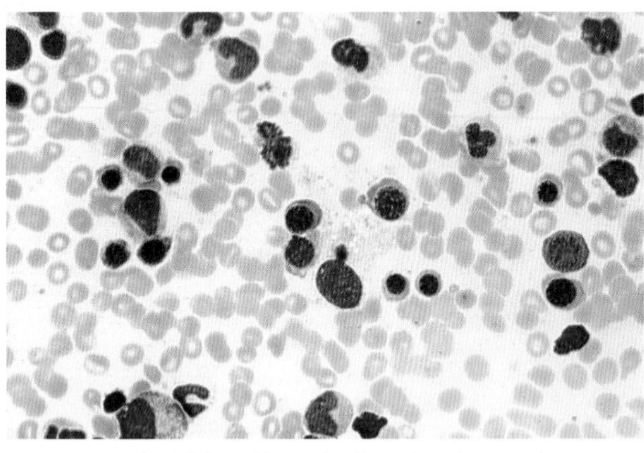

Figure 4.15 Blood infected by malarial parasites (protozoa)

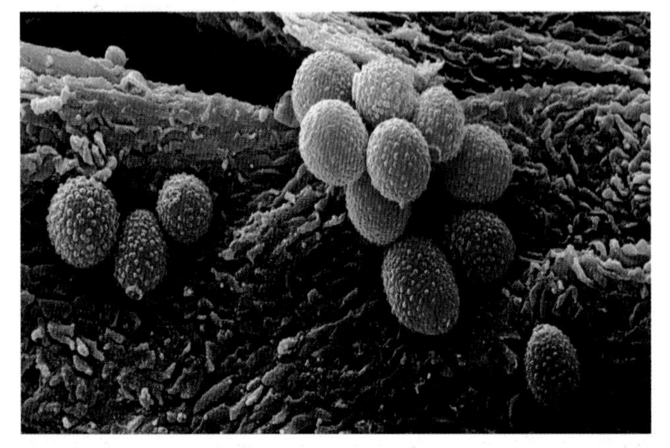

Figure 4.16 The fungus that causes athlete's foot

Disease-causing microbes

A **pathogen** is an organism that causes disease. There are four types of microorganisms (microbes) that cause disease. They are:

- viruses
- bacteria
- protozoa
- fungi.

Viruses

Viruses are tiny particles of DNA or RNA (ribonucleic acid), wrapped up in a protein covering. They are not made of cells, and they can't do anything until they are inside a living cell. Then they take over the machinery of the cell and use it to copy millions of themselves. They burst out of the cell and then infect others.

Examples of illnesses caused by viruses include flu, colds, measles, and AIDS.

Bacteria

Bacteria (singular bacterium) are single-celled organisms. They are much bigger than viruses, but much smaller than our body cells. Most bacteria are not harmful, but some can cause disease. They invade the body, where they reproduce. They often produce **toxins** which are carried all over the body and make us feel ill.

Examples of illnesses caused by bacteria include cholera, tuberculosis (TB) and septicaemia.

Protozoa

Protozoa (singular protozoan) are also single-celled organisms that are similar to our body cells. Only a few protozoa are harmful. Examples of illnesses caused by protozoa include malaria and dysentery.

Fungi

Fungi are made of microscopic threads called hyphae. Very few fungi cause illnesses, and these aren't usually dangerous. Examples of illnesses caused by fungi include athlete's foot and ringworm.

Catching diseases

Diseases caused by a pathogen can be passed on from one person to someone else. So they are **infectious** diseases. There are many different ways in which an infectious disease can be caught from someone else. Some are shown in Figure 4.17.

Figure 4.17 Ways of catching an infectious disease

Direct contact between people

A child could pick up the chickenpox virus from another child.

HIV, the virus that causes AIDS, can be passed on from one person to another during sexual intercourse, or by sharing needles to inject drugs, but not by kissing.

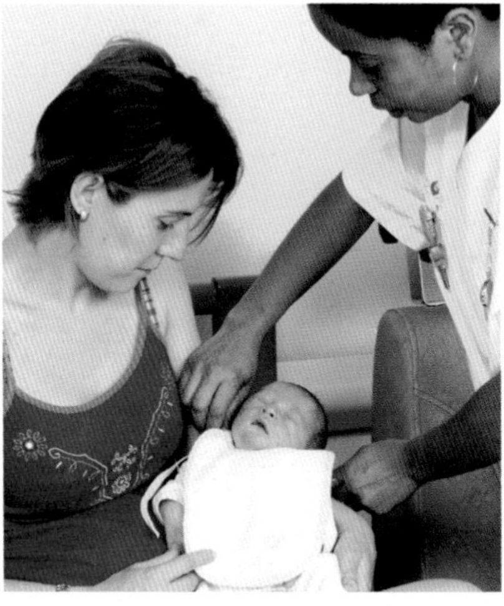

HIV can also be passed from a mother to her baby. This is sometimes called **vertical transmission**, because it passes the virus down from one generation to the next.

Indirect contact with another person
If someone who has bacteria or viruses on their hands touches an object, and then you touch it, you might pick up the pathogens on your own hands.

Droplets
When someone with a cold or flu sneezes – or even when they talk – tiny droplets of saliva spread out into the air. The droplets may contain the cold or flu viruses. If you breathe them in, you may catch the illness.

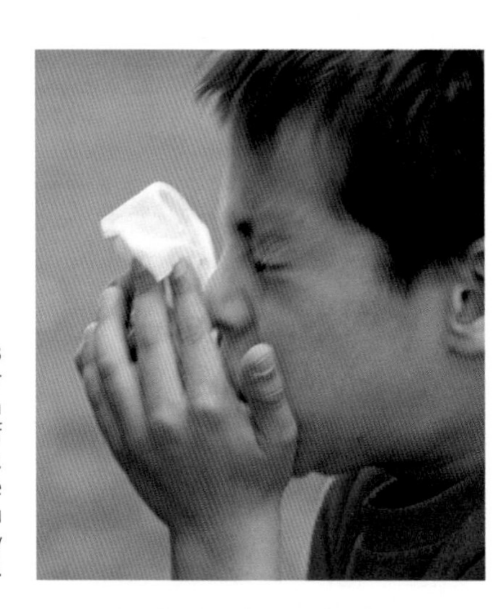

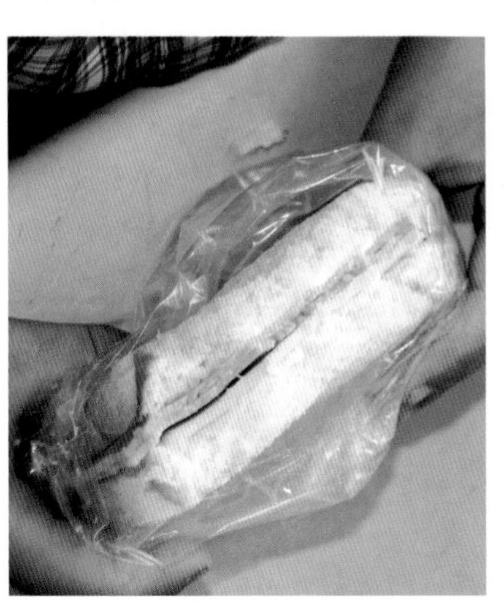

Vehicle transmission
If someone who has bacteria or viruses on their hands makes a sandwich, and then you eat it, you might get food poisoning from the pathogens that were on their hands. The sandwich is a **vehicle** that carries the infection from one person to another.

Vectors
A vector is an organism that carries a pathogen from one person to another. A mosquito may be infected with the protozoan that causes malaria. If the mosquito bites you, the protozoan may go into your blood.

Summary

- A pathogen is an organism that causes disease.
- Diseases caused by pathogens are infectious – that is, you can catch them from somebody else.
- Infectious diseases may be caught by direct contact, indirect contact, droplet infection, vehicles of infection and vectors.

Questions

4.4 Draw a table with these three headings: *Disease*, *Pathogen that causes it*, *How it is transmitted*. Use the information on pages 71–73 to complete your table.

4.5 Explain how each of these actions could reduce the chance of catching an infectious disease.

- **a** Washing your hands before you eat.
- **b** When on holiday in a country where hygiene is not very good, always peeling fruit before you eat it.
- **c** Using an insect repellent if you are in a tropical country.

C Body defences

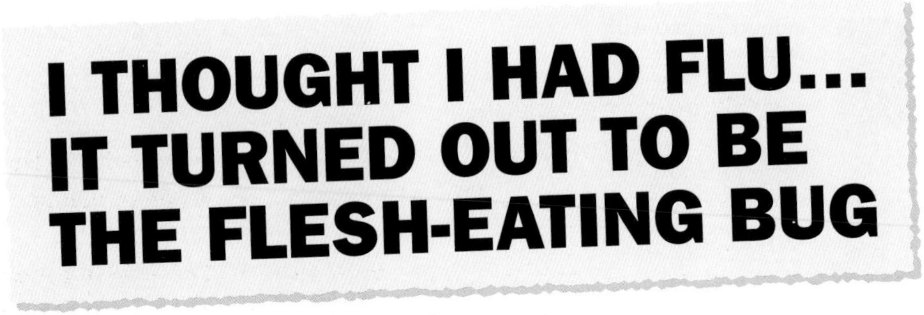

I THOUGHT I HAD FLU... IT TURNED OUT TO BE THE FLESH-EATING BUG

Figure 4.18 This story hit the headlines on 4 August 2005

Paul Cumming felt a bit unwell over the weekend. He thought he had flu. But on the Monday he developed severe pain down one arm and in his side. He took a taxi to Accident and Emergency, where he was quickly diagnosed with necrotising fasciitis – which the media preferred to call the 'flesh-eating bug'.

Necrotising fasciitis is usually caused by a bacterium called *Streptococcus pyogenes*. About one in ten of us have it living harmlessly on our skin. But if the skin is broken, it can get into the body. Sometimes – no one knows exactly why – it then starts to breed furiously. It kills cells. 'Necrotising' means 'dying', and cells all around the site of infection die. The dead tissue looks dark and angry; soon the skin disappears leaving an open wound. The infection spreads frighteningly quickly. Some people who've had it say you could almost watch it moving across and into the body. It leaves huge open wounds and can kill within days.

Paul was put on an antibiotic drip and given morphine. Almost immediately, he had an emergency operation to remove all the infected and dead tissue. Later, he needed skin grafts to cover the sites of the wounds the bacterium had caused.

The first line of defence

There are pathogens all around us. Yet most of us don't have many infectious diseases in our lifetimes. That's because the body has a whole range of defences against the pathogens.

The best way of avoiding an infectious disease is to stop the pathogen getting into the body in the first place. This is our first line of defence.

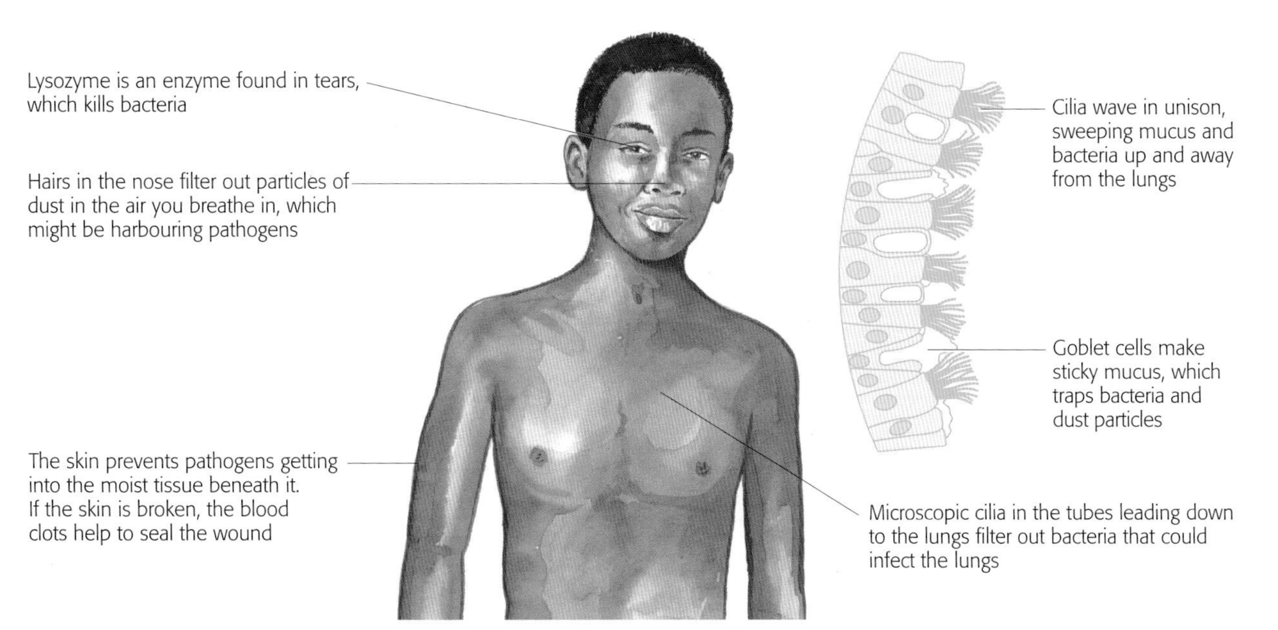

Figure 4.19 How pathogens are kept out of the body

The second line of defence

Despite the barriers that prevent pathogens entering the body, they do still sometimes get in. This is when our second line of defence comes into play.

Imagine you've pricked a finger with a thorn. There were bacteria on your skin, and now the thorn has pushed some of them through the skin and into the wound. If there are a lot of them, or if they start to breed inside the body, the cells around them begin to release special chemicals. These act like a 'help!' signal. In response, the blood capillaries carrying blood to that area get wider, so more blood flows to the scene. This extra blood supply makes the infected area look red. It feels tender and sore. This is called **inflammation**.

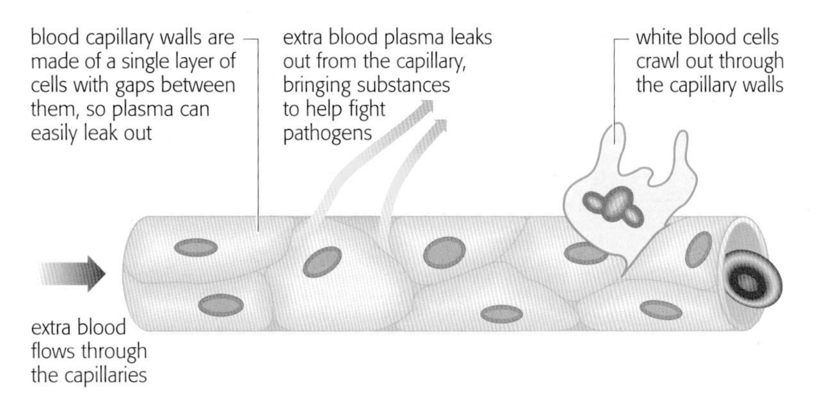

Figure 4.20 What is happening in an area of inflammation

Phagocytes are not fussy about the kinds of pathogens they destroy – they will engulf almost anything that is foreign to the body.

The extra blood supply brings extra white blood cells. Some of them are **phagocytes**. These cells are able to escape through the capillary walls and actively search out the bacteria. They put out 'arms' of cytoplasm and engulf them. The bacteria are taken into the phagocyte. Then enzymes are released onto them. The enzymes digest and kill the bacteria. This is called **phagocytosis**.

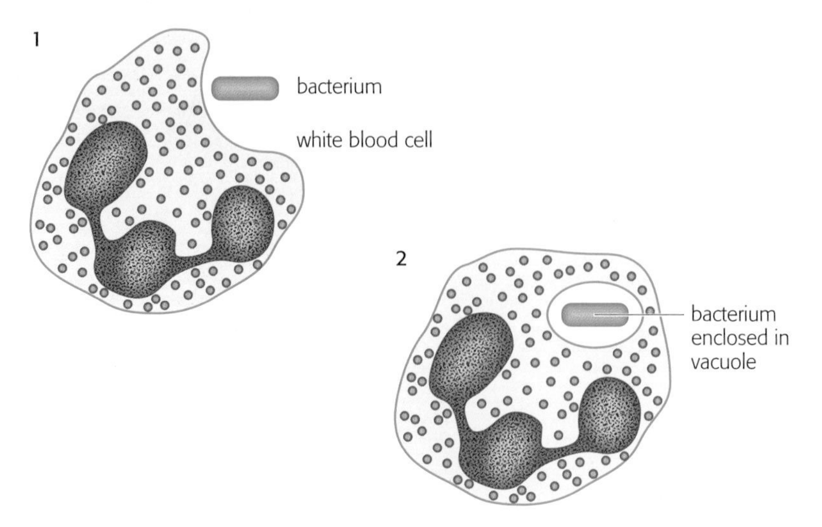

Figure 4.21 Phagocytosis

The third line of defence

If all else fails, then the third line of defence is called up for action. This involves the **immune system**.

The immune system includes white blood cells called **lymphocytes**. We have maybe a million different kinds of them, but just a few of each kind. Each kind is able to recognise one specific pathogen. It recognises the pathogen by the molecules on the outside of the pathogen's cells, called **antigens**.

Most of your lymphocytes will never have anything interesting to do; they just drift around in your blood. But if a lymphocyte does one day come across its 'own' pathogen, then it begins to fulfil its special role. First, it divides to form many more cells exactly like itself. In other words, it forms a **clone** of that specific kind of lymphocyte.

Some of these cells then begin to secrete chemicals called **antibodies**. Antibodies are proteins. The antibody molecules have a shape that exactly fits the antigens on the pathogen. They slot into them. Soon, the pathogen is surrounded by a coat of antibodies, all securely locked into the molecules on its surface. Sometimes this kills the pathogen straight away. Or it may 'label' the pathogen as being ready to be destroyed, and make it easier for phagocytes to destroy it.

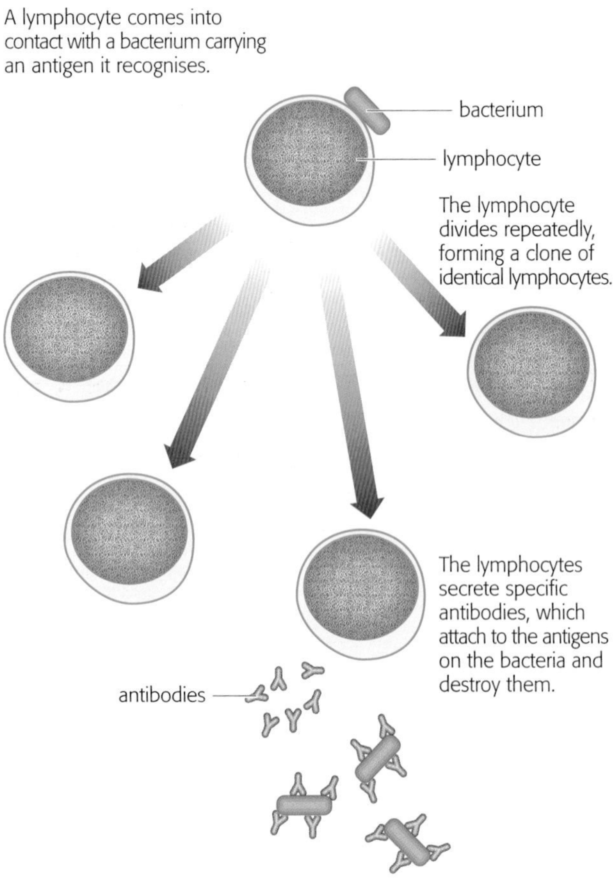

A lymphocyte comes into contact with a bacterium carrying an antigen it recognises.

bacterium

lymphocyte

The lymphocyte divides repeatedly, forming a clone of identical lymphocytes.

The lymphocytes secrete specific antibodies, which attach to the antigens on the bacteria and destroy them.

antibodies

Figure 4.22 The immune response

Summary

- Pathogens are prevented from getting into the body by physical barriers, including the skin, nasal hairs, cilia and lysozyme.
- If pathogens get into the body, inflammation may occur, bringing extra white blood cells to the scene. Some of these cells destroy pathogens by phagocytosis.
- Other white cells respond to specific pathogens, which they recognise by the antigens on their surfaces. These white cells secrete antibodies, which kill or disable the pathogens.

Questions

4.6 Explain the difference between the words in each of these pairs.

a antigen and antibody

b phagocyte and lymphocyte

4.7 This graph shows what happens to the quantities of viruses and antibodies in the blood, after a virus has infected the body.

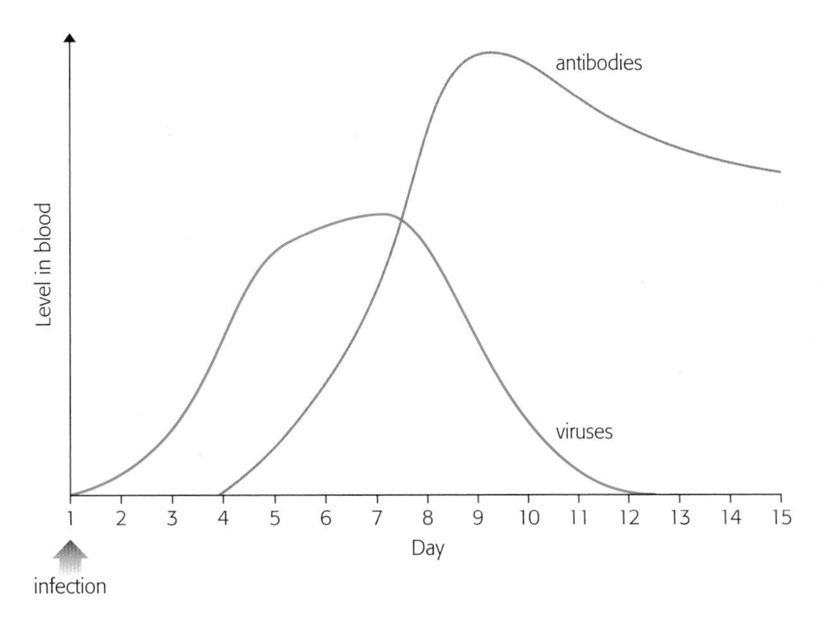

Figure 4.23 Level of viruses and antibodies in the blood after infection on Day 1

a Describe what happens to the number of viruses in the body in the first 5 days after infection.

b When does the concentration of antibodies begin to rise?

c Explain why this does not happen earlier.

d When do you think the person might start to feel better? Explain your suggestion.

D Tuberculosis

Tuberculosis is an infectious disease that has been with us for a long time. Some Egyptian mummies more than 4000 years old show signs of tuberculosis in their spines.

Tuberculosis used to be called 'consumption', because a person with this disease wasted away or was 'consumed' by the illness. In the mid-17th century, a physician called Nicholas Culpeper wrote a long book that we now call *Culpeper's Complete Herbal*. In it, he described every plant that could be used to help cure diseases. There were several that he recommended to use against consumption, which at the time was causing one in five of all deaths in London. This is one of Culpeper's entries.

Cabbages and Coleworts

The Cabbages or Coleworts ... the often eating of them wel boyled, helpeth those that are entring into a Consumption. ... Being boyled twice, and an old Cock boyled in the Broth and drunk, it helpeth the pains and obstructions of the Liver and Spleen, and the Stone in the Kidnies. ... They are much commended being eaten before meat, to keep one from surfetting, as also from being drunk with too much Wine, or quickly make a man sober again that is drunk before. ... The Ashes of Colewort Stalks mixed with old Hogs-Grease are very effectual to annoint the Sides of those that have had long pains therin, or any other place pained with Melancholly and windy humors.

Figure 4.24 Extract from Culpeper's *The English Physitian*, 1653

The TB bacterium

Tuberculosis – **TB** for short – is a disease caused by a bacterium. Up until the mid 19th century, it was one of the commonest causes of early death in Britain. It killed about 80% of the people it infected.

The bacterium usually infects the lungs. It gets actually *inside* the cells, and some people's immune systems aren't able to kill it. It breeds and spreads slowly in the lungs and other parts of the body. It is a chronic (long-term) illness. An infected person may feel weak, because their lungs are damaged and so gas exchange can't happen properly. They will begin to cough a lot, and blood will appear in the mucus they cough up. They may have chest pain and feel breathless. They may have a high temperature, and suffer from night sweats. Their weight decreases, and they gradually become weaker and weaker, until they die.

How TB spreads

You can get infected with the TB bacterium if you breathe in dust particles or droplets in the air that have been in contact with the fluid from an infected person's lungs. People seem most likely to get infected if they live in crowded conditions. For example, there have been outbreaks in prisons in Russia. People living in slum conditions are very vulnerable to getting TB.

Figure 4.25 TB outbreaks mostly occur in poor, crowded conditions

One of the reasons for the worldwide increase of TB cases is the spread of HIV (the human immunodeficiency virus). This causes AIDS, an illness in which the body's immune system is weakened. This gives the TB bacterium the chance to strike.

At the moment, it is estimated that about one third of the world population is infected with the TB bacterium. That is an amazing figure. However, only one in ten people who have the bacterium in their body actually has the disease TB. Sometimes, they never develop it in their whole life. Sometimes, they may develop it later, especially if another disease weakens them. Each year, around 3 million people in the world die from TB.

We like to think that our medical know-how and drugs keep us in control of infectious diseases like TB. But the number of cases of TB is actually getting higher and higher. And this isn't just happening in countries with poor housing and medical facilities. It is happening in the USA and even in some small areas of the UK.

Treating TB

H

Because TB is caused by a bacterium, we can treat it with antibiotics. However, the antibiotics take a long time to work. You have to go on taking them for a long time – usually at least six months. Many people are bad at doing this, and stop taking them when they feel a bit better. Then the disease comes back again. And this can also increase the risk of bacteria becoming resistant to the antibiotics. Unfortunately, there are now many strains of TB that are resistant to many of the antibiotics that are used against it.

How does this resistance arise? Anyone with TB has millions of TB bacteria in their body. Most of the bacteria will be genetically identical, but one or two might have mutations in their DNA that, just by chance, make them resistant to a particular antibiotic. If the person takes the antibiotic, it will kill most of the bacteria, but not the ones that are resistant. These resistant individuals now have the place to themselves, and there is nothing to stop them reproducing at a great rate. Their descendants are all genetically identical to them, so they are all resistant to the antibiotic.

To deal with this problem, patients with TB are often prescribed at least two antibiotics that they take at the same time. It's much harder for the bacteria to develop resistance to two drugs rather then one. The two drugs that are normally used are rifampicin and isoniazid.

Some people with TB can't be trusted to keep on taking their drugs regularly. A 'Directly Observed Therapy' scheme called DOTS has been developed, as part of which a health worker watches the patient taking their drugs each day.

A new drug for TB

Because so many people in the world suffer from TB, and because the bacterium is becoming resistant to the antibiotics that we use against it, we need to try to find new drugs if we are to win the fight to bring TB under control.

Pharmaceutical and biotechnology companies spend huge amounts of money doing research to try to find new drugs. They may have to make and test hundreds of compounds before they find something that they think will work. They may try it out to see what effect it has on bacteria, or on human cells grown in dishes in a laboratory. If it does well, then they may begin trials on humans.

Any new drug has to go through many rigorous tests before it is allowed onto the market.

1 The first stage of a drug trial involves giving the drug to around 50 volunteers. The main aim of this is to find out if the drug is safe. The volunteers will be closely monitored throughout the trial, so that the researchers can find out as much detail as possible about how it is affecting them. They will record any side effects. They will try out different doses of the drug, to find out the range of safe doses.
2 If the drug passes that stage safely, then it will be tested on people who have the illness it is meant to be helping. About 200 to 400 people are likely to be involved in this stage. The main aim of this is to find out if the drug works, and the best dose to give people. The people in the trial will again be carefully monitored by the researchers. They may include men and women, elderly people and young ones – a whole range of the kinds of people who might want to use the drug.
3 If the drug passes stage 2, then it will be tested on a much larger number of patients, usually several thousand. This time, the drug is used in the same way that it will be when it is marketed. For example, the people in the trial may simply be given the drug in a packet and asked to take say three tablets a day.

Figure 4.26 This person is taking part in the trial of a drug that may lessen some of the symptoms of cystic fibrosis

You can imagine that all of this is very expensive. What's more, many drugs will fail part way through the trials, so the drug company never makes any money out of them. They only get their money back from really useful drugs that make it through all the stages of the trials.

Because of this, drug companies aren't usually interested in finding new drugs to treat diseases that happen mostly in developing countries. They know the governments there don't have much money to spend on drugs. TB is a disease like that. Most TB cases are amongst poor people in the developing world. So, to develop new drugs against TB, some other source of funding has to be found.

In 1995, a biotechnology company called Chiron found a new drug, which they called PA-824, that looked as though it would be a good antibiotic to use against TB. They decided to work closely with an organisation called the Global Alliance for TB Drug Development. This organisation gets funds from governments and charities in many different countries. It was able to find the money and expertise that was necessary to take PA-824 through all the stages of the necessary trials.

PA-824 began the first stage of trials in 2005. By the time you read this, it may have gone further – or it may have failed. If it does get through the trials successfully, it will be the first new drug against TB for 20 years.

Summary

- Tuberculosis, TB, is caused by a bacterium.
- You can be infected with TB by breathing air that contains droplets or particles with TB bacteria in them.
- Most people with TB bacteria in their bodies don't get TB. But it still kills millions of people each year, mostly in developing countries.
- H TB is treated by drug therapy, using antibiotics. Some TB bacteria have become resistant to the antibiotics used against them. This happens most readily if people do not take the full course of antibiotics. In some countries, a scheme called DOTS is used to make sure TB patients take their antibiotics regularly.
- Developing new drugs is very expensive. They have to go through several stages of rigorous trials before they can be marketed.

Questions

4.8 Copy and complete this table to summarise what happens at each stage of a trial for a new drug.

Stage of trial	Number of people involved	Purpose of the trial	What is done
Stage 1			
Stage 2			
Stage 3			

4.9 Explain the reasons for each of the following.

a People living in crowded conditions are more likely to get TB than other people.

b As the number of cases of HIV/AIDS increases, so does the number of TB cases.

H c People with TB are usually prescribed two antibiotics at once.

5 Patterns in properties

A Identifying substances

The job of a forensic scientist is to collect evidence from the scene of a serious crime, analyse it and match it to material from a suspect, to prove that there has been contact with the victim. Wherever we go, we leave traces of ourselves behind. It may be hairs from our heads or fibres from our clothes. We also take things away with us from where we have been – perhaps soil on our trainers or paint flakes on our clothes. By painstakingly collecting tiny particles and fragments, a forensic scientist could probably tell where we have been and whom we have been in contact with.

When the evidence has been collected, it is bagged up in polythene bags and taken to a laboratory for analysis. At every stage great care must be taken to avoid contaminating the evidence through contact with other substances from other cases, or even from the forensic scientist herself. If a defence lawyer, in a courtroom, can prove contamination then the evidence has no value and the criminal could go free.

Figure 5.1 A forensic scientist at work

Flame tests

It has been known for hundreds of years that certain chemical compounds, when heated in a flame, will colour the flame. Sodium compounds will colour the flame yellow, while potassium produces a lilac flame.

Metal compounds are also called metal salts.

A **compound** contains the atoms of at least two different elements chemically bonded together. Sodium chloride is a familiar compound that contains sodium and chlorine.

To perform a flame test, a length of nichrome wire is dipped into water and then into the metal compound that you wish to test. This makes the compound stick to the wire and it is then placed in a colourless Bunsen flame. Figure 5.2 shows most of the metals that can be detected by a flame test. If a forensic scientist finds one of these metals in a powder from the crime scene then it helps in identifying the powder. We say the forensic scientist is analysing the powder. **Analysing** means finding out information – in this case discovering the identity of the metal in the powder.

Scientists often call items of information they produce **data**. This data comes from observations and measurements. The forensic scientist who performs the experiment is producing **primary data**. If other scientists then use this data, it is called **secondary data** because these scientists did not actually obtain the data themselves through an experiment.

calcium: brick red

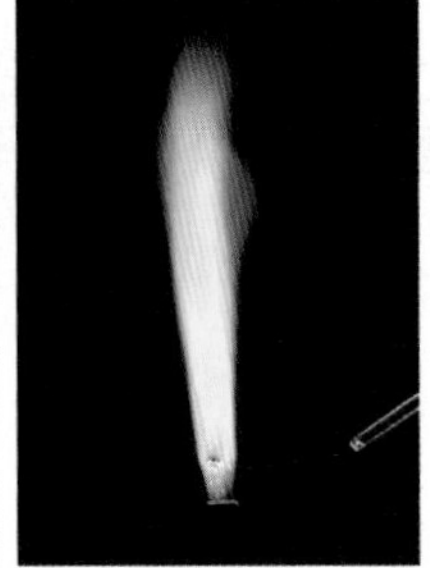
sodium: yellow

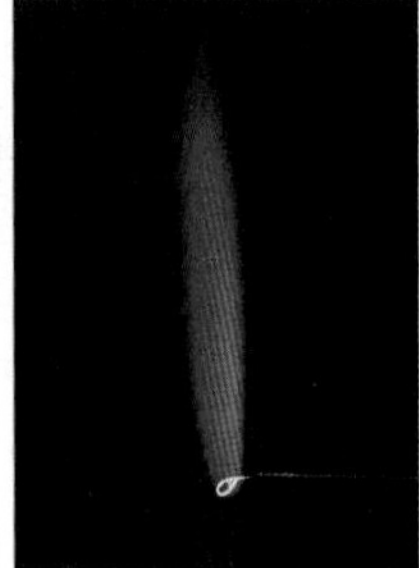
lithium: deep red

copper: blue–green

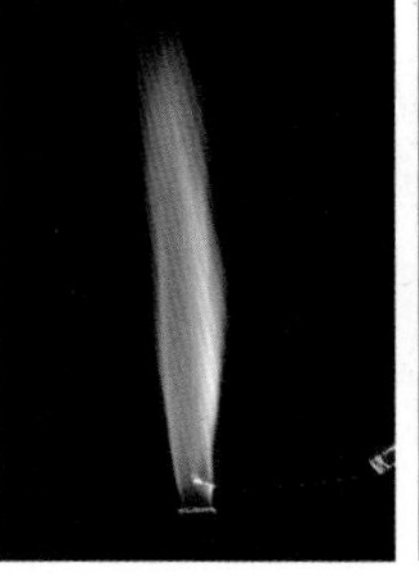
potassium: lilac

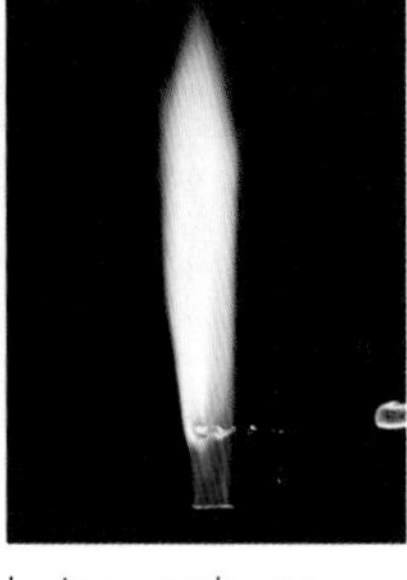
barium: apple green

Figure 5.2 The flame colours produced by different metal compounds show the presence of the metal

Forensic scientists usually collect evidence from crime scenes and take it back to analytical laboratories. To a scientist, **analytical** describes the processes used to find out information about a substance. The data produced from analysis is **analytical data**.

Another way of testing compounds

Most metals do not give a definite colour when heated in a flame, so other methods of identifying them in their compounds must be used.

The **transition metals** are a collection of metals that have some properties in common. One of these properties is that they form coloured compounds. If an unknown compound is coloured then it could well contain a transition metal.

One way to find out which metal is in a compound is to dissolve it in water and add sodium hydroxide solution. If a coloured precipitate is formed then it may be possible to identify the metal.

Copper in solution gives a jelly-like, light blue precipitate called copper hydroxide, whereas iron in solution can give one of two coloured precipitates – either green iron(II) hydroxide or brown iron(III) hydroxide. Zinc is often grouped with the transition metals but solutions of its compounds give white precipitates.

In metal compounds, the metal is in the form of ions. When metal atoms bond in a compound they form charged particles, which we call ions. Metal ions are always positive. Iron(II) in the name of a compound tells you that the iron ion has a 2+ charge. It has a symbol, Fe^{2+}.

A **solution** is formed when a solid dissolves in water.

A **precipitate** is formed when two solutions mix to give an insoluble solid.

All atoms and ions can be represented by internationally recognised **symbols**.

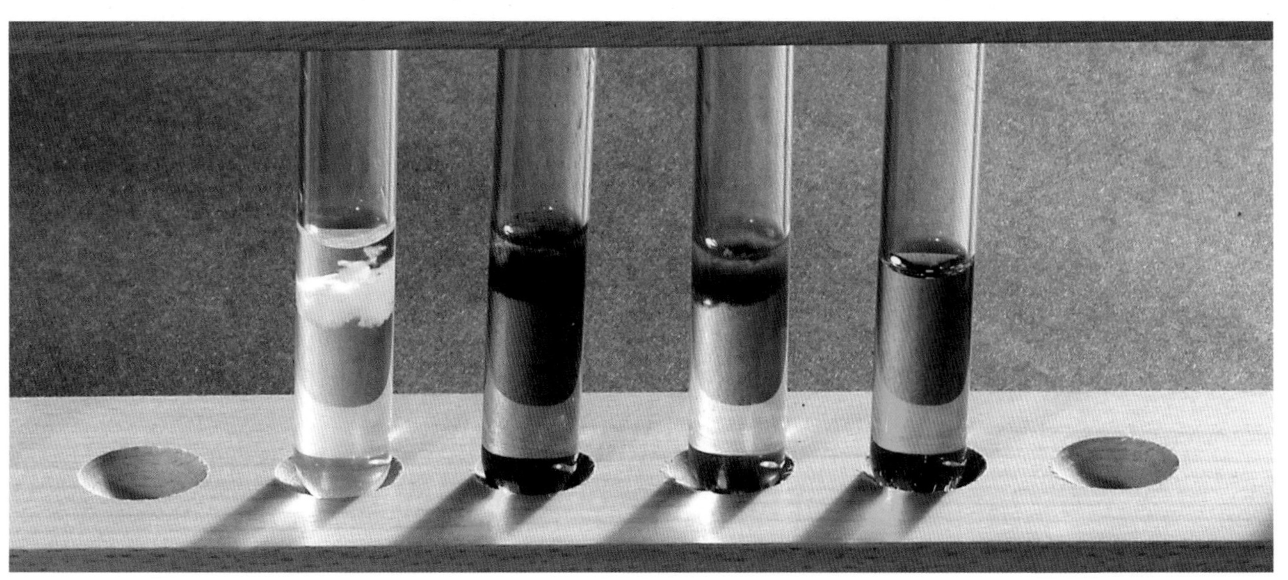
zinc: white precipitate iron(III): brown precipitate iron(II): green precipitate copper: light blue precipitate

Figure 5.3 A transition metal compound gives a coloured precipitate when its solution is reacted with sodium hydroxide

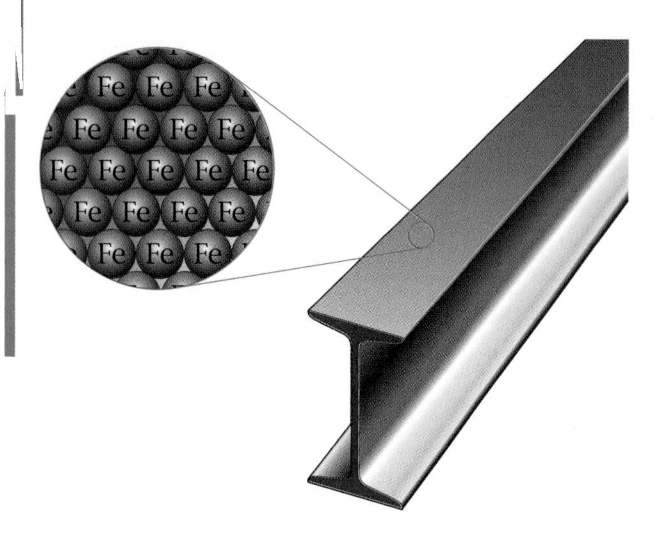

Figure 5.4 All the atoms in this iron girder are Fe atoms

Identifying elements

Elements are substances that cannot be changed into anything simpler; they are made up of atoms. Each different element is made up of only one type of atom. For example, the element iron only contains iron atoms (symbol Fe).

Just as the forensic scientist uses characteristic properties to identify substances from a crime scene, we can use the properties of elements to identify them and to understand the uses to which they are put.

Look at Table 5.1. It is full of secondary data. We can begin to see that there are patterns in some of the properties of the elements.

Element, symbol	Colour and physical state	Melting point (°C)	Boiling point (°C)	Conducts electricity when solid?	Reactivity
argon, Ar	colourless gas	−189	−186	no	Very unreactive. Until a few years ago it was thought to have no reactions with other elements. It is found uncombined. It is non-toxic
chlorine, Cl	pale green gas	−101	−38	no	Very reactive with most other elements. It reacts with metals to form chlorides. It dissolves in water to form compounds that kill bacteria and bleach material. It is highly toxic
copper, Cu	shiny brown solid	1083	2595	yes	Not very reactive and so can be found uncombined on Earth. It will react with carbon dioxide and water to form a green coating on its surface. It does not react with water on its own. It is not toxic
gold, Au	shiny yellow solid	1063	2970	yes	Very unreactive. It is found uncombined on Earth. It does not tarnish. It is non-toxic
helium, He	colourless gas	−270	−269	no	Totally unreactive. It is found on Earth uncombined. There are no known compounds. It is non-toxic
iodine, I	black solid	114	184	no	Fairly reactive. It will combine with metals to form iodides. Dissolves in alcohol to form an antiseptic that kills bacteria. It is highly toxic
iron, Fe	shiny grey solid	1535	3000	yes	Fairly reactive. It reacts with oxygen and water to form rust, which is iron oxide. It is non-toxic
neon, Ne	colourless gas	−249	−246	no	Totally unreactive. It is found on Earth uncombined. There are no known compounds. It is non-toxic
silver, Ag	shiny silvery solid	961	2210	yes	Very unreactive; non-toxic

Table 5.1 Properties of some elements

What patterns emerge? We can see that copper, silver and gold are all found uncombined on Earth. This means that they can be found as elements, rather than bonded with atoms of other elements in compounds. This is because they are very unreactive. They have high melting and boiling points.

The **reactivity** of elements means how easily they can form compounds. The most reactive elements form compounds very easily.

Can you think of another reason why copper is used for electrical cables?

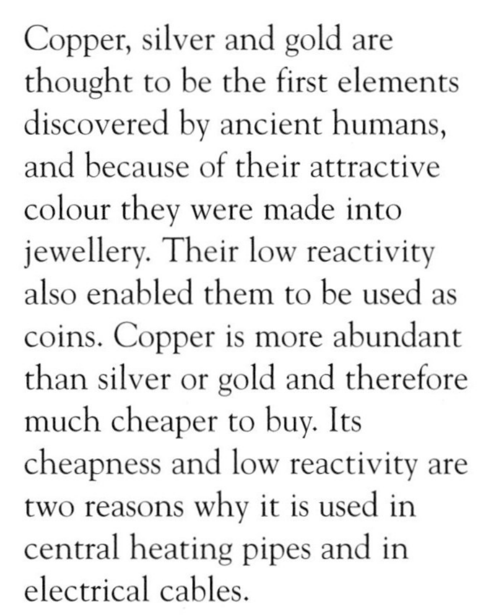

Copper, silver and gold are thought to be the first elements discovered by ancient humans, and because of their attractive colour they were made into jewellery. Their low reactivity also enabled them to be used as coins. Copper is more abundant than silver or gold and therefore much cheaper to buy. Its cheapness and low reactivity are two reasons why it is used in central heating pipes and in electrical cables.

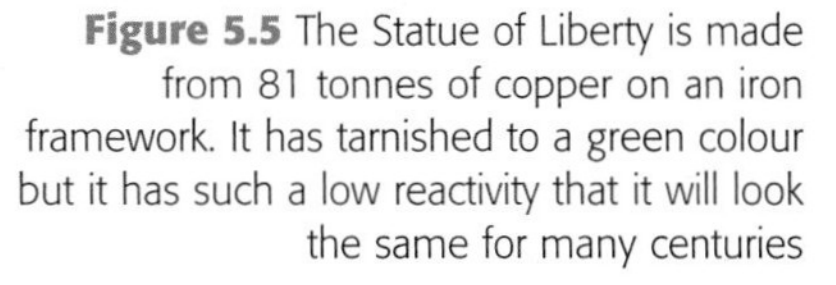

Figure 5.5 The Statue of Liberty is made from 81 tonnes of copper on an iron framework. It has tarnished to a green colour but it has such a low reactivity that it will look the same for many centuries

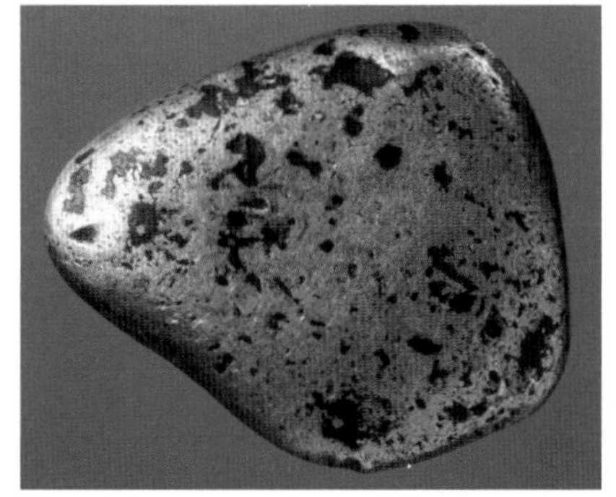

Figure 5.6 It is still possible to find gold uncombined today. This nugget was found in a Welsh stream in 1997. It weighs 21 g – this was a very rare find

Iron was discovered much later. It was chemically fairly easy to extract from iron oxide ores. It is probably the element that has most transformed the physical world we live in today. Iron is fairly easy to bend and shape, and so it is used to make chains, railings and fences.

Figure 5.7 Painting the Golden Gate Bridge, San Francisco

Iron is made stronger by combining it with a small amount of carbon to make steel. Steel is not a compound; it is an **alloy**, which means it is a mixture of more than one element, in this case iron and carbon. The strength of steel means it is used to make the girders that hold up many of our modern buildings and bridges. The major drawback with iron or steel is that it rusts. This is because the iron reacts with air and water. The usual way to prevent rusting is to paint the metal.

The other five elements in Table 5.1 are all non-metals. If you look at their properties, they form two distinct groups. Argon, neon and helium are all colourless gases that are either totally unreactive, or almost so. Iodine and chlorine are coloured elements that are reactive and highly toxic.

Some of the uses of argon, neon and helium are related to their unreactivity. Helium is less dense than air and so it is used in party balloons, weather balloons and airships. The only other gas that has this property is hydrogen, but this forms an explosive mixture with oxygen and so is too dangerous for these uses.

You probably have more argon in your home than you think. Ordinary light bulbs are filled with argon because it will not react with the tungsten filament that produces the light.

Neon is used in orange streetlights. When an electric current is passed through neon it glows orange. This phenomenon is also used in advertising signs, making our cities colourful places at night. The

Figure 5.8 Laser light decorations and neon signs in Oxford Street, London

colours can be varied by mixing neon with other atoms, such as argon, or by coating the glass tube containing the neon.

Helium, neon and argon also find uses in lasers and produce some of the fantastic effects of laser light shows.

Chlorine and iodine are the remaining two elements in the table. You will notice that both are toxic and both will kill bacteria. One of the main uses of chlorine is to kill bacteria in water supplies. Chlorine is added at water purification plants so that what comes out of your tap is bacteria-free. Iodine is dissolved in alcohol to make an antiseptic solution. This is wiped over wounds to prevent infection by bacteria.

Summary

- A flame test involves heating a small amount of substance on a nichrome wire in a Bunsen flame.
- Some metals give characteristic flame colours when they are present in compounds.
- Analysing means finding out items of information. In chemistry this is often information about a substance.
- A precipitate is formed when two solutions mix to give an insoluble solid.
- Solutions of transition metal salts form coloured precipitates with sodium hydroxide solution. This can be used to identify the transition metal.
- Elements are made up of atoms; each different element is made up of a single type of atom.
- Different elements have different properties. These properties, particularly the chemical reactivity, affect how we use the elements.
- Compounds contain the atoms of at least two different elements chemically bonded together.
- In some compounds, atoms become ions when they bond together.

Questions

5.1 What are the symbols for atoms of chlorine, iodine, helium, neon, silver and gold?

5.2 Explain the difference between an element and a compound, in terms of atoms.

5.3 Look back at the data in Table 5.1.

- **a** Give *two* reasons why iron, copper, silver and gold are classed as metals.
- **b** Give *three* reasons why copper, rather any of the other metals, is used in household electrical cables.

5.4 Think about a typical day in your life. What would you miss if iron and steel suddenly disappeared?

5.5 When collecting and testing evidence from a crime scene, it is essential to:

- accurately record where it was found
- bag each sample in a separate polythene bag
- clean down the laboratory bench thoroughly before opening an evidence bag.

Explain why each of these stages is so important.

5.6 A forensic scientist collects some powder that is stuck deep into the treads of a car tyre. In the laboratory she analyses the powder and finds it to be a mixture of two metal compounds. Here are the *analytical data* from two of the tests she carried out:

Test 1 – Flame test

The powder produced a lilac flame.

Test 2 – Some powder is dissolved in water and sodium hydroxide solution is added

A brown precipitate is formed.

a What is meant by the term *analytical data*?

b Use information from this analytical data to decide which two metals are present in the powder. Explain your reasoning.

B Bringing order to the elements

Did you know that many scientists think the atoms in your body were made in stars that exploded long ago?

Most scientists now believe that the Universe began with a gigantic explosion, called the Big Bang, about 14 billion (14 thousand million) years ago – see page 254. After the first fury of this explosion, tiny particles came together to form the first atoms, which were hydrogen atoms. Some of these hydrogen atoms were then pushed together as a result of tremendous forces present, and they fused to make helium atoms. Hydrogen still makes up 92% of the atoms in the Universe, and helium another 7%. This leaves only 1% for the atoms of all the other elements. But where did all the other different kinds of atoms come from?

After a hundred million years or so, the Universe had cooled down enough to allow clumps of atoms to come together, attracted by their own gravity. As more and more atoms came closer together, the clumps of matter became denser and hotter, until they started to glow. These were the first stars. In these stars hydrogen and helium atoms fused when they crashed into each other in the enormous heat. This produced other, heavier atoms such as carbon. Some stars became huge, and inside them atoms were fusing to make even heavier elements. These large stars eventually became unstable and exploded, blasting out the elements they had made into the Universe.

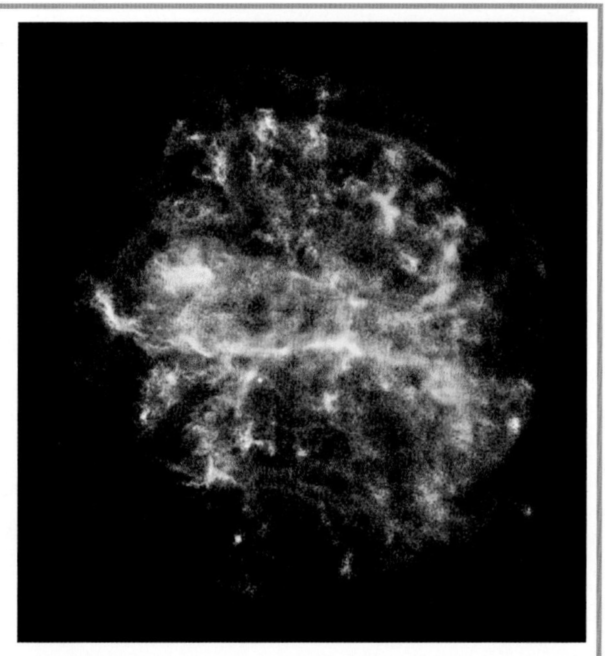

Figure 5.9 The rapidly expanding shell of gas from a 'supernova' – a star that exploded in our galaxy. The shell is 36 light years across and contains large amounts of oxygen, neon, magnesium, silicon and sulphur. These elements were produced deep within the star and were ejected by the explosion

Much later, the atoms from these exploded stars came together again to make a new generation of stars. Our Sun is one of these. The planets that surround it are the recycled remains of the very first stars, which created the many different elements that we now find on Earth.

The Periodic Table

Most chemistry laboratories have a Periodic Table on the wall. This is because chemists find it so useful. For a start, you can find the symbols of all the elements on it. But this is not its only use. The construction of this table was one of the most important scientific breakthroughs of all time.

Figure 5.10 Dimitri Mendeleev (1834–1907), who brought order to the elements in his Periodic Table

It all happened in St Petersburg over three days in 1869. Dimitri Mendeleev, a chemistry professor, had been writing a textbook and he wanted to make sure he had grouped the elements in a logical way. Other chemists had tried to do this but none had succeeded in producing any convincing order. Part of the problem was that new elements kept being discovered – 32 had been discovered in the previous 100 years, to bring the total in 1869 up to 62 elements. Unfortunately no one knew if there were any more to be found. We now know that 91 elements occur naturally on Earth.

Mendeleev became more and more frustrated, and three nights without sleep, wrestling with the problem, did not help. He put all the information about the properties of the different elements on cards. He felt that the order of the elements must have something to do with the masses of their atoms. Finally, exhausted, he fell asleep – and in his own words: 'I saw in a dream a table where all the elements fell into place as required. Awakening I immediately wrote it down on a piece of paper'. The table he produced is not very different from the one we use today.

What is so remarkable about Mendeleev's Periodic Table is that he left gaps for several elements that had not been discovered, and predicted the properties they should have. Once chemists knew what they were looking for, they quickly found the missing elements. This is an example of a scientific theory explaining not only what is known but leading to predictions of what is not known. What made scientists accept Mendeleev's ordering system was that when the missing elements were discovered, they did indeed have similar properties to the ones he had predicted.

Patterns in the Periodic Table

If you take just the first 20 elements and put them in order of the masses of their atoms, just as Mendeleev did, you get a very long row.

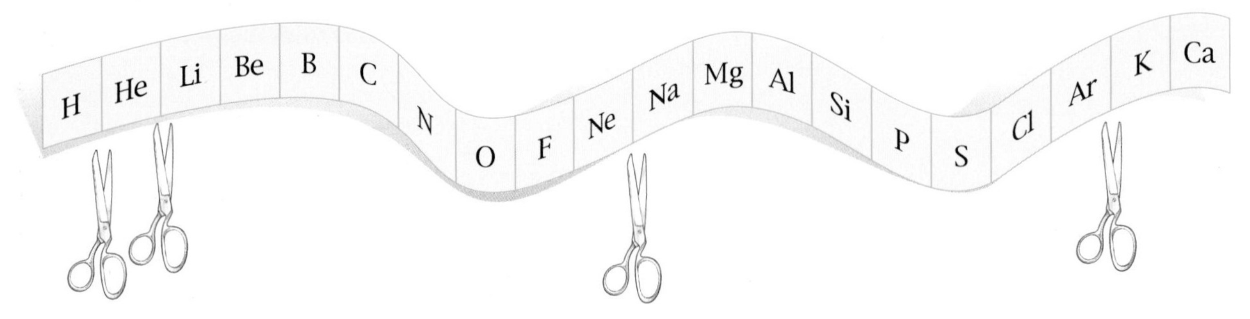

Figure 5.11 The first 20 elements in order of their masses

This long ribbon of elements can now be cut to make it more compact by putting elements with similar properties together. In the last section you read that helium, neon and argon are all very unreactive, colourless gases. (See Table 5.1 on page 84). These can be grouped together by putting them underneath each other. Another set of elements with similar properties is lithium (Li), sodium (Na) and potassium (K). These are all soft metals that are very reactive with water. These too can be put underneath each other. Look at Figure 5.12.

Figure 5.12 Beginning to group the elements

H							He
Li	Be	B	C	N	O	F	Ne
Na	Mg	Al	Si	P	S	Cl	Ar
K	Ca						

Continuing this same process of grouping elements together produces a more complete version of the Periodic Table, Figure 5.13.

Here are some things you should notice about the Periodic Table:

- Each element is given a number called the **atomic number**. This corresponds to the element's position in the table. Although most elements are in the order of their atomic masses, some are slightly out of order so that they can be grouped correctly.
- All the metals are on the left-hand side and in the middle. Most of the elements in the table are in fact metals. The non-metals are on the right-hand side.
- Rows in the Periodic Table are called **periods**. We start a new row when we need to put elements with similar properties underneath each other.
- Columns are called **groups**. Elements in a group all have similar properties.
- In the middle of the table is a large block of metals called the **transition metals**. These are dense metals that form coloured compounds. These are the metals that are mainly used to make things in the world around us.

> Anything that regularly repeats can be called a period. In school, your day is divided into periods that will be repeated tomorrow and the next day.

Figure 5.13 The Periodic Table

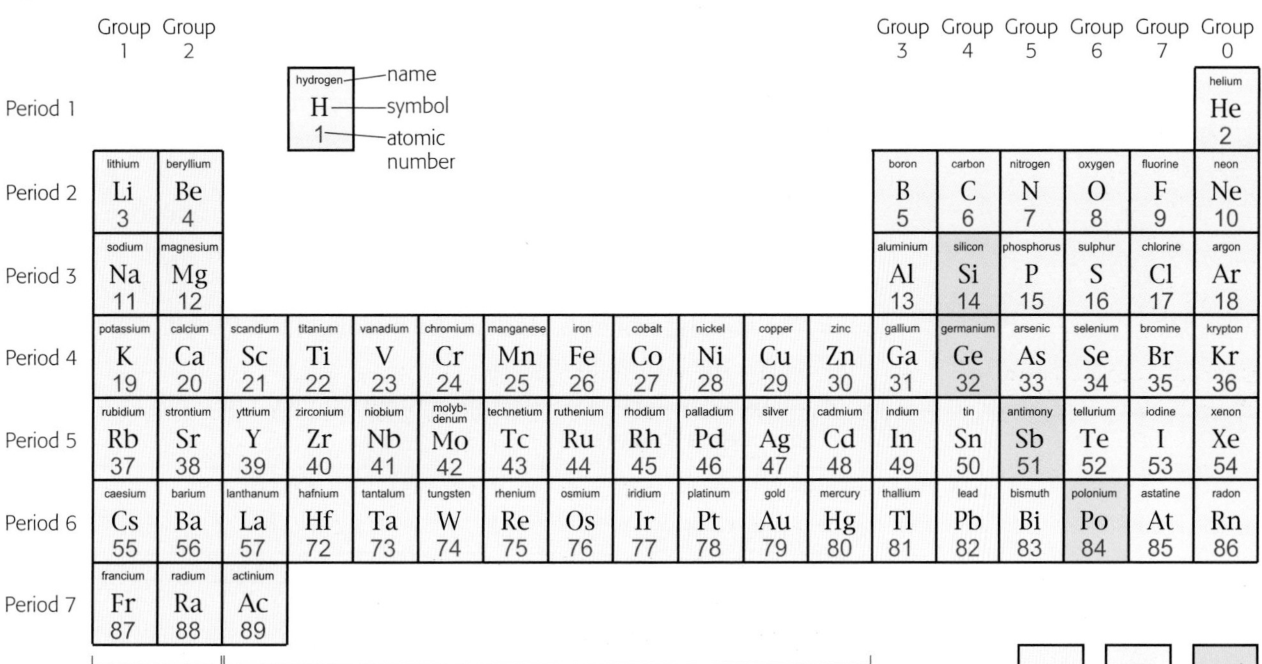

H

Using secondary data to explore how the Periodic Table was devised

Mendeleev found patterns in the properties of elements. Most of the data he used was from the observations and measurements of other scientists. This means that he used **secondary data**. From this secondary data he started to make groups of elements with similar properties. He noticed, for example, that one group of these elements were soft metals with low densities and fairly low melting points (for metals). They also had very similar reactions with water – they fizzed, produced hydrogen, and formed an alkaline solution. These metals are in Group 1 of the modern Periodic Table.

The genius of Mendeleev was that, after he put all the elements in order of atomic mass, he did three things:

- Every so often the pattern of properties of elements began to repeat, so he put these underneath one another, forming columns of elements with similar properties.
- Some elements seemed to be out of place if they were put in order of the masses of their atoms. For example, cobalt has a heavier atom than nickel but its properties suggest it should come before nickel, so he swapped cobalt and nickel. Similarly, argon atoms are heavier than potassium atoms but argon is an unreactive gas rather than a soft metal, so the order is swapped. If you look at the Periodic Table (Figure 5.13) you can see that argon is grouped with the other unreactive gases in Group 0 (called the noble gases) and that potassium is in Group 1 (the alkali metals). The placing of elements in the Periodic Table is by order of atomic number, not atomic mass.
- He left gaps for elements that had not been discovered.

None of the Group 0 elements had been discovered when Mendeleev proposed his Periodic Table. You can read about their discovery on page 97.

Summary

- In the Periodic Table, elements with similar properties are put underneath each other in columns called groups.
- The atomic number of an element corresponds to its position in the Periodic Table.
- Metal elements make up most of the Periodic Table and they occupy the left and the centre. Non-metals are found on the right.
- The rows of the Periodic Table are called periods.
- The metals forming a block in the middle of the Periodic Table are called the transition metals.
- H Mendeleev used secondary data to group elements with similar properties together.

Questions

5.7 What is the identity (name and symbol) of the following elements?

- **a** The element with an atomic number of 8.
- **b** The element in Period 3, Group 7.
- **c** The lightest element in Group 0.

5.8 What type of metal is iron?

5.9 Which Group 0 element is used to fill filament light bulbs? [Hint: Look back at page 85 to remind yourself of the uses of this element.]

5.10 Like chlorine and iodine, bromine is in Group 7. Use Table 5.1 (page 84) to predict two properties of bromine.

H 5.11 a Use secondary data from this chapter to explain why helium, neon and argon are grouped together in the noble gases group. [Hint: There is information about these elements on pages 84, 85 and 86.]

b The order of elements in the Periodic Table is by atomic number. Using argon as an example, explain why putting elements in order of atomic mass does not work.

C A closer look at Group 1

You are very unlikely to see a sample of caesium close up, because it is the second most reactive metal on our planet. The inventor of the Bunsen burner, Robert Bunsen, together with another famous scientist called Gustav Kirchhoff, discovered it in 1860 by viewing flame tests using a special piece of equipment they had invented called a spectroscope.

Caesium is a Group 1 element. It is extremely reactive with water – it explodes on contact. For this reason caesium is stored under oil. Like other members of Group 1 it is a soft metal. It doesn't have many uses but one of the most interesting is in atomic clocks. These keep very accurate time – they lose only 5 seconds in 300 years.

We said caesium is the second most reactive metal on Earth, because the most reactive metal is believed to be francium. Mendeleev left a gap for it in his Periodic Table and in 1870 predicted it would have similar properties to caesium. It was not until 1939 that it was actually discovered, by Marguerite Perey, working in Paris. It is not surprising it took so long to discover this element because it is highly unstable and there is thought to be only 20 g of it in the whole of the Earth's crust! There has never been enough collected to test its chemical properties, or to photograph its appearance. Marguerite Perey was allowed to name it francium – in honour of her country, France – by the International Union of Chemists in 1949, ten years after she had discovered it.

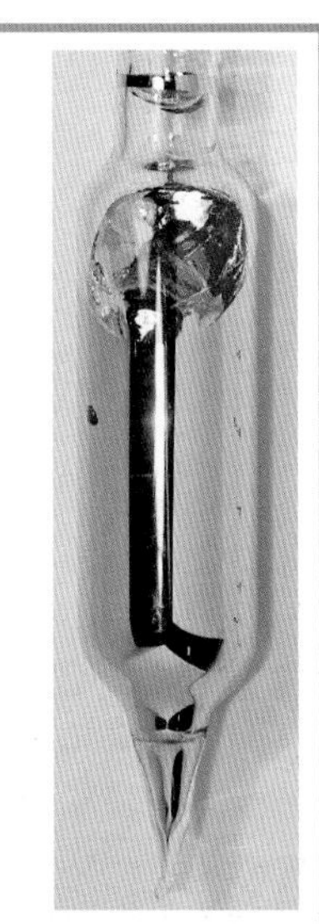

Figure 5.14 Caesium is too reactive to use in most chemical laboratories. This sample is stored in glass, away from air and water

Group 1: the alkali metals

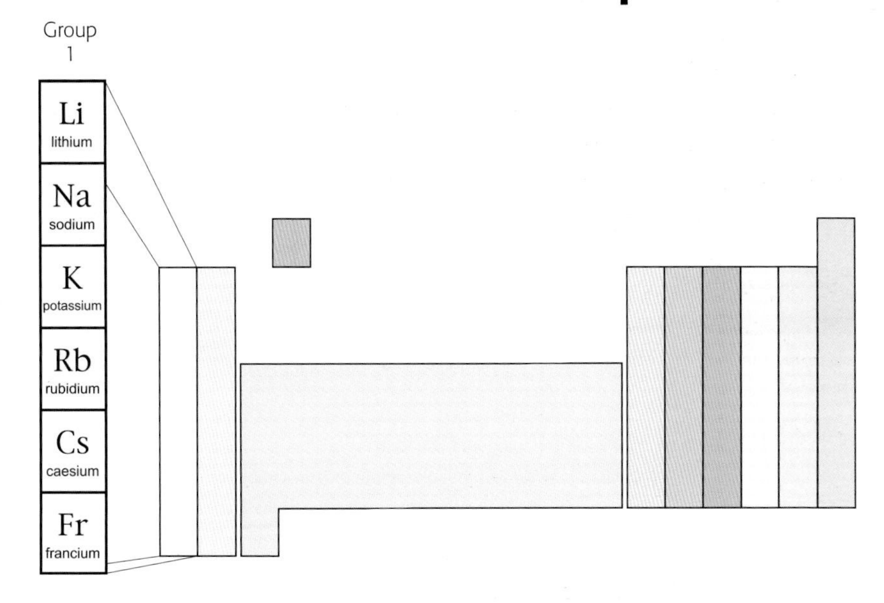

Figure 5.15 Group 1 of the Periodic Table

There are eight groups of elements in the Periodic Table. Each group contains elements with similar properties. Group 1 contains the most reactive metal in each period of the table.

To most people, metals are hard, shiny and good conductors of heat and electricity. So it comes as quite a surprise to realise that chemists call the Group 1 elements metals. For a start, they are soft and you can cut them with a knife. They also look very dull – that is until you cut them, when the inside really does have a metallic shine.

Figure 5.16 Lithium, sodium (shown here) and potassium can easily be cut with a knife to reveal shiny silvery metal

They don't stay shiny for long though – they rapidly tarnish in air and go dull again. This is because these metals are very reactive, and react with air and water. They have to be stored in oil.

However, apart from their softness they do have the familiar properties of metals. They conduct heat and electricity very well.

Reaction with water

All the Group 1 metals react vigorously with cold water to produce hydrogen gas and an alkaline solution. This is why chemists call them the **alkali metals**.

If you cut a little piece of lithium and put it in water, the first thing that becomes obvious is that it floats. All the alkali metals are less dense than water and so they float on top of it. The next thing you notice is that bubbles of gas start coming off from where the lithium touches the water. This is the hydrogen. At the end of the reaction the pH of the solution can be tested and this proves there is now an alkaline solution. Lithium hydroxide, a compound formed in the reaction, produces this alkaline solution.

Figure 5.17 Lithium is reacting with water to produce lithium hydroxide. The green colour of universal indicator is changing to purple, showing the solution is strongly alkaline

Another observation you might make is that the lithium melts and forms a ball. This is because the reaction gives out heat. Many chemical reactions give out heat – these are called **exothermic** reactions. There are also some reactions that take in heat, making the surroundings colder – these are called **endothermic** reactions.

As sodium is in the same group as lithium, we expect it to have a similar reaction with water. It does, but this time the **rate of reaction** is faster. This is because sodium is more reactive than lithium. It again produces an alkaline solution, this time of sodium hydroxide, and gives off hydrogen.

Potassium is even more reactive than sodium and so it reacts even faster with water. The reaction is quite violent and produces enough heat to set the hydrogen alight. This is why, when the potassium touches the water, it appears to burn with a lilac flame.

So you can see that as we go down the group, the reactivity of the alkali metals increases. Another way of saying this is:

The reactivity increases as the atomic number increases

Figure 5.18 Potassium reacts faster than sodium or lithium, and has a violent reaction with water. The potassium gets so hot that it ignites the hydrogen produced

H

Representing chemical reactions by equations

The reaction of lithium with water can be represented by a **word equation**:

lithium + water → lithium hydroxide + hydrogen

This shows us what has reacted – the **reactants**, and what is produced – the **products**. It is the lithium hydroxide that causes the water to become alkaline.

Using formulae equations

We can also show the reaction of lithium with water by using a **formulae equation**:

$$Li + H_2O \rightarrow LiOH + H_2$$

Look more closely at this equation. When we use **formulae** (symbols) we show which atoms are present in each substance, and in what proportions.

- In lithium there are only lithium (Li) atoms because it is an element and so made up of only one type of atom.
- Hydrogen is also an element because there are only hydrogen (H) atoms present.
- H_2 tells us there are two H atoms joined together in a **molecule** of hydrogen.
- Water (H_2O) and lithium hydroxide (LiOH) are both compounds because they have more than one type of atom in their formulae.
- H_2O has two atoms of hydrogen and one atom of oxygen bonded together.
- LiOH has one atom each of lithium, oxygen and hydrogen. If there is no small number after the atom then there is only one atom in the formula unit.
- The equation is **unbalanced**. There are two atoms of hydrogen on the left but there is an extra one on the right.

A molecule contains atoms bonded together.

Chemical reactions do not create or destroy atoms, so we need to **balance** the equation so that there is the same number of each different type of atom on each side of the equation. The easiest thing to do would be to change H_2O to H_3O, but this changes the formula of the molecule and it is not water any more. In fact H_3O does not exist. We need to change how many of each **formula unit** we have so that the atoms balance. This is how we do it:

$2Li$	+	$2H_2O$	→	$2LiOH$	+	H_2
2 × 1 = 2 lithium atoms		2 × 2 = 4 hydrogen atoms		2 × 1 = 2 lithium atoms		2 hydrogen atoms
		2 × 1 = 2 oxygen atoms		2 × 1 = 2 oxygen atoms		
				2 × 1 = 2 hydrogen atoms		

Now there are two lithium atoms, two oxygen atoms and four hydrogen atoms on both sides of the equation.

H

Sodium reacts with water to produce similar products to lithium because it is in the same group of the Periodic Table. Let's go through the steps of writing a balanced formulae equation for this reaction.

1 Write the word equation.

sodium + water → sodium hydroxide + hydrogen

2 Change the words into formulae.

$Na + H_2O \rightarrow NaOH + H_2$

Remember the formulae are fixed.

3 Balance the equation by making the number of atoms of each element the same on both sides.

$2Na + 2H_2O \rightarrow 2NaOH + H_2$

Summary

- Group 1 elements are called the alkali metals.
- Alkali metals are the most reactive metals in each period.
- Alkali metals are soft and only look shiny when cut.
- Alkali metals all react with water.
- The reactivity of alkali metals increases with increasing atomic number, so potassium is more reactive than sodium, and sodium is more reactive than lithium.
- Chemical reactions (such as the reactions of alkali metals with water) happen at different rates. The rate of reaction of potassium with water is much faster than that of lithium.
- Some chemical reactions give out heat. These reactions are called exothermic reactions. In the reaction of the alkali metals with water, the heat is enough to melt the metal.
- Some chemical reactions take in heat. These reactions are called endothermic reactions.
- A molecule contains atoms bonded together. It may be an element (e.g. H_2) or a compound (e.g. H_2O).
- Reactions can be represented by word equations.
- H In a balanced formulae equation, the number of atoms of each element must be the same on both sides of the equation.

Questions

5.12 Which of potassium, lithium or sodium has the highest atomic number?

5.13 Why are Group 1 metals called the alkali metals?

5.14 **a** Which properties of the alkali metals make them unusual metals?

b Which properties of the alkali metals suggest they really are metals?

c Why are sodium, lithium and potassium stored under oil?

5.15 Write the *word equation* for the reaction of potassium with water.

H **5.16** Work out the *balanced formulae equation* for the reaction of potassium with water.

5.17 Rubidium is another Group 1 element (see Figure 5.15).

a Find out who discovered the element rubidium (Rb).

b Predict three properties of the element rubidium.

c Will rubidium be more or less reactive than potassium? Explain your reasoning.

Two new superheavy elements weigh in

On 1 February 2004, a team of researchers from Russia and the USA working in a Russian laboratory made an announcement to the scientific world – they had discovered two new elements, with atomic numbers of 113 and 115. This was big news, reported to the public the following day by *New Scientist*.

The last naturally occurring element in the Periodic Table is uranium, with an atomic number of 92. However, for the past 50 years scientists have been able to make 'artificial' elements by crashing different particles into atoms.

The two elements that were created in the Russian particle accelerator were called Ununtrium (meaning one-one-three) and Ununpentium (one-one-five). They have symbols Uut and Uup. These may seem very uninspiring names for two newly discovered elements, but naming elements can give rise to bitter disputes between different groups of scientists. Eventually there will be agreement about a name but it can take years. When the element with an atomic number of 101 was named there was no dispute. It was named Mendelevium in honour of Mendeleev, the creator of the Periodic Table.

Group 7: the halogens

The name halogen comes from two Greek words, *hals* for salt and *gen* for producer. The Group 7 elements were named in this way because they all occur in salts found dissolved in seawater. They were discovered about 200 years ago and are all called after Greek words. Chlorine was named after the Greek word *chloros* meaning green, as it is a pale green gas. *Bromos* is a Greek word meaning stench and gives us the name bromine. In fact all the halogen are very pungent smelling when they are gases. It is very dangerous to breathe in the vapour of any of them, as they are all highly toxic. Chlorine was used as a war gas in World War 1. Iodine has a violet-coloured vapour when heated – *iodos* is the Greek word for violet.

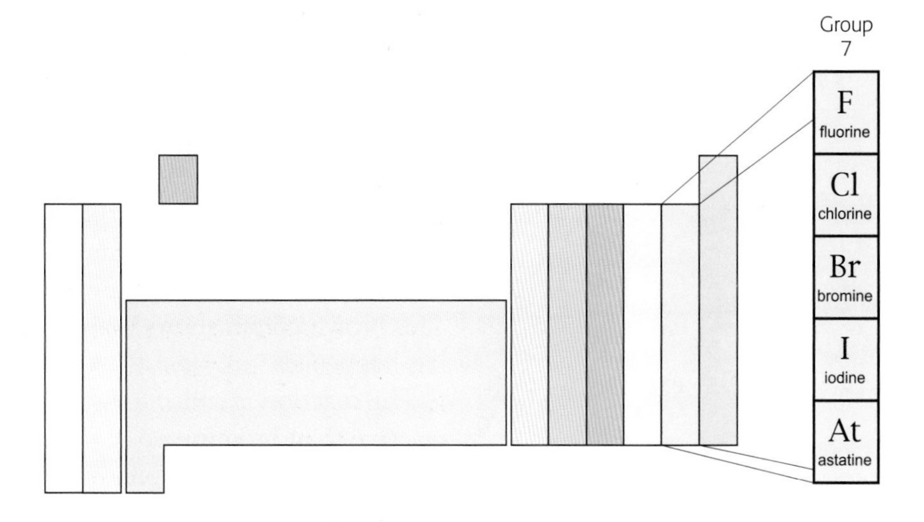

Figure 5.19 Group 7 of the Periodic Table

We have already met chlorine and iodine in the first section of this chapter, and considered some of their properties and uses (see page 86). Now we are going to look at them together with bromine to see why they have all been put in the same group of the Periodic Table. We shall not be considering fluorine because it is too reactive to be used in school, nor astatine as it is radioactive.

Figure 5.20 Chlorine, bromine and iodine

Comparing chlorine, bromine and iodine

All the halogens are reactive non-metals. Let's look first at some of their physical properties, shown in Table 5.2.

Table 5.2 Physical properties of some halogens

Property	Chlorine, Cl_2	Bromine, Br_2	Iodine, I_2
Colour	pale green	brown	black
Physical state at room temperature	gas	liquid	solid
Boiling point (°C)	–35	59	187

The table gives us the following information about these three halogens:

- Their molecules are **diatomic**. This means that they have two atoms bonded together in each molecule.
- They are all coloured elements.
- Their boiling points are fairly low and they increase down Group 7 with increasing atomic number.

The reactivity of the halogens

We saw that the alkali metals of Group 1 get more reactive as you go down the group. With the halogens, they become more reactive going *up* Group 7. So chlorine is more reactive than bromine, and bromine is more reactive than iodine.

It is easy to show this in the laboratory. If you add some chlorine water to a solution of a halide such as potassium bromide or sodium bromide, you will immediately see a brown colour (Figure 5.21). This brown colour shows that bromine has been produced. So what has happened?

The reaction is called a **displacement reaction**. Chlorine is more reactive than bromine, so it will displace (push out) bromine from its compound potassium bromide. This is easier to understand if we look at the word and formulae equations for this reaction:

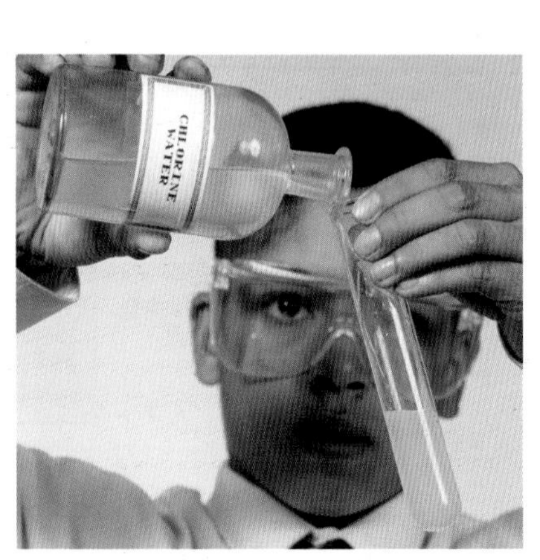

Figure 5.21 Bromine has been displaced by chlorine

chlorine + potassium bromide → potassium chloride + bromine

$$Cl_2 + 2KBr \rightarrow 2KCl + Br_2$$

The chlorine is now in a compound, potassium chloride, instead of the bromine.

The displacement of bromine from sodium bromide is similar, and is commercially very important because it is used to manufacture bromine. Sea water contains sodium bromide, and chlorine is added to sea water to displace the bromine. Sodium chloride is also produced.

H

Compounds of halogens are called **halides**. The ending *-ide* tells you that there are only two elements in a compound, for example ZnO is zinc ox*ide* and NaI is sodium iod*ide*.

Another common ending of compound names is *-ate*, and this usually tells you that oxygen is present – for example sodium carbon*ate* has the formula Na_2CO_3.

Chlorine is also more reactive than iodine, so you can see iodine forming when chlorine water is added to potassium iodide solution. Although iodine is a black solid, it looks dark brown in solution. Again we can see what is going on if we look at the word and formulae equations:

chlorine + potassium iodide → potassium chloride + iodine

$$Cl_2 + 2KI \rightarrow 2KCl + I_2$$

See if you can predict what will happen when bromine water is added to a solution of sodium iodide.

Chlorine displaces iodine from the potassium iodide solution because it is more reactive than iodine. The chlorine ends up in the compound potassium chloride.

Group 0: the noble gases

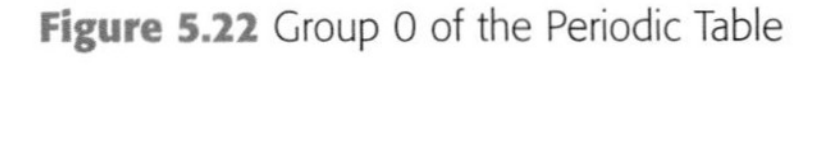

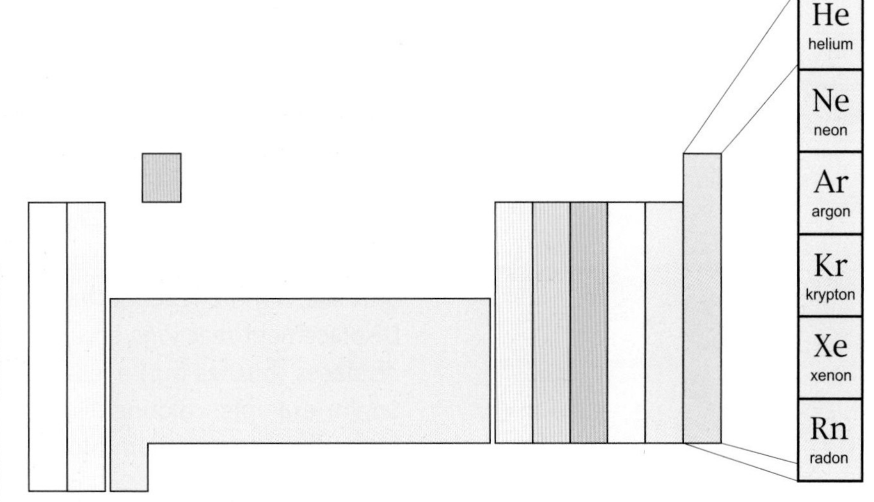

Figure 5.22 Group 0 of the Periodic Table

Figure 5.23 William Ramsey received the Nobel Prize for Chemistry in 1904 for his discovery of argon, helium and neon in air

When Mendeleev devised the first Periodic Table in 1869, Group 0 elements had not been discovered on Earth. The first noble gas to be discovered was argon in 1894, by Sir William Ramsey. He found it in the air – in fact we now know it makes up nearly 1% of the atmosphere. At first Mendeleev would not accept that argon was an element. This is because he had left no gap for it in his Periodic Table. Ramsey realised that there must be a whole group of elements that were still to be discovered, between Group 7 and Group 1. He went on to discover helium in 1895 and neon in 1898.

The discovery of Group 0 was a brilliant use of Mendeleev's Periodic Table because, once argon had been discovered, it was obvious that there must be other elements of Group 0 to be found. Far from causing scientists to reject Mendeleev's work it showed just how useful the concept of the Periodic Table was.

Naming the noble gases

Why are Group 0 elements called the noble gases?

When they were first discovered, these elements were found to be so unreactive that chemists did not believe they would react with anything. There were certainly no compounds of helium, neon or argon that had been discovered, so they did not appear to combine with any other element. Ramsey named argon from the Greek *argos*, which means 'idle', because it was so unreactive. Until 1962 no one could make a Group 0 element react. The group was called the 'inert gases' because a substance that doesn't react is said to be **inert**.

In 1962 the scientific world was stunned when Neil Bartlett, a British chemist, made the first 'inert' gas compound – a compound of xenon. Now only two members of Group 0 have not yet been made to form any compounds – neon and helium. However, the Group 0 gases are still chemically almost inert compared with other elements. The name of the group was changed to the 'noble gases', because they were likened to the very unreactive precious metals gold and platinum, which are sometimes called 'noble' metals.

Some of the properties and uses of argon, helium and neon were described on pages 84, 85 and 86.

Summary

- Group 7 elements are known as the halogens.
- Halogens are members of the same group of the Periodic Table, so they have similar properties.
- The halogens are coloured elements. Chlorine is green, bromine is brown and iodine is a black solid with a violet vapour.
- Boiling points of the halogens increase going down Group 7. At room temperature chlorine is a gas, bromine is a liquid and iodine is a solid.
- The reactivity of the halogens increases as you go up the group. So chlorine is more reactive than both bromine and iodine, and bromine is more reactive than iodine.
- Displacement reactions occur when a more reactive element displaces (pushes out) a less reactive element from its compound. So, for example, chlorine displaces bromine from a solution of a bromide such as sodium bromide.
- The noble gases are in Group 0. They include argon, neon and helium.
- The noble gases are chemically inert compared with other elements.

H

- The ending *-ide* in the name of a chemical compound tells you that there are only two elements bonded together, e.g. sodium bromide has the elements sodium and bromine bonded together.
- The ending *-ate* in the name of a compound usually tells you that oxygen is present in the compound.

Questions

5.18 Predict three properties fluorine might have as a member of Group 7.

5.19 Predict whether or not there will be a reaction when the following solutions are mixed:

- **a** iodine solution and sodium bromide solution
- **b** bromine water and sodium chloride solution
- **c** chlorine water and sodium iodide.

5.20 As the atomic number of the halogens increases, what happens to their reactivity?

5.21 **a** Write a word equation for the reaction of chlorine with sodium bromide.

- **b** Why is this reaction commercially important?
- H **c** Write a balanced formulae equation for the reaction.

5.22 What elements do you predict are in chemical compounds with these names:

- **a** calcium chloride
- **b** magnesium carbonate
- **c** sodium sulphate?

E The structure of the atom

Atoms are much too small to be seen by the naked eye. Even the most powerful light microscopes don't come anywhere near magnifying an atom enough for us to see it. But 'light' is not all visible – there are waves of different wavelengths that we can't see, for example X-rays. Using waves of tiny wavelengths it is now possible for computers to capture data and produce highly magnified images of even the smallest atoms, so that we can 'see' what they are like. The name of one type of instrument that does this is a scanning tunnelling microscope, which was invented in 1986. In Figure 5.24 you can see that scientists were even able to move individual molecules around to create the world's first piece of atomic art. Techniques that were used to do this are now helping scientists to make new materials (see Chapter 8, page 160).

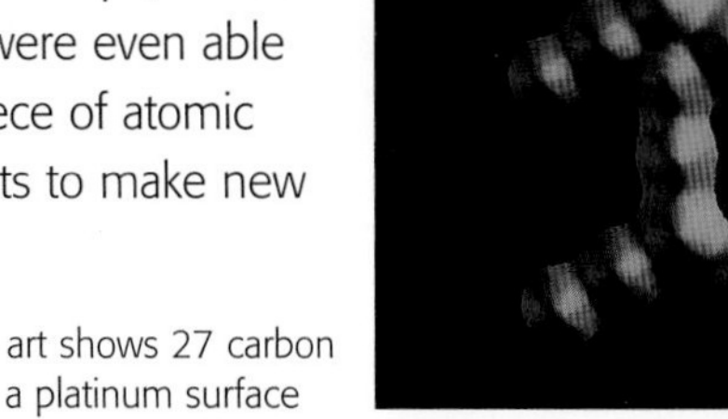

Figure 5.24 This piece of atomic art shows 27 carbon monoxide molecules arranged on a platinum surface

Inside the atom

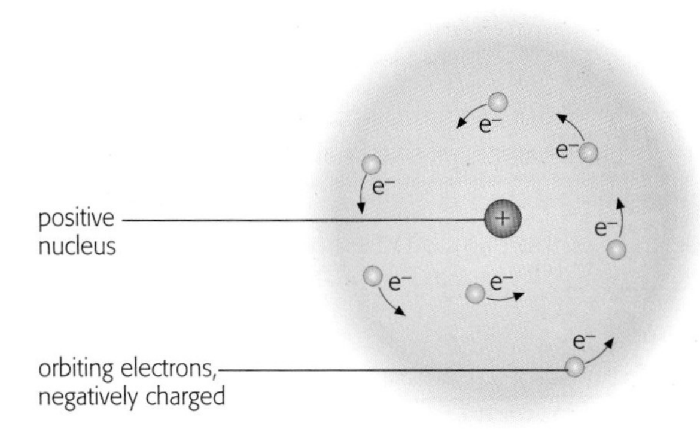

Figure 5.25 Our modern model of the atom

For a long time scientists believed that atoms were the smallest particles that made up all the matter in the Universe. They were predicted to be small round objects that could not be split. The observations that scientists made of the way matter behaved fitted this model. A **model** is the scientist's attempt to explain observations.

We now know that there are two parts to an atom – a tiny nucleus that is in the middle and electrons that orbit this, a bit like the planets orbit our Sun.

Electrons are much lighter than the nucleus and they are negatively charged. The nucleus is

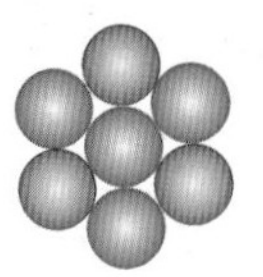
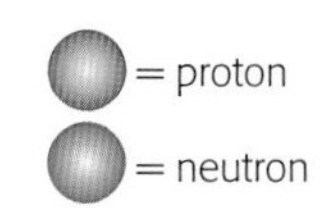

Figure 5.26 Inside a lithium nucleus

positively charged, and if we look more closely at it we find that there are two particles that make it up – **protons** and **neutrons**.

Protons and neutrons have the same mass. They are much heavier than electrons. The mass of an electron is 1/1840 of the mass of a proton or neutron. Neutrons, as the name suggests, are neutral particles. Protons are positively charged and give the nucleus its positive charge. The positive charge on the proton is the same as the negative charge on the electron. As atoms are neutral, this means that each kind of atom must have the same number of protons as it does electrons. For example, chlorine has 17 electrons and 17 protons in each atom.

It is the electrons that cause chemical reactions. Each element has a different number of electrons and this gives the element its characteristic properties. Sodium has 11 electrons and is very different from chlorine.

We have said that in the Periodic Table the elements are arranged in order of their **atomic number**. When Mendeleev first constructed his Periodic Table, no one even knew that protons and electrons existed. We now know that all atoms of the same element have the same number of protons and electrons. The atomic number tells you how many protons – or how many electrons – are in an atom.

atomic number = number of protons = number of electrons

Summary

- Atoms are made up of protons, neutrons and electrons.
- Protons and neutrons are found at the centre of the atom in the nucleus; electrons orbit the nucleus.
- An atom's nucleus is positively charged because the protons are positively charged.
- A proton has a positive charge that is equal to the negative charge on an electron.
- Neutrons have no charge; they are neutral particles.
- All atoms of the same element contain the same number of protons.
- The number of protons in any atom is equal to the number of electrons.
- The order of elements in the Periodic Table is due to an element's atomic number.
- The atomic number is the number of electrons, or protons, in the atom.

Questions

5.23 Draw a table to show the three particles that make up an atom, where they are found in the atom, and their charges.

5.24 Using the atomic numbers of the elements below, state how many protons and electrons each atom contains.

a neon – atomic number 10

b potassium – atomic number 19

5.25 Explain why an atom must contain the same number of protons as electrons.

5.26 Find out who first discovered the electron.

6 Making changes

A Making salts

Big celebrations often involve firework displays. They are exciting, colourful and often quite spectacular. No one knows who invented the first firework but it probably happened in China more than a thousand years ago. Bamboo canes were hollowed out and tightly packed with gunpowder. Gunpowder is a mixture of two fuels, charcoal and sulphur, with another chemical that provides the oxygen to allow the fuels to burn. This other chemical is a metal salt called potassium nitrate.

Compared with today's fireworks, those first fireworks would seem rather dull. They only burnt with one colour – gold. It was in the 19th century that chemists started to add the chemicals that make the colourful effects we enjoy today.

Figure 6.1 A firework display is an essential part of many celebrations

The reason it took so long to develop the modern-day firework is that the substances used to make the special effects were just not available. For example, magnesium and aluminium were only extracted from their compounds in the 1800s. When added to fireworks as powders they produce brilliant white and silver sparks and flames. The yellows, reds, greens and blues we now take for granted at firework displays are produced by metal salts in much the same way as flame test colours appear. (You can read about flame tests at the beginning of Chapter 5.) The firework burns and heats up these chemicals and they produce the colours.

Neutralisation

If you have ever had indigestion you will know that it can be extremely uncomfortable. It is caused by too much hydrochloric acid in your stomach. The treatment is to neutralise the acid with a base. A **base** is a substance that neutralises an acid to make a **salt** and water only. Apart from acids and bases, the most common chemicals you are likely to meet in the laboratory are salts.

Salts are compounds made up of two parts – a metal or ammonium part, and a part that has come from an acid. It is easy to work out which acid has made a salt by looking at the name of the salt:

- **chlorides** are produced from hydro**chlor**ic acid
- **sulphates** are produced from **sulph**uric acid
- **nitrates** are produced from **nitr**ic acid.

Salts are very useful compounds. Potassium nitrate is a salt. As you may have read in the box above, this salt provides the oxygen that aids the combustion of the fuels in fireworks. There is, of course, oxygen in the air, but the amount of air that is in the tightly packed gunpowder is just not enough for the very fast burning of a firework. Other metal salts are added to fireworks to produce the beautiful colours that shoot out when the fuel heats up.

The yellow colours in a firework display are produced by a sodium salt; the greens by a barium salt. To find out more about the colours of different metal salts, see the section on flame tests on pages 82–83 of Chapter 5.

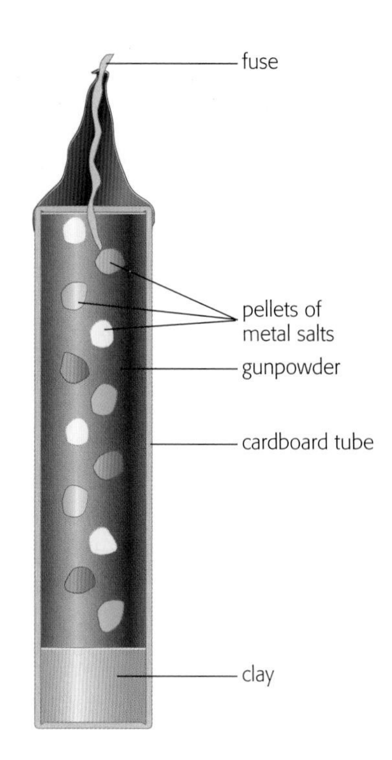

Figure 6.2 Inside a Roman Candle. The 'stars' are pellets of different metal salts which shoot out, get hot as they burn and make the beautiful colours we enjoy

The neutralisation of an acid by a base to produce a salt can be written as an equation:

base + acid → salt + water

Potassium nitrate can be made by neutralising nitric acid with potassium hydroxide solution:

potassium hydroxide + nitric acid → potassium nitrate + water

If a substance dissolves it is **soluble**. Alkalis are soluble bases.

Potassium hydroxide is the base in this reaction. It is also called an **alkali** because it is a base that dissolves in water.

We can write a formulae equation for this reaction:

You can remind yourself about balancing formulae equations by reading page 93 in Chapter 5.

$$KOH + HNO_3 \rightarrow KNO_3 + H_2O$$

You will notice that this equation balances – there is the same number of each type of atom on both sides of the equation.

Using salts to improve crop growth

Plants need nitrogen from the soil to grow. Although nitrogen gas is all around us in the air, plants can't take it up and use it because it is very unreactive and it won't dissolve in water. The nitrogen needs to be in a compound that dissolves in the water around the roots. When a crop is harvested from a field it takes its nitrogen with it.

This means that farmers have to add it back to the soil. This is where fertilisers come in. Fertilisers can be natural, like manure or compost, or they can be manufactured. Manufactured fertilisers that provide nitrogen contain salts of ammonia. When you look at the formula of ammonia you can tell it contains nitrogen – the formula of ammonia is NH_3.

Figure 6.3 When this crop is harvested it will take its nitrogen with it

Figure 6.4 The farmer is adding nitrogen-containing fertiliser to the soil to replace the nitrogen removed by the harvested crops

Making ammonium salts

Ammonia is an alkali (a soluble base) so it will react with acids to make ammonium salts.

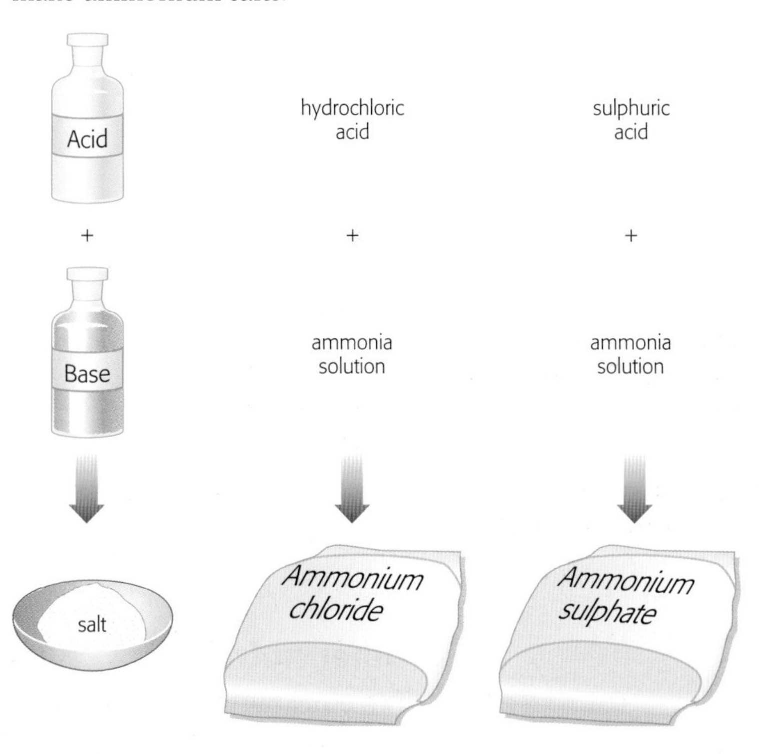

Figure 6.5 Neutralising ammonia solution with acids produces ammonium salts

Table 6.1 shows the acids that make three important ammonium salts. You can see from the table that all the ammonium salts contain nitrogen.

Table 6.1 Ammonium salts are manufactured by reacting ammonia with different acids

Acid	Formula	Name of ammonium salt	Formula
hydrochloric acid	HCl	ammonium chloride	NH_4Cl
sulphuric acid	H_2SO_4	ammonium sulphate	$(NH_4)_2SO_4$
nitric acid	HNO_3	ammonium nitrate	NH_4NO_3

Ammonia can be neutralised in the laboratory to produce a salt as shown in Figure 6.6. Using this method any acid and any alkali can be reacted together to produce a salt.

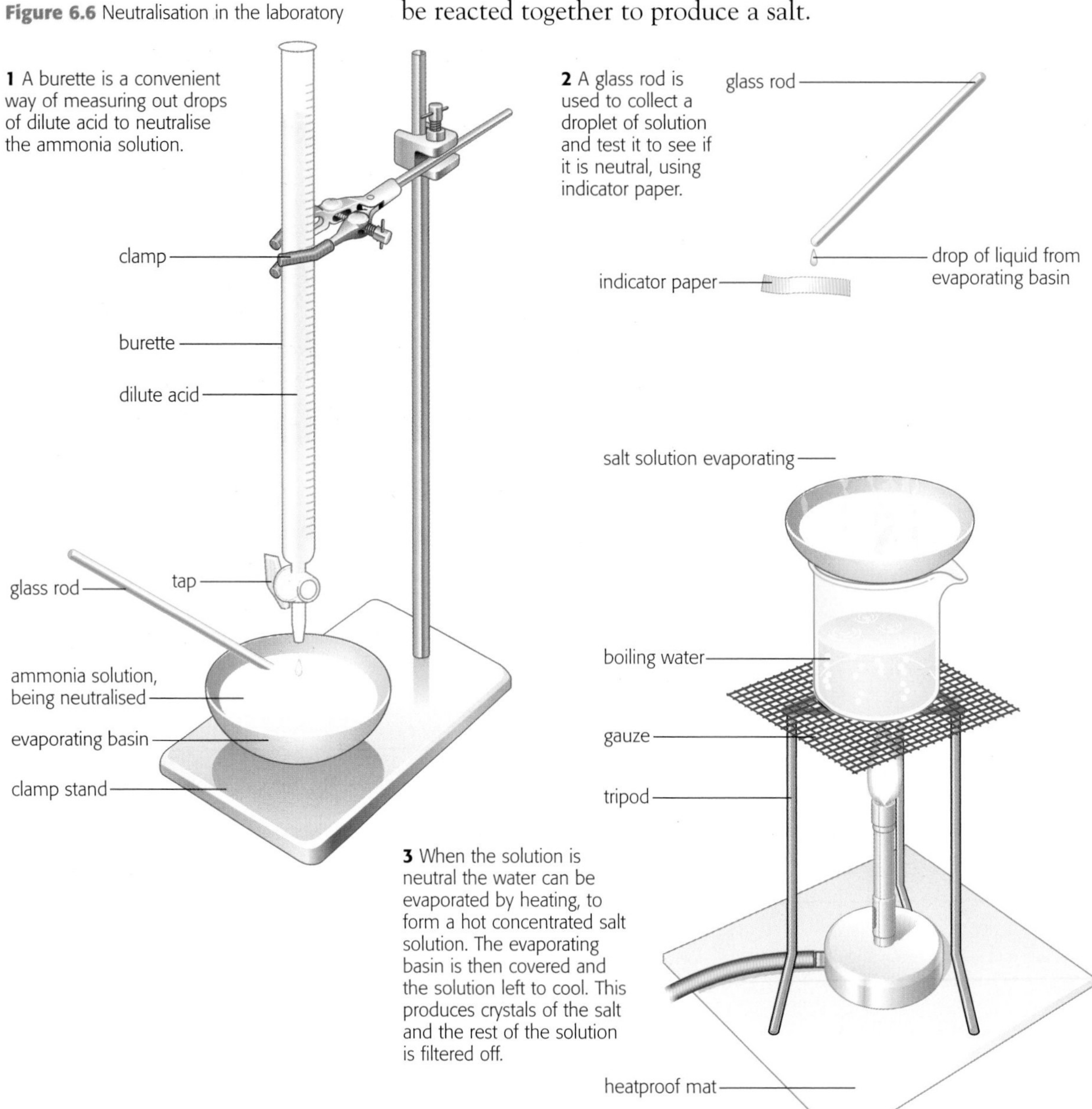

Figure 6.6 Neutralisation in the laboratory

Other nutrients for plant growth

Two other nutrients needed by plants for healthy growth are potassium and phosphorus. Potassium nitrate, KNO_3, contains both potassium and nitrogen, so it is a very useful salt to add to an artificial fertiliser mix.

Ways of preparing salts

Reacting dilute acids with metal oxides

Metal oxides contain metal atoms bonded to oxygen atoms. Metal oxides are usually bases and most of them are **insoluble**, which means they do not dissolve in water. If the metal oxide is a base then it will react with an acid to make a salt:

metal oxide (base) + acid → salt + water

Copper chloride is the salt that produces a blue–green colour in fireworks. It can be made by reacting copper oxide with dilute hydrochloric acid, using the method shown in Figure 6.7 below.

copper oxide + hydrochloric acid → copper chloride + water

The formula equation is:

$$CuO + 2HCl \rightarrow CuCl_2 + H_2O$$

Notice that to balance this equation we have put two formula units of hydrochloric acid.

H

The ending *-ide*, as in oxide, tells you that there are atoms of only two different elements in the compound.

Reacting dilute acids with metal hydroxides

Metal hydroxides are also bases. If they are soluble in water, for example sodium hydroxide or potassium hydroxide, they are also called alkalis. These can be neutralised by acids to produce salts. The salt sodium sulphate can be prepared by neutralising sodium hydroxide with sulphuric acid:

metal hydroxide (base) + acid → salt + water

sodium hydroxide + sulphuric acid → sodium sulphate + water

$$2NaOH + H_2SO_4 \rightarrow Na_2SO_4 + 2H_2O$$

The method is then the same as that shown in Figure 6.6 on page 104. Many metal hydroxides are insoluble, however, so salts from these bases can be prepared by using the method shown in Figure 6.7.

Figure 6.7 Preparing a salt from a solid base in the laboratory

solid base

1 The solid is reacted until some is left at the bottom of the beaker. This is called an **excess**.
Note: sometimes the mixture will need warming to speed up the reaction.

glass stirring rod

dilute acid

substance reacting

2 The excess solid base that has not reacted is filtered off to leave a clear salt solution.

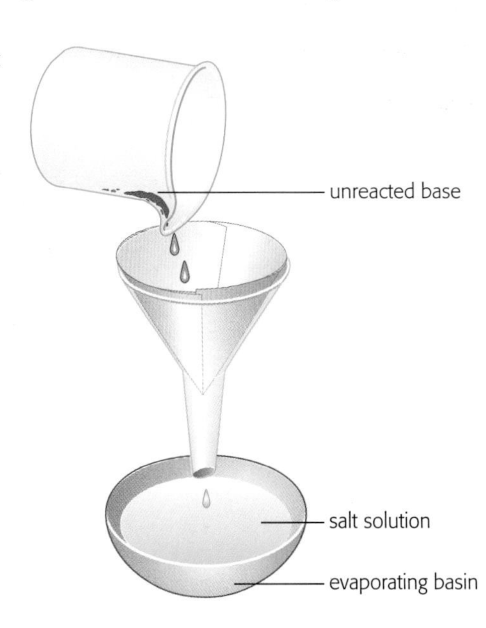

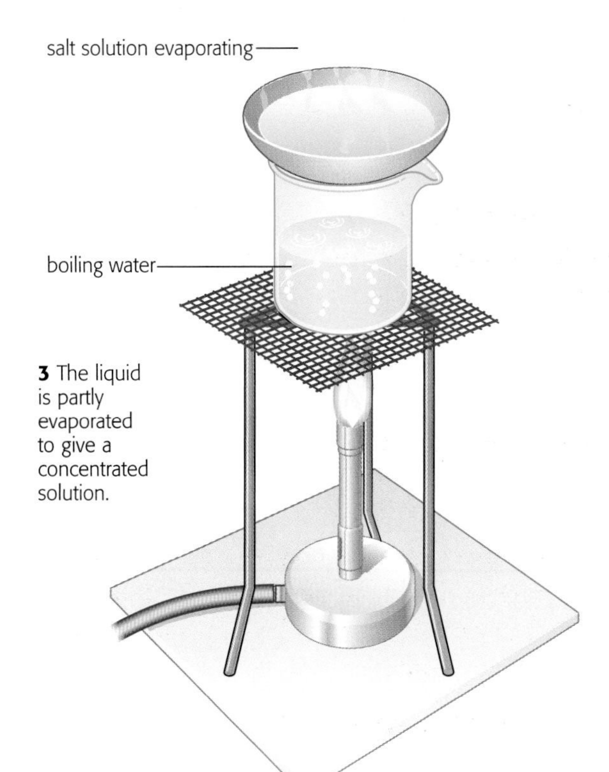

paper covering evaporating basin

4 The concentrated solution is left to cool and crystallise.

As the concentrated solution cools in step 4, the salt becomes less soluble. This is why it forms crystals.

Reacting dilute acid with metal carbonates

Any metal carbonate reacting with an acid will produce a salt, carbon dioxide and water:

metal carbonate + acid → salt + water + carbon dioxide

Copper carbonate, for example, will react with dilute sulphuric acid to give copper sulphate:

copper carbonate + sulphuric acid → copper sulphate + water + carbon dioxide

$$CuCO_3 + H_2SO_4 \rightarrow CuSO_4 + H_2O + CO_2$$

The test for carbon dioxide is to bubble it through limewater. The limewater produces a milky white precipitate.

As carbon dioxide is a gas, the reaction will fizz. Nearly all metal carbonates are insoluble, so the salt can be prepared in exactly the same way as in Figure 6.7. When all the fizzing has stopped, you know that all the acid in the beaker has reacted.

Preparing insoluble salts

So far in this section we have only considered reactions that produce soluble salts. These are salts that dissolve in water. If you look again at Figures 6.6 and 6.7, you will see that once the salt solution is produced, the water is evaporated to concentrate the solution and then it is left to crystallise. This method will not work if the salt is insoluble.

A **precipitate** is formed when two solutions mix to give an insoluble solid.

To make an insoluble salt you need to mix two salt solutions and form a **precipitate** of the salt you want to make. The only lead salt that is soluble is lead nitrate. If you want to make lead iodide, you could mix lead nitrate solution with sodium iodide solution:

lead nitrate (solution) + sodium iodide (solution) → lead iodide (solid) + sodium nitrate (solution)

$$Pb(NO_3)_2 + 2NaI \rightarrow PbI_2 + 2NaNO_3$$

Distilled water is pure water with no dissolved substances in it.

A precipitate forms that can be filtered off from the sodium nitrate solution, washed with distilled water and allowed to dry to produce pure lead iodide.

1 Sodium iodide solution is added to lead nitrate solution to produce a yellow precipitate of lead iodide.

2 The solid precipitate of lead iodide is filtered off.

3 The precipitate is washed several times with distilled water to remove any sodium nitrate or other soluble impurities.

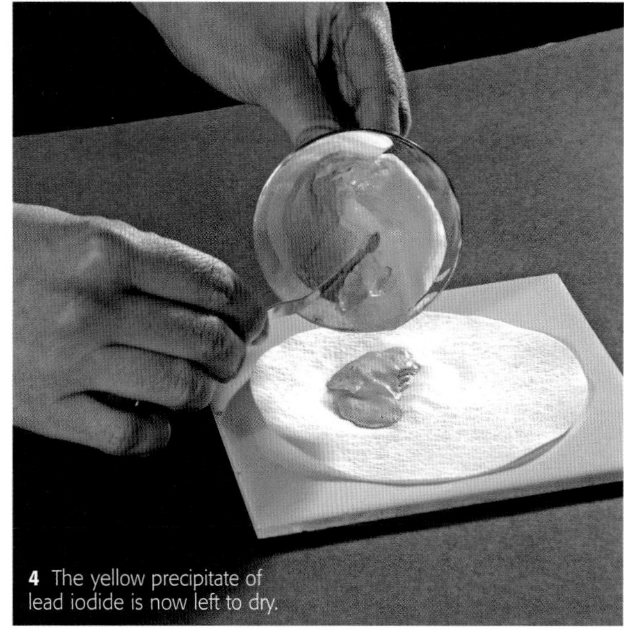
4 The yellow precipitate of lead iodide is now left to dry.

Figure 6.12 Preparing an insoluble salt: lead iodide

Summary

- Neutralisation is when a base reacts with an acid to produce a salt and water:

  base + acid → salt + water

- Alkalis are soluble bases.
- Some salts are used to create colours in burning fireworks.
- It is easy to work out which acid has made a salt by looking at the salt's name:
  - chlorides are produced from hydrochloric acid
  - sulphates are produced from sulphuric acid
  - nitrates are produced from nitric acid.
- Potassium nitrate provides the oxygen to aid combustion of the fuels in gunpowder.
- Manufactured fertilisers are made from salts which provide essential nutrients for healthy plant growth.
- Salts can also be produced by these reactions:

  metal oxide + acid → salt + water

  metal hydroxide + acid → salt + water

  metal carbonate + acid → salt + water + carbon dioxide

- Soluble salts are crystallised from their hot, concentrated solutions.
- Insoluble salts are made by mixing two salt solutions and filtering the precipitate formed, then washing it and leaving it to dry.

Questions

6.1 Which acid and which metal oxide can be reacted together to make:
  a calcium chloride **b** zinc sulphate?

6.2 Which salt is produced from reacting the following?
  a sodium hydroxide and hydrochloric acid
  b barium carbonate and hydrochloric acid

6.3 What is the name given to any soluble base?

6.4 Why is potassium nitrate an important ingredient in gunpowder?

6.5 Write the word equation and (Higher level) the formulae equation for the reaction of:
  a zinc oxide and sulphuric acid
  b magnesium carbonate and hydrochloric acid.

6.6 Lead sulphate is an insoluble salt. Sodium sulphate and lead nitrate are soluble salts.
  a Explain, using diagrams, how you would prepare a pure, dry sample of lead sulphate.
  b Write the word equation and (Higher level) the formulae equation for this reaction.

B Extracting metals from the Earth's crust

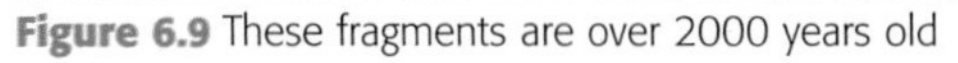

Figure 6.9 These fragments are over 2000 years old

Figure 6.10 The famous Copper Scroll may contain clues to a fabulous treasure

In 1947, in a cave near the Dead Sea, a Bedouin shepherd made a remarkable discovery. He was throwing rocks into the cave when he heard what sounded like a dull cracking sound. He went to investigate and found his rock had hit a clay jar. Inside the jar was a scroll of animal parchment with writing on it. This was the first of what we now call the Dead Sea Scrolls, which are more than 2000 years old. Over the next 10 years, fragments of between 800 and 900 manuscripts were found in eleven caves. They give fascinating insights into the times of the Old Testament of the Bible and Jewish history.

At the back of one of the caves, archaeologists discovered a different scroll. It was made of very thin copper, with writing engraved into the metal. The copper had reacted with oxygen in the air and it was very brittle. To unroll it would destroy it, and its secrets would be lost. After 4 years a way was found – and it was gently sawn into 27 pieces that could be read. This was done in Manchester at the College of Technology (now incorporated into the University of Manchester).

When it was translated, the writing on the Copper Scroll appeared to describe a hidden treasure of gold and silver – coins and vessels worth an absolute fortune. Tantalising clues were given about the sites where the treasure was hidden, such as *'In the sepulchral monument, in the third course of stones: 100 bars of gold'*. Various ideas have been put forward about the fate of the treasure, but the truth is no one knows if it even exists or whether the scroll is a work of fiction. People have gone looking for the treasure, but the clues are so old that the places described are now unknown. The last entry on the scroll suggests that treasure seekers must first find a second copper scroll, with more clues, before they can locate the places where the treasure is hidden.

Metal ores

The box on the previous page shows the importance of gold, silver and copper to humankind. These metals were discovered so long ago that we have no record of exactly when people began to use them.

The first to be discovered was almost certainly gold; gold articles dating back almost 10 000 years have been found. Ancient people probably found gold glinting on the ground and in the beds of streams. Gold is very unreactive so it can be found uncombined with other elements (see pages 84–85). It is a soft metal so it was only used for ornaments and jewellery, and because it was rare it was prized for its beauty and value. Silver and copper were also discovered very early on in human history, because again they could be found uncombined, as elements. Silver too is soft and its main use was again decoration.

You can see a picture of a gold nugget found in a Welsh stream on page 85.

Extracting copper from copper ore

Copper is more reactive than gold, and nearly all of the copper in the Earth's crust has formed compounds. These compounds are found in rocks. The rocks that contain metal compounds are called **ores**.

Metal compounds found in rocks are called **minerals**.

Copper was probably first extracted from its ores around 7000 years ago. It almost certainly happened by accident, in the fires of potters who used copper minerals to produce beautiful coloured glazes on their pots. When the fires in their kilns died down, pieces of copper metal were found in the ashes. So what had happened in the fire to produce the copper?

Many metal ores are **oxides** – compounds containing oxygen. Copper oxide is a copper ore. In the wood fires that heated the potters' kilns, carbon was formed and this took the oxygen way from the copper oxide to leave copper:

copper oxide + carbon → copper + carbon dioxide

The formulae equation is:

$$2CuO + C \rightarrow 2Cu + CO_2$$

H This formulae equation is balanced. The number of each type of atom is the same on both sides of the equation. See page 93 of Chapter 5.

When copper oxide loses its oxygen we say that it has been **reduced**. At the same time, carbon has gained oxygen to become carbon dioxide. We say that the carbon has been **oxidised**.

Adding oxygen to a substance is **oxidation**
Removing oxygen from a substance is **reduction**

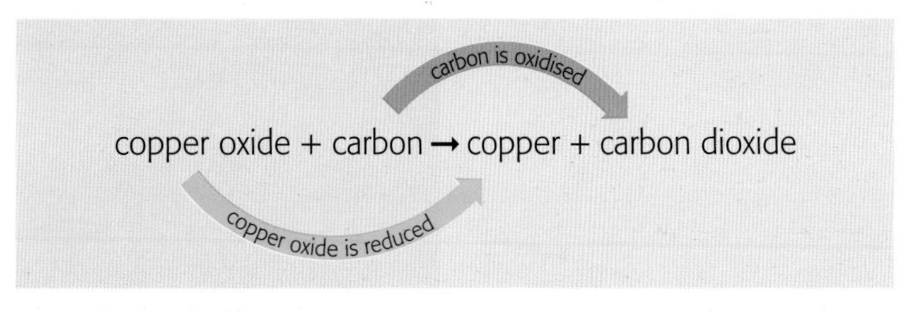

Figure 6.11 Oxidation and reduction

Figure 6.12 In this reaction magnesium is being oxidised – oxygen is being added to it

You have almost certainly carried out an oxidation reaction in the laboratory. When magnesium burns in oxygen, it does so with a very bright, white light, leaving behind a white powder called magnesium oxide:

magnesium + oxygen → magnesium oxide

$2Mg + O_2 \rightarrow 2MgO$

Extracting lead and iron by reduction with carbon

We think that lead was the next metal to be discovered, about 5000 years ago. It too can be extracted by using carbon. In modern industry we still use carbon but instead of heating wood to provide it, we heat coal to produce coke.

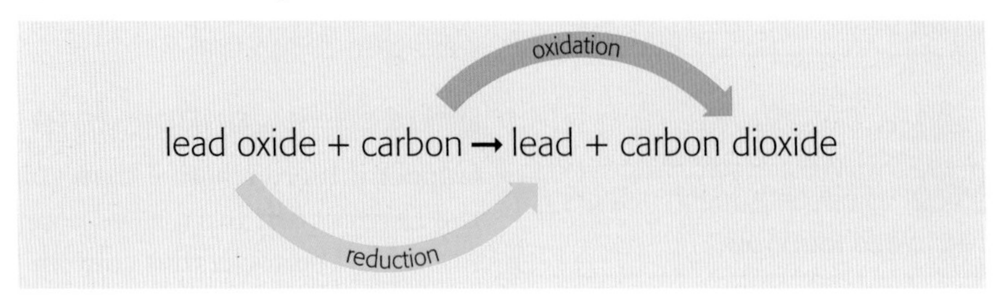

Figure 6.13 Reducing lead oxide to produce lead

H

The balanced formulae equation is:

$2PbO + C \rightarrow 2Pb + CO_2$

Lead soon found its way into making lead pipes to carry water. It is easy to shape and it melts at the relatively high temperature of 327 °C. It also does not corrode easily. Lead pipes bearing the marks of Roman emperors still exist today.

However, no metal has transformed the world we live in like iron, and its alloy steel. Iron makes up 5% of the Earth's crust. When iron began to be extracted in quantity it could be forged into agricultural implements, revolutionising agriculture. It was also hard enough for making weapons. Today you do not have to go far to appreciate how much we rely on the strength and hardness of iron, from bridges and car bodies to the building of homes and offices.

An **alloy** is a mixture of two or more elements; in this case iron with a tiny amount of carbon to make steel, which is stronger.

Again, the oxygen from iron oxide can be removed using carbon:

iron oxide + carbon → iron + carbon dioxide

$2Fe_2O_3 + 3C \rightarrow 4Fe + 3CO_2$

Figure 6.14 Iron and steel find uses in all areas of life, from knives to building structures and pins in broken bones

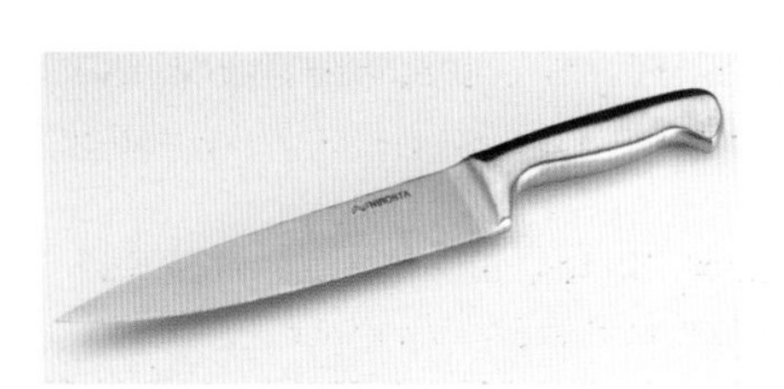

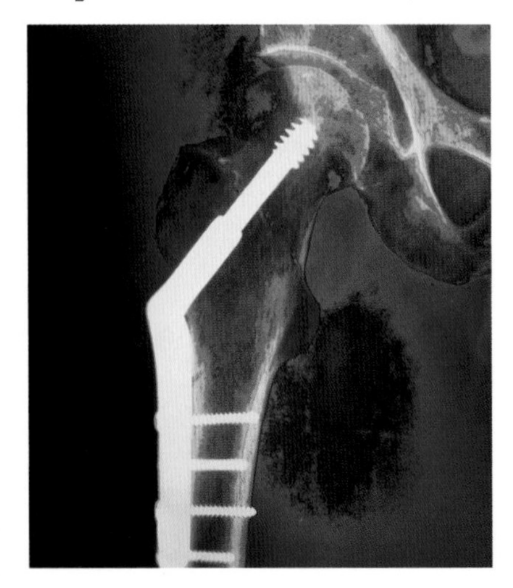

The order of reactivity of metals

As we have seen, the first metals to be discovered were the ones that were found uncombined – these had not reacted to form compounds. The next to be discovered could be fairly easily extracted using carbon. But how exactly is carbon able to remove the oxygen from the metal oxide?

The answer lies in the fact that carbon is more reactive than copper, lead or iron. It combines with the oxygen from these metals, taking it away from them.

Other metals are more reactive than carbon, so cannot be extracted in this way. We can build up a metal league table of reactivity, called the **Reactivity Series** (Figure 6.15).

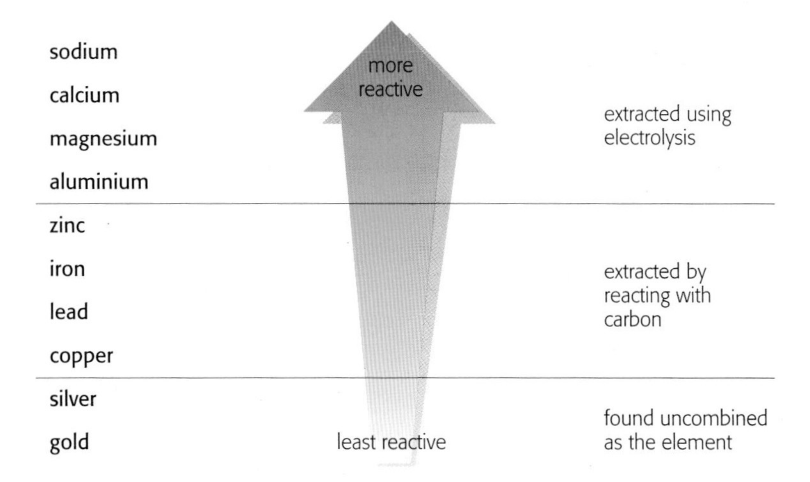

Figure 6.15 Some of the metals in the Reactivity Series. Notice that the very unreactive metals were those discovered uncombined. Other metals can be extracted using carbon. The most reactive metals require a different method called electrolysis

The most reactive metals are strongly bonded in their compounds. We say their ores are stable. It was not until the 19th century that these metals were extracted from their compounds. They were extracted using a powerful method called **electrolysis**. This is the splitting of a compound using electricity. You can read more about electrolysis in Chapter 7 (page 150).

Summary

- Most metals are not found uncombined as elements but are extracted from their ores, which are found in the Earth's crust.
- Many metal ores are oxides. Oxides are compounds containing oxygen.
- Iron, copper and lead can be extracted from their metal oxide by reaction with carbon.
- When oxygen is added to a substance it is oxidised.
- When oxygen is removed from a substance it is reduced.
- A metal league table called the Reactivity Series puts metals in order of their reactivity.
- The most reactive metals in the Reactivity Series cannot be extracted from their ores using carbon. The ores are too stable and the metals must be extracted by electrolysis.
- Metals at the bottom of the Reactivity Series, such as copper, gold and silver, can be found uncombined in the Earth's crust.

Questions

6.7 What do we call rocks that contain metal compounds?

6.8 Zinc can be found as zinc oxide (ZnO) in the Earth's crust. Zinc oxide can be *reduced* to zinc.

a What is meant by the term *reduced*?

b What substance will reduce zinc oxide?

6.9 In the box on page 108 we said that the Copper Scroll had reacted with oxygen. Explain why this reaction is called an oxidation reaction.

6.10 Write a word equation and (Higher level) a formulae equation for each of the reactions in questions 6.8 and 6.9.

6.11 Potassium is a very reactive metal. What method must be used to extract it from its ore?

C Chemicals and food

This was one of many headlines in 2002. A group of Swedish scientists had discovered a chemical in food, produced by cooking. This chemical, called acrylamide, was a known carcinogen (a cancer causer) but no one had realised that it was in most cooked foods – from crisps to bread. It was also found in coffee.

Scientists around the world set out to discover if the amount produced by cooking was really a cancer risk. In 2005 a study of 43 000 women in the USA could find no link to breast cancer. In fact, to date the small amounts of acrylamide found in nearly all cooked foods do not appear to give any increased risk of cancer.

Food additives

Everyone has to eat food to live. So how food is produced and what goes into it can cause fierce debate.

All food is made up of chemicals. When food is cooked, chemical reactions take place and produce new products. Most of these chemical reactions make food more palatable and more enjoyable to eat. Cooking also starts to break down some of the chemicals in food and make them easier to digest. Some reactions, however, may be harmful, such as the production of acrylamide – see the box above. But even if food is just left on the shelf, chemical reactions are happening in it all the time, usually making the food deteriorate, and bacteria are growing which could cause food poisoning.

It may surprise you that chemicals have been added to food for centuries. In the Middle Ages meat could be kept for longer when saltpetre was added to it. Saltpetre is an old name for potassium nitrate, which is still used in preserving cooked meats today. Sugar has long been used to preserve cooked fruit in jams, and also of course makes foods taste sweeter. Putting cooked vegetables such as onions into vinegar is traditional way of preserving them, called pickling.

But in recent decades – since food production was industrialised – many more chemicals have been added to our food. These are called food additives.

In the UK the Food Standards Agency is the body given responsibility to monitor and maintain food safety.

Figure 6.16 There are many additives in these sweets. They are mostly colouring agents

Food additives in UK food today are carefully regulated. If you read a food label you will often see numbers with the letter E before them. The E tells you that the particular additive has been approved for use in the European Union, of which the UK is a member. The number after the E gives you an idea of what type of food additive has been added. See Table 6.2.

E numbers	Type of additive
E100s	colouring agents
E200s	preservatives
E300s	antioxidants
E400s	emulsifiers, gelling agents and thickeners
E500s	anti-caking agents and salts
E600s	flavourings
E900s	sweeteners, surface coating agents and gases

Table 6.2 What the E numbers mean

Natural versus artificial additives

Natural additives are chemicals that are already present in plants and other foods that occur in nature. This seems a quite simple definition. For example, curcumin (E100) is a yellow food colour that is extracted from turmeric roots and added to custard to make it yellow. However, some chemicals that occur naturally, such as the preservative benzoic acid (E210) which is found in fruit, are actually made in a factory because it is cheaper to do this than to extract them from natural foods. Does this make the additive non-natural? The manufactured benzoic acid is exactly the same compound as the naturally occurring one, so has exactly the same properties.

There is a further dilemma. Just because an additive is 'natural', it does not necessarily mean it is safe to eat. Some natural additives are thought to have health impacts. For example, caffeine – found in coffee and tea, and added to cola drinks – is linked to high blood pressure in some people. Similarly, 'artificial' additives may not be bad for us – they may have no negative health impact and may help to keep food safe for longer.

Preservatives give food a longer shelf life. Antioxidants act as preservatives by stopping oxygen reacting with certain ingredients. Emulsifiers allow different ingredients such as oil and water to mix (see page 172).

Artificial additives are those that are not found in natural plants or foodstuffs. Good examples of these chemicals are the sweeteners aspartame (E951) and saccharin (E954). They are made in a factory. They are very much sweeter than sugar and so only tiny amounts are needed, and they have the advantage that they contain no calories.

When sugar is eaten, it is used by our bodies to produce energy. This energy is sometimes measured in 'calories'. If this energy is not needed for activity, our bodies make fat from it.

Baking powder

Baking powder is added to cake mixture to make it rise when it is baked in the oven. Sometimes it is added as a separate ingredient, but often it is in the flour already. The type of flour that already contains baking powder is called 'self-raising' flour. Baking powder works by producing carbon dioxide in the cake mixture. In the oven the tiny bubbles of carbon dioxide expand with the heat and the cake mixture rises.

Figure 6.17 This sponge cake has a texture that is full of holes because carbon dioxide was released during baking

The simplest type of baking powder is sodium hydrogencarbonate, $NaHCO_3$. It is also known as 'bicarbonate of soda'. When sodium hydrogencarbonate is heated it breaks down, releasing carbon dioxide:

$$\text{sodium hydrogencarbonate} \xrightarrow{\text{heat}} \text{sodium carbonate} + \text{water} + \text{carbon dioxide}$$

$$2NaHCO_3 \xrightarrow{\text{heat}} Na_2CO_3 + H_2O + CO_2$$

When a substance breaks down during heating, the reaction is called **thermal decomposition**.

Heating carbonates and hydrogencarbonates

Sodium hydrogencarbonate is not the only salt to decompose when heated. Other carbonates and hydrogencarbonates will also undergo thermal decomposition.

A very important decomposition reaction is the heating of limestone, which is calcium carbonate. This again produces carbon dioxide, leaving calcium oxide, also called quicklime:

$$\text{calcium carbonate} \xrightarrow{\text{heat}} \text{calcium oxide} + \text{carbon dioxide}$$

$$CaCO_3 \xrightarrow{\text{heat}} CaO + CO_2$$

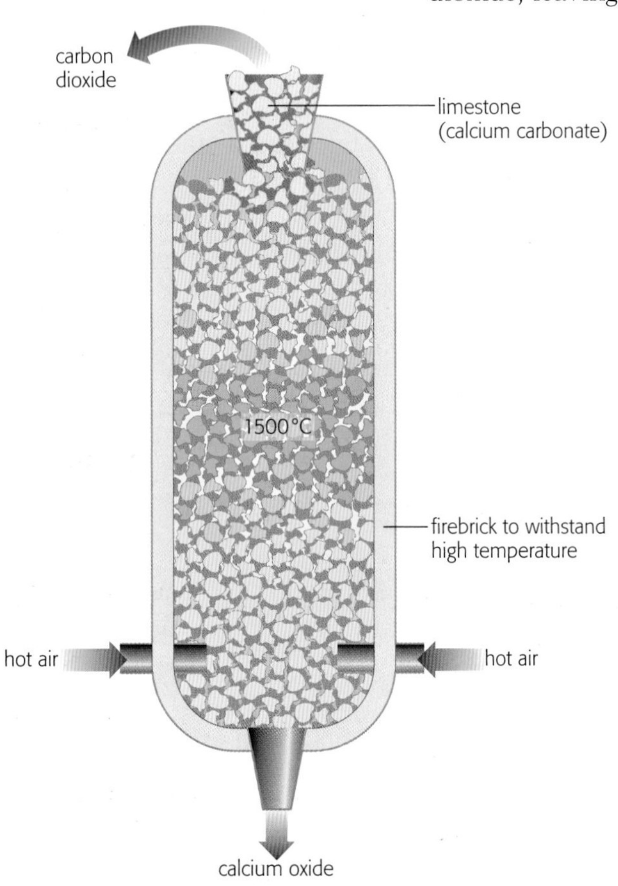

Figure 6.18 The reaction occurring in this lime kiln is thermal decomposition

In industry this reaction is carried out in a lime kiln (Figure 6.18). The product, calcium oxide, is used to make the alkali calcium hydroxide, which is used by farmers to neutralise acidic soils.

Some other carbonates will also thermally decompose in the same way.

$$\text{zinc carbonate} \xrightarrow{\text{heat}} \text{zinc oxide} + \text{carbon dioxide}$$

$$ZnCO_3 \xrightarrow{\text{heat}} ZnO + CO_2$$

But not all carbonates will decompose when heated. If you look back at the baking powder reaction, you'll see that sodium carbonate is produced by heating the sodium hydrogencarbonate. This does not decompose further.

Calcium hydrogencarbonate is soluble in water. It is often present in tap water. When the tap water is heated, calcium hydrogencarbonate decomposes to produce calcium carbonate, which is not soluble in water and forms the 'fur' on the heating element of kettles.

$$\text{calcium hydrogencarbonate} \xrightarrow{\text{heat}} \text{calcium carbonate} + \text{water} + \text{carbon dioxide}$$

$$Ca(HCO_3)_2 \xrightarrow{\text{heat}} CaCO_3 + H_2O + CO_2$$

H

This formulae equation balances. Check that you understand why it balances.

Hydration and dehydration

Bacteria and other microorganisms need water to grow and reproduce. If food is **dehydrated** then the water is taken out of it and bacteria can no longer grow. You may think that dried coffee and dried pasta are recent innovations, but this method of preserving food is far from new. People have been drying fish and meat in the sun for centuries. Nowadays most foodstuffs can be dehydrated. Not only does this preserve them, it also makes them much lighter to transport, and they take up less room.

Dehydrated foods become really important when disasters strike. Then food needs to be got to people quickly and in quantity. The food can be hydrated again by adding water back.

In Section A we prepared different metal salts by allowing them to crystallise out from hot concentrated solutions of the salts as they cooled down. These crystals contain water that is chemically bonded to the metal salt particles. For example, copper sulphate is blue when it is hydrated. If you heat the blue crystals they become dehydrated and form white anhydrous copper sulphate. (**Anhydrous** means without water.) This is another thermal decomposition.

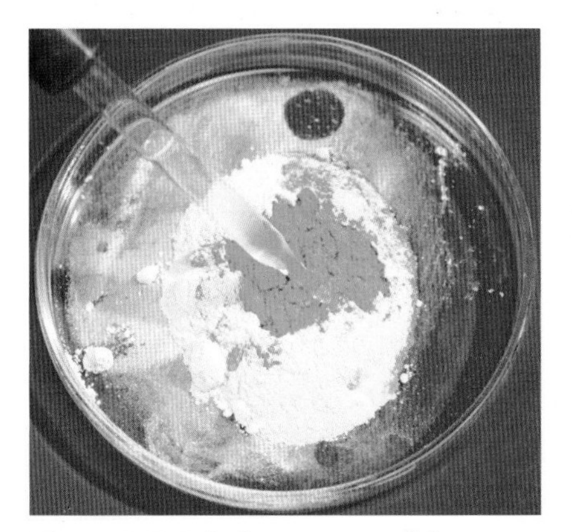

Figure 6.19 Anhydrous copper sulphate being hydrated

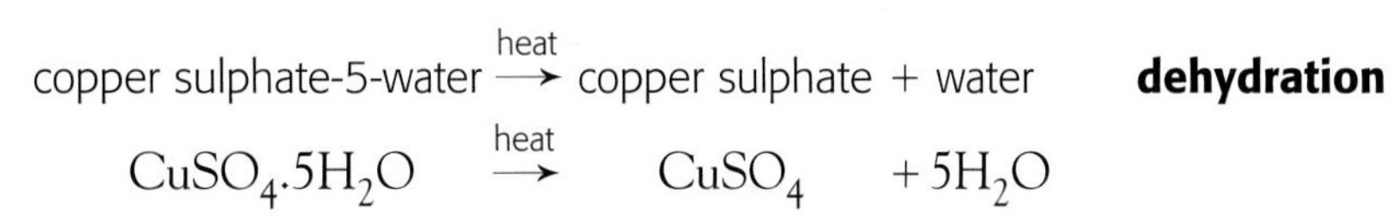

copper sulphate-5-water $\xrightarrow{\text{heat}}$ copper sulphate + water **dehydration**

$$CuSO_4.5H_2O \xrightarrow{\text{heat}} CuSO_4 + 5H_2O$$

Notice that we show water of crystallisation in a formula by using a dot before the water. As this reaction takes in heat to make it happen, it is an endothermic reaction (see page 92). When water is added to anhydrous copper sulphate, an exothermic reaction occurs. This reaction is a hydration reaction.

copper sulphate + water $\xrightarrow{\text{heat}}$ copper sulphate-5-water **hydration**

$$CuSO_4 + 5H_2O \xrightarrow{\text{heat}} CuSO_4.5H_2O$$

Do food additives cause health problems?

H

This is a very difficult question to answer. The Food Standards Agency is constantly monitoring the safety of food additives and looking for any evidence that suggests there might be a link between a food additive and a health impact.

One recent example is aspartame (E951). This sweetener is 200 times as sweet as sugar and is added to many prepared foods in minute quantities, as well as being used as a sugar substitute in tea and coffee. It can be found in some 'diet' drinks because it does not have the calories that sugar contains (see page 113). Since the introduction of aspartame as a food additive in the 1980s, some people believe that it is having an effect on their health. Aspartame has been claimed to cause headaches, memory loss, mood swings, even brain tumours and epilepsy, and this list continues to grow.

Figure 6.20 These sweeteners contain aspartame, which is 200 times sweeter than sugar

Scientists have tried to discover if there really is a link between aspartame in foods and health problems. To do this they must produce **data**. Data is obtained from observations and measurements.

H

But data is not necessarily evidence. To become evidence, scientists must decide if the data is **reliable** (can be trusted) and if it is **valid** (relevant to the effects they are trying to investigate).

Let's examine the data for one claim about aspartame – that it causes headaches. One study asked patients who had been referred to hospital because they were having frequent headaches to fill in a questionnaire about the foods they were eating. There did appear to be a **correlation** between when they had eaten aspartame-containing foods and the onset of their headaches. However, just because there seems to be a relationship this does not mean that aspartame is the cause of the headaches. The cause might be linked to something completely different in the foods they had eaten, or it could just be a coincidence. This is why scientists always try to repeat findings to ensure their data is valid and reliable.

Correlation means that there appears to be a relationship or connection between two factors.

Another study involved 40 people who complained of getting headaches after consuming food and drink containing aspartame. In the study, on some days they were given a drink that contained aspartame while on other days they were given a drink that looked the same but did not contain any aspartame. Neither the 40 volunteers in this experiment nor the people giving the drink knew whether it contained aspartame or not. When the results were analysed, 35% of the volunteers had headaches following the aspartame-containing drink but 45% complained of getting headaches after drinking the drink that contained no aspartame. This experiment suggests that aspartame is not the cause of headaches, but even this is not conclusive. All we can say is that for these 40 patients there is no reliable data that provides evidence of a link.

Summary

- Food is made up of chemicals. When food is cooked, chemical reactions take place to produce new chemicals.
- Substances that are added to food are called additives. These are often labelled by E numbers.
- 'Natural' additives are chemicals that are already present in some foodstuffs.
- 'Natural' additives can be manufactured. They have exactly the same chemical composition and properties as the chemicals that occur in nature. They cannot be distinguished from 'natural' substances.
- 'Artificial' additives are not found in natural foodstuffs.
- Sodium hydrogencarbonate ($NaHCO_3$) can be used as baking powder.
- When carbonates and hydrogencarbonates are heated they produce carbon dioxide gas; this is called thermal decomposition.
- When water is added to a substance and chemically combines with it, the reaction is called hydration. Dehydration reactions remove water from substances.
- Some food additives are said by some people to have an impact on their health.
- H Data must be interpreted carefully to decide if a chemical in food has a health impact; even if there is a correlation this does not provide evidence of cause.

Questions

6.12 A 'natural' food additive can be manufactured.

a Give one example of a manufactured 'natural' food additive.

b Explain why it is not possible to distinguish a manufactured 'natural' additive from one that is already present in nature.

c Explain what is meant by the term 'artificial additive'.

6.13 Sodium hydrogencarbonate is also called baking powder.

a How does baking powder make cakes rise?

b What is thermal decomposition?

c Write a word equation and (Higher level) a balanced formulae equation for the thermal decomposition of sodium hydrogencarbonate.

H

6.14 Hydrated sodium carbonate is sometimes known as washing soda. It has the formula $Na_2CO_3.10H_2O$. Write a balanced formulae equation for the *dehydration* of washing soda.

6.15 Tartrazine is a yellow food-colouring additive. See if you can find any data linking this food additive to a health impact.

D Collecting and testing gases

There is a cave in Italy called *Grotto del Cane*, which literally means 'cave of the dog'. It got its name because dogs and other small animals that went into it soon became unconscious and died. It is not a large cave – it's about the size of a garden shed – but its unusual properties have been known for at least 2000 years. Travellers used to come from far and wide to see for themselves this remarkable cavity. The keeper of the grotto would, for a small fee, tie his dog by the legs and place the animal in the middle of the cave. Within about 30 seconds the dog was unconscious. The dog would then be removed from the cave and allowed to recover.

Figure 6.21 The dog placed into the Grotto del Cane quickly became unconscious

The explanation for this phenomenon is that carbon dioxide gas seeps up into the cave through cracks in its floor. Carbon dioxide is denser than air and so it forms a layer at the bottom of the cave. Any small animal that ventures into the cave is immersed in this layer of carbon dioxide and becomes unconscious through lack of oxygen.

Collecting gases in the laboratory

Density is the mass of matter in a particular volume, under particular conditions of temperature and pressure.

Air is mainly a mixture of oxygen (20%) and nitrogen (79%). If a gas is less dense than these gases it will move upwards in air. You will have seen this if you've ever had a party balloon filled with helium. When you let it go the balloon floats upwards and stays on the ceiling.

Hydrogen is the least dense gas of all and this is why it was used in airships and balloons.

Figure 6.22 Ballooning pioneers Jacques Charles and Noel Robert landing successfully in 1783 after the first hydrogen-filled balloon flight. They had flown for 27 miles

Can you think of the reason why hydrogen is not used any more to fill balloons or airships? (You can find the answer on page 85 of Chapter 5.)

Knowing how dense a gas is helps scientists to decide how it can be collected in the laboratory after it has been made.

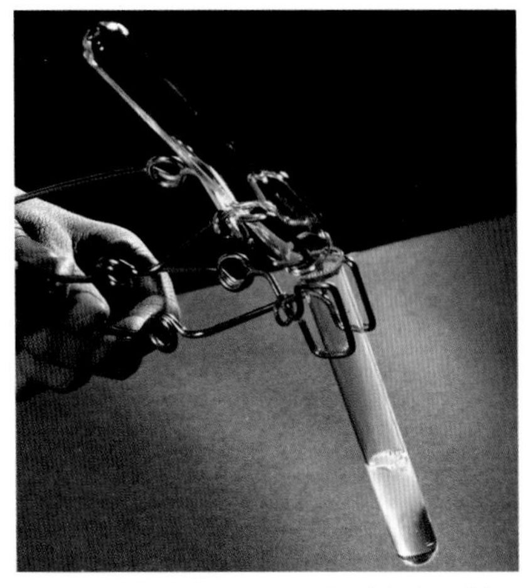

Figure 6.23 Magnesium and sulphuric acid react to produce hydrogen. The hydrogen is easy to collect because it is less dense than air

Collecting gases by upward delivery

You may remember from earlier work that acids react with many metals to produce hydrogen. For example:

magnesium + sulphuric acid → magnesium sulphate + hydrogen

$$Mg + H_2SO_4 \rightarrow MgSO_4 + H_2$$

A simple way to collect the hydrogen given off is to hold a test tube above the one being used to produce hydrogen. This method of collecting gas, shown in Figure 6.23, is called 'upward delivery'. Hydrogen can be collected this way because of its low density – it displaces the air from the test tube.

Collecting gases over water

Hydrogen has another property that is useful in collecting the gas – it is insoluble in water. In Figure 6.24 you can see the arrangement for collecting hydrogen over water into a gas jar. This time we are using the reaction of zinc with sulphuric acid:

zinc + sulphuric acid → zinc sulphate + hydrogen

$$Zn + H_2SO_4 \rightarrow ZnSO_4 + H_2$$

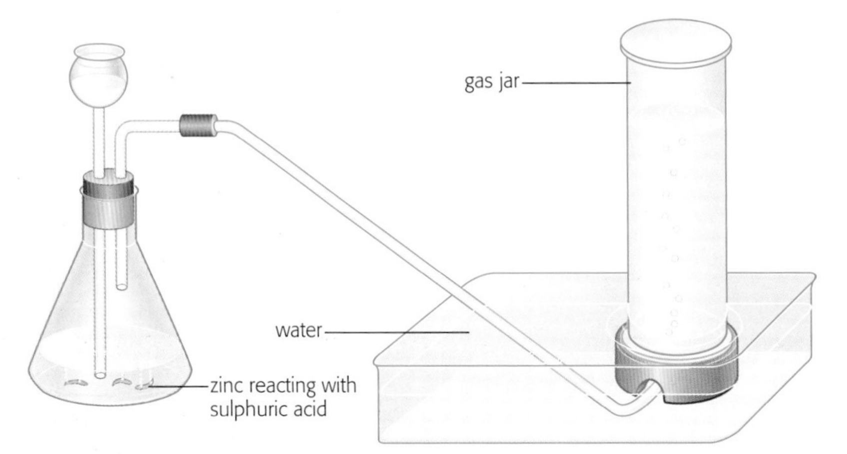

Figure 6.24 This method of collecting hydrogen is called 'collecting over water'

Any gas that is insoluble, or does not dissolve very much in water (that is, is **slightly soluble**), can be collected in this way. Oxygen is slightly soluble in water (otherwise fish would not be able to breathe) and so is carbon dioxide, but both of these gases can be collected over water.

Collecting gases by downward delivery

If a gas is denser than air then the gas can be collected by downward delivery. Carbon dioxide is one such gas – as we saw in the section-opening box (page 117). One method used to prepare carbon dioxide is the reaction of a metal carbonate with acid (you can read about this reaction on page 106). In Figure 6.25 carbon dioxide is made by reacting calcium carbonate (marble chips) with dilute hydrochloric acid, and is collected by downward delivery. The carbon dioxide sinks to the bottom of the test tube, gradually pushing all the air out.

calcium carbonate + hydrochloric acid → calcium chloride + water + carbon dioxide

$$CaCO_3 + 2HCl \rightarrow CaCl_2 + H_2O + CO_2$$

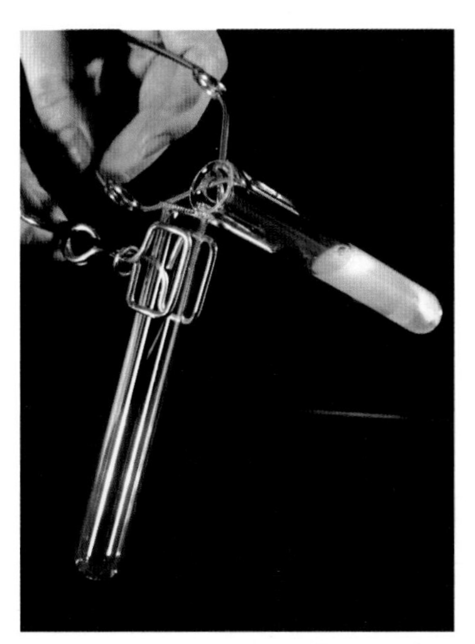

Figure 6.25 This method of collecting carbon dioxide is called 'downward delivery'

Figure 6.26 Chlorine gas being collected by downward delivery. This is done in a fume cupboard because chlorine is toxic

You can read more about chlorine in Chapter 7 (page 151).

Another gas that can be collected in this way is chlorine – it is much denser than air. In Figure 6.26 you can see a gas jar being filled with chlorine.

Collecting soluble gases

Ammonia can be collected by upward delivery because it is less dense than air.

Ammonia gas is extremely soluble in water. 100 cm^3 of water will dissolve 100 litres of ammonia. This gas cannot be collected over water. In fact, if the apparatus in Figure 6.24 were used to produce ammonia, the water in the trough would dissolve the ammonia so fast that it would create a partial vacuum in the delivery tube and the water would rush into the conical flask, stopping the reaction. Ammonia is collected by upward delivery, as in Figure 6.23.

Chlorine is not as soluble as ammonia but 100 cm^3 of water will still dissolve 300 cm^3 of chlorine, so collecting chlorine over water is not feasible either. Chlorine is collected by downward delivery – see above.

Using a gas syringe

A gas syringe can be used to collect any gas, provided the gas does not react with the material of the syringe. A glass gas syringe is shown collecting carbon dioxide gas in Figure 6.27.

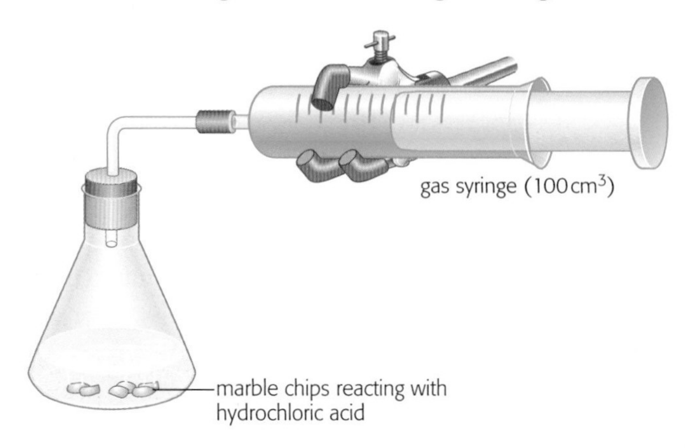

Figure 6.27 This method can be used to collect any gas

Tests for some gases

Table 6.3 shows tests for the gases we have discussed in this section.

Table 6.3 Tests for some gases

Gas	Test
hydrogen, H_2	A lighted splint gives a squeaky pop
oxygen, O_2	A glowing splint is relit
carbon dioxide, CO_2	Limewater gives a milky white precipitate when the gas is bubbled through it
ammonia, NH_3	White smoke is produced when the gas is exposed to fumes from concentrated hydrochloric acid
chlorine, Cl_2	Moist litmus or moist universal indicator paper is bleached by the gas and turns a whitish colour

Hazard labels in the chemistry laboratory

In a chemistry laboratory we need to know which substances are hazardous so that we can ensure our safety when carrying out experiments. Figure 6.28 shows the common hazard labels you are likely to see.

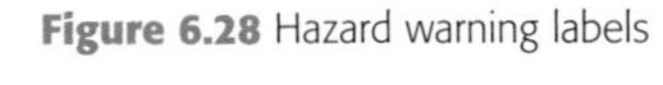

Figure 6.28 Hazard warning labels

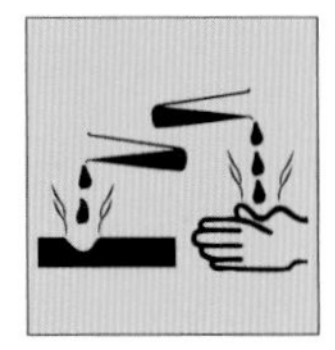

Corrosive: substances with this hazard label will attack and destroy living tissue, including eyes and skin. Even quite dilute solutions of sodium hydroxide can be corrosive; concentrated acids are also corrosive.

Irritant: substances with this hazard label are not corrosive but they can cause reddening or blistering of the skin. Dilute laboratory acids are often irritant but are usually too dilute to be called corrosive; copper oxide is also an irritant.

Toxic: this label warns you that these substances can cause death. They may have their poisonous effect when they are swallowed, or breathed in, or absorbed through the skin. Chlorine gas is toxic when breathed in; so is ammonia gas.

Harmful: these substances have similar effects to toxic substances but are less dangerous. The salts ammonium chloride and copper sulphate are both harmful.

Highly flammable: these substances easily catch fire. Ethanol and petrol both come into this category.

Oxidising: this tells you that these substances provide oxygen, which allows other materials to burn more fiercely. Hydrogen peroxide and potassium manganate(VII) are both oxidising.

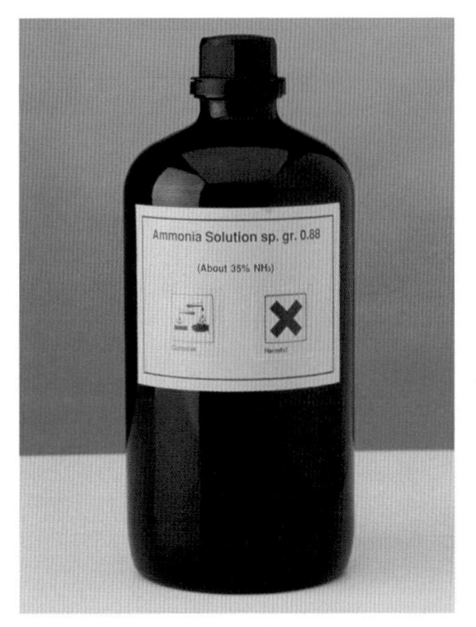

Figure 6.29 Concentrated ammonia solution is highly corrosive and harmful if swallowed

Many chemicals present more than one hazard and so you will often see more than one hazard symbol on a bottle, as in Figure 6.29.

Figure 6.30 Hazard warning sign on a tanker

Transporting chemicals

The chemistry laboratory is not the only place you will see hazard labels. The same labels can be seen on the backs of tankers, as in Figure 6.30. These labels let the emergency services know what the hazard is in the event of an accident.

Summary

- Gases can be collected in the laboratory in different ways, depending on their properties.
- Gases that are insoluble, or only slightly soluble, can be collected over water.
- Gases that are less dense than air can be collected by upward delivery.
- Gases that are denser than air are collected by downward delivery.
- All gases can be collected in a gas syringe, provided they do not react with the substances making up the syringe.
- Hazard labels on containers warn you of the hazards of particular chemicals.

Questions

6.16 Hydrogen chloride gas is very soluble in water. Which method is definitely not suitable for collecting this gas?

6.17 Sulphur dioxide gas has a density much greater than air. It is soluble in water. Draw *two* possible methods of collecting this gas.

6.18 Name a gas that bleaches litmus paper.

6.19 Why should oxidisers and highly flammable chemicals not be stored together?

6.20 Find out what the hazard symbol is for an explosive chemical, and draw it.

E Uses of some compounds

The acid that gives cola drinks their tangy flavour is phosphoric acid, which is quite a strong acid. It is very dilute but manufacturers of cola drinks such as Coca-Cola and Pepsi are frequently required to defend the use of this artificial food additive. Some people claim it is so acidic that it reacts with tooth enamel and starts to dissolve it away.

When we look at this claim we must be aware that many foods and drinks contain acids. Manufacturers of cola drinks point out that acids in drinks are essential to prevent bacteria from growing. All citrus fruits such as lemons and oranges contain citric acid. Drinks made using citrus fruit juices are just as likely to react with tooth enamel as phosphoric acid.

Figure 6.31 Does your cola drink rot your teeth?

Cola manufacturers are quick to say that drinks are in less contact with teeth than are acidic solid foods, such as those that are pickled in vinegar. Vinegar is also an acid, called ethanoic acid or acetic acid.

Ten common compounds

The box above mentions three common chemicals – phosphoric acid, citric acid and ethanoic acid. The names of these chemicals may not be familiar to you, but it is likely that these acids are present in many homes. Many other chemicals are common in our homes. Here we look at the three acids just mentioned and seven other compounds. One or two common uses are given. You should use the internet to investigate other uses.

Figure 6.32 Chemicals in our homes

carbon dioxide

CO_2

This is used to make fizzy drinks fizzy. It is dissolved under pressure into the drink.

phosphoric acid

H_3PO_4

Apart from its use in cola drinks, when in its concentrated form it can be used to remove rust from cars.

ammonia

NH_3

You can find this compound in household cleaners. It is also responsible for the smell produced when bacteria break down urine in babies' nappies.

water

H_2O

This is what comes out of our taps. We all need to drink water to live.

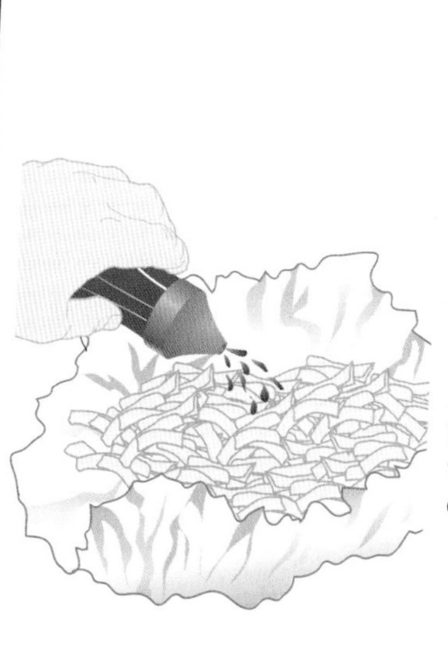

ethanoic acid or acetic acid

(vinegar)

CH_3COOH

Apart from pickling foods to preserve them, this is used to flavour foods.

Acetic acid is the old name. The acid in vinegar is now called ethanoic acid.

sodium chloride

(common table salt)

NaCl

Salt has been used for centuries as a preservative in foods and as a flavouring agent. Too much is bad for you but without any you could die.

caustic soda

sodium hydroxide

NaOH

This is used as a drain cleaner.

Caustic soda is an old name for sodium hydroxide. *Caustic* means to burn or corrode organic matter, such as flesh.

hydrochloric acid

HCl

This chemical is in your stomach. It helps to break down and digest your food.

carbohydrates

e.g. sugar $C_6H_{22}O_{11}$

Carbohydrates come in many different forms. You will find them in starchy foods such as potatoes and in sugar. They provide us with energy.

citric acid

Apart from its natural use in making lemons and oranges taste sharp, citric acid may be found in the kitchen in powder form and is used by amateur winemakers. It is also used in making lemonade and boiled sweets.

Summary

- **You should be able to identify uses of the common compounds in this section.**

Question

6.21 Choose *five* of the compounds mentioned in this section. Use the internet to research a use outside of the home for each of your chosen chemicals.

There's one Earth

A Global warming and climate change

It's the stuff of science fiction. The UK, instead of getting warmer as a result of global warming, suddenly starts to freeze over. You may have seen *The Day After Tomorrow*, a film based on the United States experiencing a massive drop in temperature. Some scientists think that this is much more likely to happen here than on the other side of the Atlantic. We would see icebergs floating off the coast, prolonged snowfalls for much of the winter and our ports freezing over during the winter months.

Figure 7.1 Is this scene likely to become commonplace in this country every winter?

We may moan about our climate as it is now, but other countries on our latitude in the northern hemisphere have very much cooler climates than we experience. What keeps us warm is an ocean current called the Gulf Stream. It starts in the southern hemisphere and passes across the equator along the Gulf of Mexico, absorbing heat all the time. It then bathes the UK coastline with much warmer waters than would otherwise be there. This has been our good fortune, but some scientists believe that the Gulf Stream could suddenly stop flowing in our direction.

So what could stop such a massive current, that moves billions of tonnes of water every second? The Gulf Stream cools as it makes its journey northward, and when it reaches Greenland it becomes so dense that it sinks and makes the return journey back to the southern hemisphere. It's like a gigantic conveyor belt. But global warming is causing huge amounts of fresh water to pour off the melting ice sheets of Greenland. This makes the cooling Gulf Stream less dense and less likely to sink to make its return journey. This could effectively halt the Gulf Stream – which would mean a dramatic change in climate for the UK and western Europe.

The theory of global warming

The idea that increasing amounts of carbon dioxide in the atmosphere could cause warming of the Earth is not a new one. The first person to attempt a calculation of this effect was a Swedish scientist called Svante Arrhenius over 100 years ago. He worked out that carbon dioxide must have been increasing in the atmosphere due to increased coal burning.

Figure 7.2 During the Industrial Revolution vast amounts of coal were burnt to manufacture things we now take for granted. In this picture from 1835, coal is being burnt to extract iron from iron ore

Coal is mainly carbon and when it burns it produces carbon dioxide:

carbon + oxygen → carbon dioxide

$$C + O_2 \rightarrow CO_2$$

Coal burning greatly increased in the 19th century because it was the main source of energy that powered the Industrial Revolution. Arrhenius calculated that if the level of carbon dioxide in the atmosphere were to double, then it would lead to a 6 °C rise in temperature. Unfortunately, it took a long time to go from one scientist's idea to a theory that is now widely accepted.

Scientists now think the resulting rise in temperature will be less than Arrhenius predicted, but whereas he believed that the doubling in carbon dioxide concentration would take about 3000 years, at current rates of increase it will take less than 200 years. Arrhenius also believed that the temperature increase would be largely beneficial to the world.

Although Arrhenius didn't get it all right, his work influenced other people to investigate climate change.

For a scientist's idea to become widely accepted by other scientists, there must be evidence to back it up. A British engineer called Guy Callendar put forward the first piece of evidence for the Earth's warming in 1938. He used to study the climate as a hobby. He discovered that data collected over 50 years from 200 weather stations around the world showed a definite rise in average temperatures. He also calculated that during these 50 years carbon dioxide concentrations in the atmosphere had steadily risen. He concluded that this was evidence for global warming. However, hardly anyone took any notice.

It was almost another 50 years before scientists had produced enough evidence of global warming for many governments across the world to decide it was time to act. The United Nations set up a panel of scientists to report on climate change and suggest ways to combat global warming. In 1997 an international agreement was proposed, at Kyoto in Japan, to reduce the amount of carbon dioxide that countries release into the atmosphere. Many countries have now signed up to this agreement. One country that still hasn't is the United States, which produces a quarter of the carbon dioxide that goes into the atmosphere each year. The US Government points out that not all scientists agree that the Earth is warming up solely as a result of increasing amounts of carbon dioxide in the atmosphere. It is argued by some that there has always been natural variation in the global climate and that the period we are in now could just be one of those warmer times. So it is important to realise that, although the vast majority of climate scientists accept the theory of global warming, it is not accepted by all.

The greenhouse effect

So how does global warming occur? The Earth has an atmosphere that acts a bit like a duvet on your bed, trapping some of the heat that comes from the Earth. The Sun's rays warm the Earth. The land and the sea then radiate this energy back into space but the atmosphere absorbs some of this heat radiation. This stops our planet cooling too much at night, when half the Earth has no warming sunshine.

Even in winter if you step into a greenhouse when the sun has been on it, it feels warm inside. This is because the sunshine has warmed everything inside the greenhouse and all these things start radiating the heat back out again. However, the greenhouse glass doesn't let this heat escape so the inside warms up. This is similar to the way our atmosphere works in trapping the warmth of the Earth. Hence we call this effect the **greenhouse effect**.

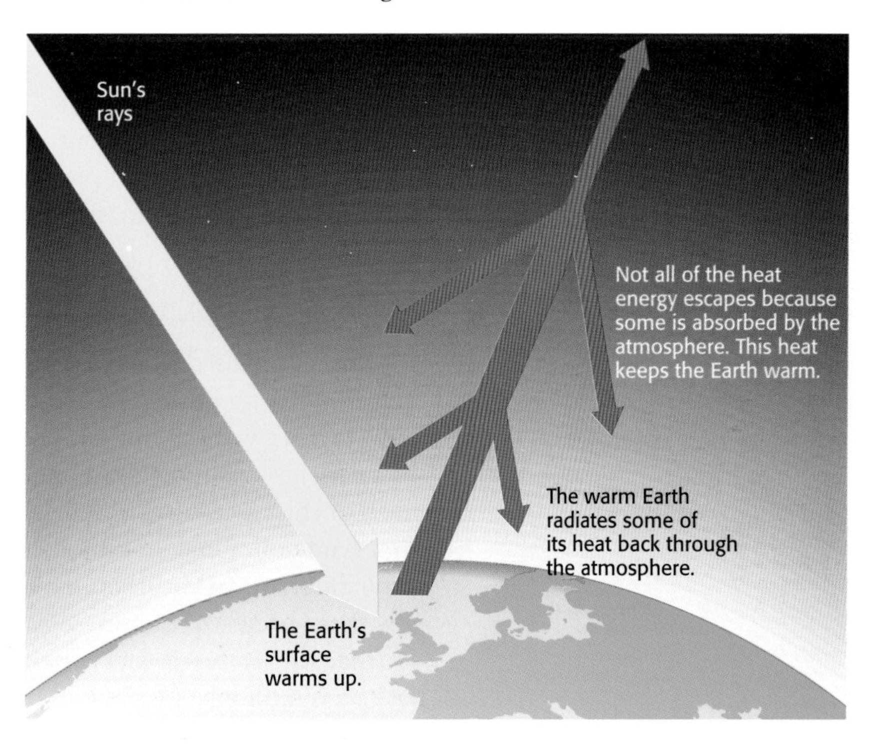

Figure 7.3 The Sun's rays warm the surface of the Earth. The Earth then radiates this heat energy into the atmosphere. Some of this heat energy goes straight back into space but some is trapped by the atmosphere and keeps our planet warm

Nitrogen makes up 79% of the Earth's atmosphere and oxygen 20%. Carbon dioxide makes up only about 0.038%.

Carbon dioxide is not the only greenhouse gas in the Earth's atmosphere. Two others are methane and water vapour.

Our atmosphere is made up of a mixture of many gases. The main two that make up 99% of the atmosphere are nitrogen and oxygen. These two gases are not 'greenhouse gases' – they do not absorb heat from the Earth. Carbon dioxide makes up only a tiny percentage of the atmosphere but it is a greenhouse gas.

Why we need the greenhouse effect

If there were no carbon dioxide in the atmosphere, our planet would be a very different place. It is the small amount of carbon dioxide that keeps the temperature of the Earth moderate and reasonably constant.

Burning hydrocarbon fuels

Combustion means burning. **Complete combustion** means that there is enough oxygen present for all the carbon atoms in the fuel to burn to form carbon dioxide.

Every time you make a car journey or take a bus ride, the fuel being used to transport you is being burnt. Buses run on diesel fuel, while most cars use petrol. These fuels are made up of molecules called **hydrocarbons**. Hydrocarbon molecules contain only hydrogen and carbon atoms, bonded together. When these molecules burn, they react with oxygen in the air and release energy. This is an **exothermic** reaction. The products of the **complete combustion** of hydrocarbons are carbon dioxide and water.

Figure 7.4 Most of the vehicles we use to transport us run on hydrocarbon fuels

Methane is the gas that comes out of gas taps in the chemistry laboratory and in gas cookers at home.

The simplest hydrocarbon is **methane**. It has the formula CH_4. This formula tells you that in a molecule of methane there are four hydrogen atoms bonded to one carbon atom.

When methane burns it produces carbon dioxide and water. The word and formulae equations are:

$$\text{methane} + \text{oxygen} \rightarrow \text{carbon dioxide} + \text{water}$$
$$CH_4 + 2O_2 \rightarrow CO_2 + 2H_2O$$

The energy released from burning this hydrocarbon means that the temperature of a Bunsen burner flame can reach almost 900 °C.

Every year we burn billions of tonnes of hydrocarbons to provide the energy to transport us, to manufacture our goods and to produce much of the electricity we use. We get most of our hydrocarbon fuels from **crude oil**. Crude oil is a natural resource that is a mixture of many different hydrocarbon molecules. On page 142 you will see how these hydrocarbons are separated from the crude oil mixture.

Crude oil is a **fossil fuel**. The term 'fossil' tells you that the fuel has taken millions of years to form from dead animals and plants. The crude oil we use today is formed from the remains of tiny sea creatures and plants that died about 400 million years ago, sank to the bottom of the oceans and got buried by mud. Many chemical reactions occurred, over a huge span of time, to make the crude oil that is now piped out of the ground. Coal is also a fossil fuel and was formed from trees that died millions of years ago.

Burning fossil fuels and global warming

Fossil fuels all contain carbon atoms in their molecules. When carbon burns it produces carbon dioxide, and most of this carbon dioxide ends up in the atmosphere. Since carbon dioxide is a greenhouse gas, if there is more of it in the atmosphere then more heat energy will be trapped. This in turn may make the planet warmer – hence the theory of global warming.

Nearly every climate scientist believes there is a link between burning fossil fuels and global warming. But how do we know that carbon dioxide in the atmosphere is increasing?

Even though the amount of carbon dioxide in the air is very tiny, it is possible to measure it. The problem with measuring it is that you need to find a place that is far away from towns and cities, so that direct emissions of carbon dioxide from car exhausts do not interfere with the readings. The measuring station should not be near growing crops, either, as these are constantly removing carbon dioxide from the air for photosynthesis and adding it back when they respire.

The choice of a site to measure carbon dioxide concentrations in the atmosphere is an example of how scientists attempt to ensure that their results are valid.

The most famous measuring station is in Hawaii at the Mauna Loa Observatory (Figure 7.5). Scientists have been taking measurements of carbon dioxide in the air there, using the same method for nearly 50 years. The graph of their results is one of the most important in history. See Figure 7.6. It clearly shows that the concentration of carbon dioxide is steadily increasing every year.

Figure 7.5 This monitoring station at Mauna Loa, Hawaii, is 3400 metres above sea level on volcanic rock. Scientists there have been taking monthly measurements of carbon dioxide concentrations in the atmosphere since 1958

Repeating the measurements every month ensures that the results are reliable.

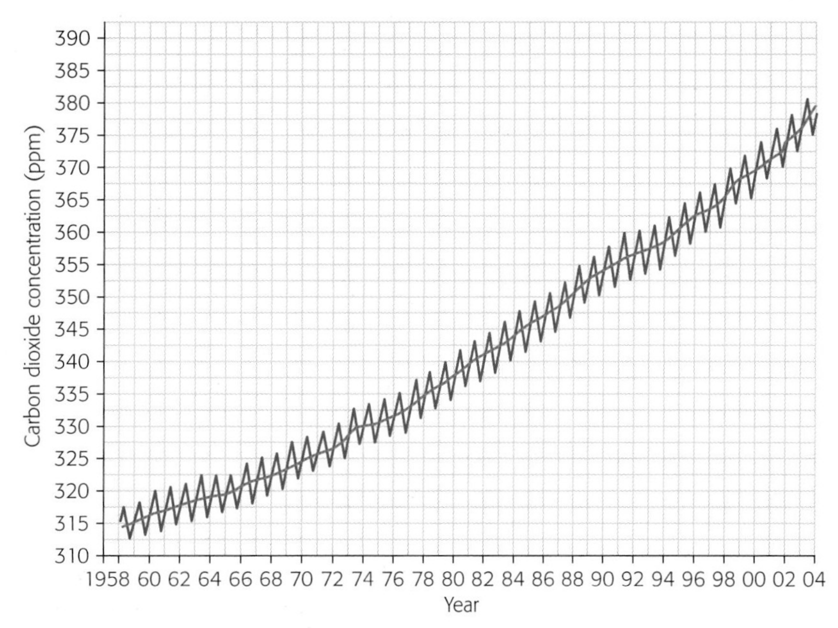

Figure 7.6 The Mauna Loa Observatory measurements show the increase in carbon dioxide concentration in the atmosphere. The zigzag line tells us that the concentration of carbon dioxide varies with each month, but the trend (red) is very definitely upwards

The concentration of carbon dioxide is measured in parts per million (ppm). One of the monthly readings in from Mauna Loa 2005 was 380 ppm. This means that in a million cm^3 of air, 380 cm^3 is carbon dioxide. As a percentage this is 0.038%.

Figure 7.7 Slicing an ice core for analysis of carbon dioxide content

Evidence from ice cores

The ice sheets in Greenland and the Antarctic started to form about 900 000 years ago. As the water froze it trapped tiny bubbles of the atmosphere at the time. Scientists can drill into this ice and find out the history of our atmosphere. The deeper the drill, the older the ice. The ice comes out in long cylinders called ice cores (Figure 7.7). These cores are melted and the air trapped inside is released. The carbon dioxide concentration can then be measured.

These studies of ice cores have shown that over the last 400 000 years the carbon dioxide concentration in the atmosphere has been as low as 180 ppm (0.018%) and as high as 280 ppm (0.028%). When it was 180 ppm, the Earth was in an ice age. When it rose to 280 ppm, the planet was much warmer. It was 280 ppm just before the Industrial Revolution started 250 years ago. During the whole of the previous 400 000 years the carbon dioxide concentration had never risen above 280 ppm. But in the 1980s, when the evidence from ice cores started to be published, the measured concentration of carbon dioxide stood at 350 ppm – much higher than the highest concentration before the Industrial Revolution.

The data from ice core research showed a clear link between carbon dioxide concentration in the atmosphere and the temperature of the planet. This was critical in making governments across the world aware of a looming crisis.

Predicting the future

In 1896 Svante Arrhenius first made his prediction that a doubling of carbon dioxide concentration would produce a 6 °C rise in average global temperature (see page 126). This was based on his limited knowledge of the atmosphere at the time. We now think that the rise will be 2–3 °C. But we realise now that this seemingly moderate temperature increase will produce many different weather patterns across the world, as well as the average warming. The effect of our atmosphere on the world's climate is incredibly complex and we now use computers to predict the future.

Scientists use the information they have about the climate and atmosphere to make a computer model. There are many factors that affect the average global temperature. For example:

- a huge volcanic eruption would send clouds of dust particles into the air, which would reflect sunlight away from Earth
- if the atmosphere gets warmer, more clouds will form and these will also reflect more sunlight back into space
- the planet's icecaps act as giant reflectors, reflecting the Sun's warmth away, but these are beginning to melt as the Earth warms up, so this effect will increase global warming.

These factors and many thousands more are used to attempt to model our changing climate. This is why it is impossible to say by exactly how much the Earth will warm and what the effect will be on our climate. However, each year more information is used to make more sophisticated climate models, reducing the level of uncertainty in predicting the future.

So what is to be done?

By 2005, 140 countries had signed the international agreement called the Kyoto Protocol. This commits these countries to trying to reduce the amounts of carbon dioxide they put into our atmosphere. The key to reducing this carbon dioxide is to burn less fossil fuel. But this will lead to a change in lifestyle for many. That is why we are investing in wind power, solar power and other alternatives to fossil fuels (see Chapter 10). We can all make a contribution. Just switching off lights when they are not needed reduces the amount of fossil fuel burnt in power stations. Insulating homes so that less energy is used to keep them warm in winter is another important thing to do. Walking or cycling, instead of using a car for short journeys, is another way of reducing fossil fuel use.

Acting now to combat global warming means taking precautions before our climate changes as dramatically as suggested in the box on page 125. This is sometimes called the **precautionary principle**. Many scientists today think we are leaving it much too late.

Using the internet to research information

In your lifetime the internet has revolutionised the way we can find out information. When we were doing research for this book we used the internet quite often to discover the latest data from research scientists and different organisations. With all such information, there are issues about its reliability. Can we trust it to be correct?

Here are some questions you might ask to evaluate information that you find, to help you decide whether it consists of well-founded scientific facts.

- Is it data or opinion? If it is opinion it may be subject to bias. Do many people agree with the conclusions, or is it just a few people?
- If it is data, can you find the same data anywhere else? In other words, have other scientists come up with similar results?
- Do other scientists agree with the data? Do other websites quote the data?
- Is it real data from experiments, or is it based on a computer simulation?
- Is the source reliable? Does the information come from an acknowledged expert in the field? Were the findings published by a respected scientific journal? Does the website belong to an official body with responsibility in the area you are researching? Or does it come from an organisation with an interest in a particular conclusion?

Often, in researching a particular topic, you will come across opposing views from different scientists. Some of the questions above may help you decide which side of the argument you favour.

Summary

- It has taken more than 100 years for one single scientist's ideas about global warming to become a widely accepted theory. However, some scientists still argue that increasing global temperatures may be part of a natural cycle.
- Carbon dioxide is a greenhouse gas. Greenhouse gases trap some of the heat that the Earth radiates into space.
- Most climate scientists now accept the theory of global warming because of evidence that carbon dioxide is increasing in our atmosphere and that the average global temperature has increased during the last century.
- Hydrocarbons are compounds containing only atoms of hydrogen and carbon joined together in molecules.
- When hydrocarbons are completely burnt they release energy and produce carbon dioxide and water.
- In this country we have come to rely on burning hydrocarbons to produce the energy we need for transport, for manufacturing and for the production of electricity.
- Oil is a hydrocarbon fuel, also called a fossil fuel. It took millions of years to form from sea creatures that died long ago. Coal is another fossil fuel that formed from trees that died millions of years ago.
- From studying ice cores we know that the composition of our atmosphere has changed over the last 400 000 years. The carbon dioxide concentration varied between 180 ppm and 280 ppm until the start of the Industrial Revolution, when it began to rise steadily. Readings of atmospheric carbon dioxide in 2005 were as high as 380 ppm.
- Climate scientists use computer models to predict the effect of rising carbon dioxide levels on our climate. But no computer model is perfect. There will always be uncertainties.
- Countries and individuals can take action to combat global warming. The precautionary principle means taking action just in case the worst predictions of climate change become reality.
- The internet is a good place to research up-to-date information, provided you check its reliability and lack of bias.

Questions

7.1 **a** What is a hydrocarbon?

b When hydrocarbons burn and completely combust, what products are formed?

c Propane is a hydrocarbon, with the formula C_3H_8. It is used in a fuel called Calor Gas. Write the word equation and (Higher level) the formulae equation for propane burning completely in oxygen.

7.2 **a** Why is the Mauna Loa Observatory a good place to measure the composition of the atmosphere every month?

b Look at the graph of carbon dioxide concentration measured each year since 1958, Figure 7.6. Using the trend line (red), work out the increase in carbon dioxide concentration from 1990 to 2000.

c Every year the concentration of carbon dioxide goes up and down to produce the zigzag line on the graph. Why do you think the carbon dioxide concentration varies during each year?

7.3 What evidence are scientists able to collect from ice cores in Antarctica to support the theory of global warming?

7.4 Computer models are only as good as the information that is used to construct them. Recently some scientists have come up with a theory called Global Dimming. Use the internet to find out what this term means, and what effect this might have on the Earth's average temperature.

H **7.5** Imagine you are part of the United Nations scientific panel advising on climate change. Use the precautionary principle to argue for five steps the UK could take to combat global warming.

B Wasting away

The title of this chapter is *There's one Earth*. There may be no other planet like ours anywhere in the Universe. It has abundant resources and, most important of all, it is teeming with life. The Earth, as we know it, is a complex balance of billions of interconnecting systems.

When we use Earth's resources we have to be careful because we are always in danger of disturbing the balance that exists. When we mine metal ores we produce vast heaps of waste material, and we use huge quantities of fossil fuels to convert the metal ore into the metals we require. The burning of fossil fuels produces carbon dioxide and, as you read in the last section, most scientists believe this is a major cause of global warming. Also, the crude oil on which our modern life depends will soon run out.

It is the same with paper production. We cut down trees, destroying natural habitats, and the energy required to convert wood into paper comes mainly from burning fossil fuels.

Faced with this information, why do we throw 30 million tonnes of rubbish away each year in dustbins? This is one tonne (1000 kg) for every household in the UK. We are not good at recycling in this country. Up to 70% of the things we throw away could be recycled but only about 15% is. We cannot keep manufacturing new things and using them only once. If we do, then we will run out of many of Earth's natural resources, and we will destroy the delicate balance on which life on our planet depends.

Figure 7.8 Of all the planets, the Earth is unique because it supports life

Reducing our rubbish

Every year, on average, each one of us produces half a tonne of waste that has to go somewhere. Nearly three-quarters of it goes into landfill sites. These are holes in the ground, often left over from quarrying. The rubbish is then buried. This creates two major issues.

The first is that there are not enough landfill sites for the rubbish we now produce. The second issue is that bacteria in the ground break down any organic waste, such as paper or potato peelings, to produce methane gas. This is a greenhouse gas that absorbs 20 times more energy in the atmosphere than carbon dioxide, so it makes a big contribution to global warming.

Every 2 hours we could fill the Royal Albert Hall with what we throw away.

Figure 7.9 Most of our rubbish goes into landfill sites but very soon we will run out of these sites

We can all do something about this, in three ways. We can:

- ***reduce*** the amount of rubbish we create by buying things that are not over-packaged
- ***reuse*** things rather than throwing them away
- ***recycle*** as much as possible.

If something is **recycled** it means that the material it is made of is used again.

Recycling materials makes a lot of sense. It not only saves the Earth's precious resources but it is better for our economy and the environment. We'll look at the recycling of three materials.

Figure 7.10 Huge quantities of paper are produced every day

Recycling paper

Think about the number of articles you come across in one day made from paper. To satisfy our demand for paper products in this country, a forest the size of Wales is cut down each year. It is true that most of our paper comes from forests that are grown for this purpose in Scandinavia, so for each tree cut down another one is grown. But these forests have replaced older forests with established wildlife habitats. They do not support the range of animals and plants that older forests do. And even these 'renewable' forests do not match our need for paper.

The contents of the average dustbin contain about 20% paper or card. Every year 5 million tonnes of this goes into landfill sites. We are getting better at recycling paper. In England alone in 2004, 1.3 million tonnes of paper were recycled, but this is nowhere near what it could be.

The advantages of recycling paper are:

1 It reduces the number of trees cut down.
2 It means we import less from Scandinavia, which is better for our economy.
3 It uses less energy, depending on the grade of paper being produced – this means that less fossil fuel is burnt and less carbon dioxide goes into the atmosphere.

The argument about saving energy is always a complicated one. To calculate whether recycling saves energy you must look at all the energy used in all the processes. To produce brand new paper from trees it takes energy to:

- plant the seeds
- manage the forests
- cut down the timber and transport it to the paper mill
- pulp the wood to make paper
- transport the paper to where it will be used.

To recycle paper it takes energy to:

- transport the paper from homes and offices to the recycling paper mill
- clean the old paper
- convert it into usable paper again
- transport the paper to where it will be used.

When all of this energy is added up, the paper industry calculates that between 30% and 70% less energy is used in recycling paper – a large saving in energy and cost. It is also much better for the environment.

Figure 7.11 It is an environmental tragedy when glass ends up in household refuse

Recycling glass

We still occasionally dig up glass bottles used by the Egyptians as long ago as 3000BC. Glass is very slow to break down in the ground. Each year 1 400 000 tonnes of glass is put into landfill sites and it will still be there in thousands of years' time.

The raw materials for making glass are sand, limestone and soda ash. Although these may be relatively cheap and readily available, it takes a lot of energy to turn them into glass. Most of the energy required to do this comes from fossil fuels.

At the moment we recycle about one-third of the glass we use, but *all* of it could be recycled. There are real advantages in recycling it:

- It uses much less energy. For every two bottles recycled you save enough energy to boil a kettle to make five cups of tea.
- It significantly reduces the amount of carbon dioxide going into the atmosphere. The saving is about 225 kg of carbon dioxide for every tonne of glass recycled.
- It saves natural resources – 1.2 tonnes of quarried raw material are needed to make one tonne of glass.
- One tonne recycled is one tonne that is not dumped in landfill sites.

Recycling metal

Of the rubbish we throw into bins, 8% is metal and most of this is in the form of cans. There are two main materials used to make these cans – aluminium and steel.

Most drink cans are made from aluminium. Recycling them saves a huge amount of energy, and therefore money. It takes 95% less energy to melt the cans back down to liquid aluminium than it does to extract the aluminium from its ore. Each can recycled saves enough energy to run your television for 3 hours.

It takes 8 tonnes of ore to make 1 tonne of aluminium.

Figure 7.12 Steel is quite easy to extract from unsorted household rubbish using magnets

Drink cans are the most recycled packaging containers worldwide, but in the UK we recycle only 42% of the 5 billion drink cans we consume each year. This is one of the lowest recycling rates in Europe and well behind Switzerland and Finland, where they recycle 91% of their aluminium cans.

Recycling steel cans saves 74% of the energy required to make new steel. Most of the steel cans you have in your house will be food cans. These are often called 'tin cans' or 'tins' because of the very thin layer of tin that coats them and stops them rusting.

As steel is magnetic, it is the most easy material to sort from household waste. However this only recovers 85% of the steel thrown away. Many local authorities are asking us to put steel cans into recycling containers so that even more can be recycled.

Making more of the planet's water usable

At first sight it seems ridiculous that water is quite a scarce resource. After all, most of the surface of the planet is water. However, 97% of this water is salt water and 2% is frozen in glaciers. In many hot countries there are real problems with the supply of fresh water for irrigating crops, for animals to drink and for human consumption.

One solution to the problem of fresh water supply is to remove the dissolved salts in sea water. This is called **desalinisation** (or **desalination**).

Figure 7.13 Desalinisation plant in the USA

There are two main ways of doing this. Either the sea water can be boiled, and fresh water distilled off by collecting the steam, or the sea water can be pushed under pressure through a membrane that has very fine pores in it. These pores are so small that they only allow water molecules through, so separating them from the dissolved salts.

There are economic and environmental considerations for desalinisation, whichever process is used. A lot of energy is required. If this energy comes from burning fossil fuels then the process is adding to the amount of carbon dioxide in the atmosphere, which most climate scientists believe will increase global warming. Some desalinisation plants use solar energy, instead of burning fossil fuels. As there are no emissions from solar energy production, this is a much more environmentally friendly option. Other plants are sited next to power stations. In this way the steam produced in the power station is used to provide the energy required for desalinisation, making it a much more energy-efficient process.

Another environmental issue is the effluent that is left over after most of the water has been removed. This has a very high concentration of salts. Disposing of this is very difficult as it will kill both marine and land life.

From an economic point of view, as desalinisation is energy intensive it will also be expensive. Putting the desalinisation plant next to a power station will offset these costs but they are still considerable. Also, if the membrane process is used, the membranes do not last very long and they are expensive to make.

There are also positive environmental reasons for desalinisation. For example, if too much fresh water is removed from rivers, habitats both in the rivers and along their banks may be destroyed.

Summary

- **By using Earth's resources we may disturb the balance that sustains life on the planet. Also we may run out of some resources, such as crude oil.**
- **Most of the UK's rubbish goes into landfill sites; suitable places for landfill are running out.**
- **Recycling is usually good for the economy. It is often cheaper to recycle than to manufacture or import new materials.**
- **Recycling of metal, glass and paper saves energy. This energy usually comes from burning fossil fuel, so recycling means less carbon dioxide is released into the atmosphere.**
- **In hot countries, where water is scarce, desalinisation can provide much needed fresh water. However desalinisation is expensive and uses much energy. Also the waste is highly concentrated in salts and requires careful disposal.**

Questions

7.6 Recycling usually uses less energy than making the material from scratch. This often means less fossil fuel is burnt. Give two reasons why this is better for the environment.

7.7 Most of our paper comes from managed forests in Scandinavia.
  a What does the term *managed forest* mean?
  b Give one positive reason for using wood from a managed forest.
  c Explain why there are environmental disadvantages to using managed forests.

7.8 Aluminium is extracted from an ore called bauxite. It takes 8 tonnes of bauxite to make 1 tonne of aluminium. What are the environmental issues behind these figures?

7.9 **a** What does *desalinisation* mean?
  b There are two major environmental issues concerning desalinisation. What are they?

7.10 Keep a 'dustbin diary' for a week. Estimate what mass in kilograms your family throws away in the ordinary household waste and how much of it is paper, metal or glass.

C Fuelling the future

Over the last few decades we have become very dependent on the private car for getting us around. People often travel quite long distances to work, to the shops, or to take children to school. But some of the distances we travel by car are very short. One in ten of our car journeys is under a mile long – a distance that most of us could walk in 20 minutes. But why does it matter?

For a start, we are beginning to run out of the crude oil that fuels much of our modern life in this country. We produce most of our own petrol and diesel from oil that we pump out of the North Sea, but by 2010 there will not be enough to allow us to do this.

Then there is the problem of global warming. Many people now believe this is the biggest crisis to face humankind. Much global warming is caused by the greenhouse gas carbon dioxide, which is produced when fossil fuels burn. Petrol and diesel are both fossil fuels. Every year the average motorist drives 10 000 miles, producing 2 tonnes of carbon dioxide.

Figure 7.14 Our modern way of life depends on the car

Clearly one way of reducing carbon dioxide emissions is to make fewer car journeys. Another way is to make the car engine more efficient so that it produces less carbon dioxide per mile. A third way is to consider alternatives to petrol and diesel – and there are some interesting developments in this area.

What makes a useful fuel?

A fuel is a substance that releases energy that can do work. Most fuels release this energy during combustion reactions (see page 128). If we are to consider alternative fuels for the car, the lorry and the bus, we must first consider what properties these fuels should have.

The properties of an ideal fuel are that it:

- is widely available and reasonably cheap to produce
- is safe to store and easy to transport
- releases a lot of heat energy for its mass
- burns easily
- does not give off dangerous fumes
- does not leave any solid residue, such as ash, smoke or soot.

Residue is any solid left over after the fuel is burnt.

Now we have listed these properties we can see that petrol is not an ideal fuel, but it is still very useful. What makes it so popular is that it is still quite a cheap fuel, and it is safe to transport. For its mass, it produces lots of energy and, if it burns completely, it does not produce solid waste that has to be dumped. However, it does produce some harmful gases. Even if these gases were just carbon dioxide and water, the fact that carbon dioxide is a greenhouse gas means that it is not an ideal fuel. What are the alternatives?

Biofuels

Biofuels are fuels that are made from the products of living things, usually from plants. There is increasing interest in the use of biofuels in vehicles. An important reason for this is that the carbon dioxide given out by these fuels when they are burnt was originally taken in from the atmosphere by the plants when they grew. So, in theory, using a biofuel does not increase the carbon dioxide in the atmosphere. Also, using biofuels conserves our precious crude oil.

The use of two biofuels is set to increase significantly in this country. One of these is biodiesel. It is made from vegetable oil, much of it produced from growing oilseed rape. (You may have seen huge fields of this crop – the flowers are a very bright yellow.) Biodiesel has almost the same properties as ordinary diesel. In the month of May 2005, biodiesel sales were 3 million litres, an increase of 1 million litres from May 2004. Many local authorities are now using a mixture of 95% fossil fuel diesel to 5% biodiesel for all their vehicles, and the Government has decided that all petrol and diesel on sale in the UK from 2010 will contain 5% biofuel.

The second biofuel set to play a significant role in the UK is alcohol – the same alcohol that is found in beers, wines and spirits. Brazil has been using alcohol in many of its cars since 1985. New Brazilian cars can run on pure alcohol, or a mixture of alcohol and petrol. Whatever the mixture of the fuel, a computer chip inside the car analyses its composition and adjusts the engine accordingly.

If fossil fuels are used to provide the energy to plant, harvest and process crops to make biofuels, then the carbon dioxide added to the atmosphere will be more than the plants took out in the first place.

Brazil is ideally placed to make use of alcohol as a biofuel. It grows millions of hectares of sugar cane and this is fermented into alcohol. Of course it requires energy to crush the cane, ferment the sugar and distil the alcohol, but this energy comes from burning the crushed cane stems. This means the amount of carbon dioxide produced from processing the sugar cane is still only what was taken out of the atmosphere in the first place.

To produce alcohol in this country for a biofuel without importing large quantities of sugar cane would mean growing more sugar beet. This would take up fertile land that could otherwise be used to grow crops for food. At the moment we grow more food than we can eat, so there is in theory some agricultural land available for biofuel crops, but this will only provide so much. This is one reason why biodiesel and alcohol can never completely replace fossil fuels – there simply is not enough fertile land to spare.

Figure 7.15 Brazilians use Alcool as a vehicle fuel. It is made from sugar cane

Hydrogen fuel

Hydrogen has been called the 'pollution-free fuel'. When hydrogen combusts it produces only water:

$$\text{hydrogen} + \text{oxygen} \rightarrow \text{water}$$
$$2H_2 + O_2 \rightarrow 2H_2O$$

No carbon dioxide is produced because there is only hydrogen in the fuel. There are experimental cars in operation now that run on hydrogen fuel (Figure 7.16), but it will be some time before these go into commercial production.

Figure 7.16 This car has a new design of engine that uses hydrogen for fuel

There are drawbacks to using hydrogen as a fuel. First of all we have to have a way to make the hydrogen that does not involve using a fossil fuel. Otherwise any saving on carbon dioxide emissions by burning hydrogen will be cancelled out. Secondly, hydrogen is a gas, which means it occupies a large volume. One way to make hydrogen's volume smaller is to cool it to a liquid. This is not very practical, as it must be cooled to below –253 °C. Another way is to keep the gas in pressurised tanks, but these are bulky and heavy. Chemists are researching a third way, which involves storing the gas on the surface of transition metals.

The other way of using hydrogen as a fuel is in a **fuel cell**. A fuel cell transfers the energy from reacting hydrogen and oxygen directly into electricity. The fuel cell too is in the experimental phase.

There are already cars in operation that run on electricity – see page 199 of Chapter 10.

Sustainable development

We live in a world of more than 6 billion people. Almost 1 billion live in the most terrible poverty, 800 million do not get enough food and 2.5 billion do not have access to adequate sanitation. If everyone on the planet were to live as we do in the UK, the WorldWide Fund for Nature (WWF) estimates that we would probably need three planets' worth of resources. Yet, as this chapter's title tells you, there's only one Earth.

Many scientists believe that we cannot **sustain** (keep up) the way we live now. Already in this chapter we have looked at how carbon dioxide emissions could bring about climate change, and how over-use of natural resources could disrupt the balance of the planet. By considering alternatives to fossil fuels, we are taking a small step on the road to **sustainability**. We can take another step by reusing and recycling materials – the more we reuse and recycle, the less we use up precious resources.

Many say that sustainable development is about fairness. The standard of living we enjoy should be available to the rest of the world, yet as a nation we live wastefully. Other countries have a right to develop so that their economies can feed, house and provide a healthy lifestyle for their citizens.

Sustainable development involves balancing:

- the world's need for economic development
- the desire for standards of living that are fair not only to everyone now but also to future generations
- respect for the environment.

Summary

- A useful fuel has the following properties:
  - fairly cheap to produce and readily available
  - safe to store and easy to transport
  - releases a lot of heat energy for its mass
  - does not give off dangerous fumes or leave solid residues.
- Biofuels are mostly produced from plants grown as crops.
- Biodiesel is made from vegetable oil; alcohol is made by fermenting sugar from sugar cane or sugar beet.
- The advantages of biofuels are that they reduce the demand for petrol and diesel, and that they do not add to the carbon dioxide in the atmosphere, as burning them only puts back into the atmosphere what was taken out to grow the crops.
- The disadvantage of biofuels is that they take up fertile agricultural land that could be used for growing food crops.
- Hydrogen is a fuel that produces only water when it burns – it is pollution free.
- The drawbacks to using hydrogen instead of petrol are storage and economic production that does not involve burning fossil fuels.
- Sustainable development involves balancing the need for economic development, standards of living and respect for the environment.

Questions

7.11 **a** What is a biofuel?
b Give two examples of a biofuel.
c Explain why wood is called a biofuel.

7.12 When biofuels burn they are said not to add to the amount of carbon dioxide in the atmosphere. Why is this?

7.13 **a** Find out how hydrogen can be manufactured without using fossil fuels.
b Why is it important not to use fossil fuels in the manufacture of hydrogen fuel?

7.14 Use the principles of sustainable development to suggest three actions your school could take to make its daily life more sustainable.

D Making crude oil useful

Sometimes the ancient Chinese would come across crude oil when digging for salt. They used it as a fuel as long as 2000 years ago. In the Middle East it was often found on the ground, but all the components of it that would burn had long since evaporated away, leaving a sticky tar called pitch. The pitch was used for waterproofing boats. But no one could have predicted how the first oil well was to change our world. It was sunk in 1859 by Edwin Drake in the USA.

We look back through history and give names to certain periods, often based on the material that was used at the time. We have named the Stone Age, the Bronze Age and the Iron Age. Perhaps historians will look back at this period and call it the Oil Age. If they do, then it will probably only have lasted a brief 200 years from the sinking of the first well to supplies running out. You are likely to see the end of the Oil Age in your lifetime.

Although we burn 90% of the crude oil that comes out of the ground, the other 10% is used to make many things that we take for granted – such as plastics, detergents, medicines, dyes and paints.

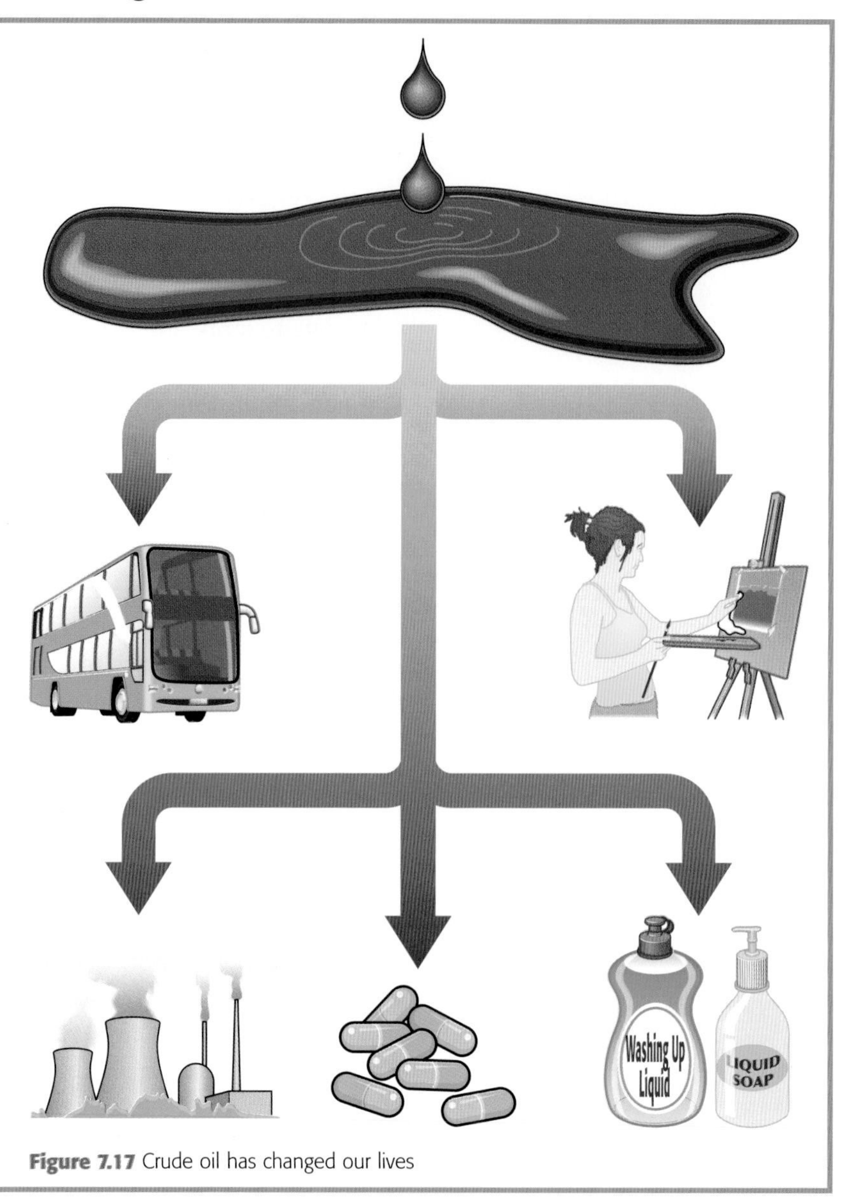

Figure 7.17 Crude oil has changed our lives

Fractional distillation

When crude oil is pumped out of the ground it's not of much use. It's a black, smelly mixture of hundreds of different compounds, nearly all of them hydrocarbons. To make it useful it must first be separated into its components. This is where **fractional distillation** comes in.

When the crude oil arrives at the refinery, it is heated to about 400 °C, which turns it into gases. The gaseous crude oil then enters a huge **fractionating column**, which can be up to 60 metres high. As the gases rise up the column they cool and turn back into liquids. Liquids with the highest boiling points condense first. The liquid mixtures that are condensed off at different heights are called **fractions**. Each fraction is a mixture of different compounds with a similar boiling point range. See Figure 7.18.

Figure 7.18 The fractionating column separates crude oil into fractions with different boiling points. All the fractions have particular uses

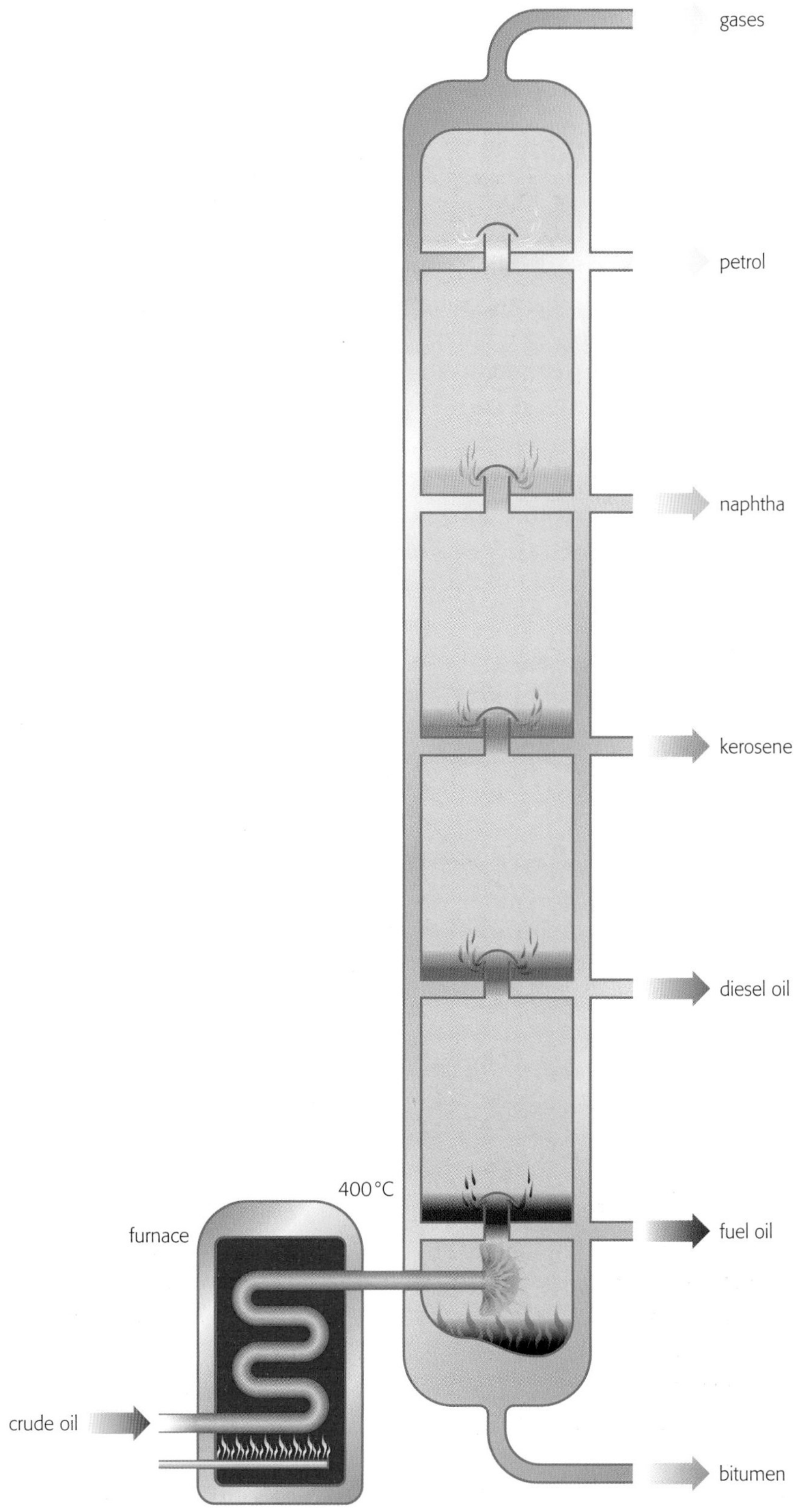

gases
petrol
naphtha
kerosene
diesel oil
fuel oil
bitumen
400 °C
furnace
crude oil

bottled gases
car fuel
manufacturing chemicals, e.g. detergents
aeroplane fuel
lorry fuel
ship and power station fuel
road building

How a fractionating column works

The fractionating column is very tall. It is hot at the bottom where the gaseous mixture of crude oil enters at 400 °C, and cools down as the gases go upwards. The bitumen fraction has such a high boiling point (above 350 °C) that it condenses as soon as it cools on entering the column, so bitumen is piped off at the bottom. As the gaseous mixture rises it cools further, so compounds with lower and lower boiling points condense as different fractions.

It is important to realise that a fraction is not a pure liquid. It is a mixture of different compounds.

Some compounds in crude oil do not condense in the fractionating column, because their boiling point is very low. These come out of the top of the fractionating column as a mixture of gases.

The boiling points of the fractions depend on the size of the molecules that they contain. Bitumen, with a very high boiling point range, is made up of very large molecules. Typically bitumen molecules may have more than 70 carbon atoms in them. Petrol, in comparison, has molecules that are much smaller. These molecules contain only five to ten carbon atoms. Petrol boils in the range 20 °C–70 °C and so is collected almost at the top of the column. The gases that come out of the top have molecules with only one to four carbon atoms in them.

Table 7.1 Properties of the fractions

Fraction	Boiling point range (°C)	Number of carbon atoms in the molecules	Viscosity	Ease of ignition	Uses
gases	less than 20	C_1–C_4	Viscosity increases as molecules get larger		bottled gas for camping stoves and barbecues
petrol	20–70	C_5–C_{10}			car fuel
naphtha	70–120	C_8–C_{12}			to make thousands of different chemicals from plastics to medicines
kerosene	120–240	C_{10}–C_{16}		Ignition gets easier as the molecules get smaller	aeroplane fuel
diesel oil	220–250	C_{15}–C_{25}			fuel for buses, lorries and trains
fuel oil	250–350	C_{30}–C_{60}			fuel for ships and power stations
bitumen	above 350	C_{60} and above			road surfacing and waterproofing roofs

If a liquid is **viscous** it means it is not very runny and does not pour well. Cold treacle is a viscous liquid.

The size of the molecule not only affects the boiling point. Heavy fuel oil, with large molecules of 30 to 60 carbon atoms, is **viscous**.

Going up the fractionating column, as the molecules get smaller the **viscosity** gets less. This means the liquids are much more runny. Petrol has a much lower viscosity than fuel oil.

Compounds with small molecules also **ignite** (catch fire) much more easily. On bonfire nights you sometimes hear of people who have put petrol on their bonfires and been badly burnt. This is because petrol ignites easily. As the size of the molecules gets larger, ignition becomes more difficult. The fuel oil for ships and power stations must be sprayed into fine droplets in a combustion chamber to make it ignite.

Burning fuels sensibly

At first sight this may seem a silly title. We know very well that when we light a Bunsen burner we must be careful. If we allow too much gas out before we ignite it then we can expect it to be very dangerous. But here we are going to concentrate on two issues where an understanding of the science of combustion is very important to our health.

Incomplete combustion

On page 128 we looked at what happens when natural gas (methane) completely combusts. The reaction is:

methane + oxygen → carbon dioxide + water

$$CH_4 + 2O_2 \rightarrow CO_2 + 2H_2O$$

Complete combustion means that there is enough oxygen present to react with all the atoms of carbon in the fuel and produce carbon dioxide. However, if there is not enough oxygen to do this you get incomplete combustion. This is very dangerous because **incomplete combustion** produces **carbon monoxide**, a highly toxic gas:

Toxic substances are poisonous and can cause death.

carbon + oxygen → carbon monoxide

$$2C + O_2 \rightarrow 2CO$$

The term **sootiness** is sometimes used to describe a flame when it is burning – incomplete combustion leads to a high sootiness.

Carbon can also be produced. This is the black soot that you find on the bottom of the gauze if you have been heating with a yellow Bunsen flame.

Carbon monoxide combines with haemoglobin in the blood and lowers the ability of the blood to carry oxygen round the body. It does not take much carbon monoxide in the air you breathe to have very serious effects. Unfortunately, it has no smell. In the UK, about 30 people a year die of carbon monoxide poisoning because their bodies are starved of oxygen.

Figure 7.19 A yellow flame is evidence of incomplete combustion and there is likely to be carbon monoxide present. A blue flame shows complete combustion and it is unlikely that carbon monoxide is being produced

In the home it is gas appliances, such as cookers and fires, that are frequently the cause of carbon monoxide poisoning. When these appliances are well serviced and able to take in enough air to completely combust the gas then there is no problem. It is when they are old, or faulty, that problems can occur. A tell-tale sign is that the gas burns with a yellow, sooty flame, rather like a Bunsen flame with the air hole closed.

Any fuel you burn in the home can produce carbon monoxide if there is not enough oxygen present, so oil- and coal-fired boilers also need regular servicing.

Another major source of carbon monoxide is the car engine. Only a small amount of air is drawn into the engine and this is not enough to burn all the carbon in the hydrocarbon fuel to carbon dioxide. Catalytic converters are an important way of turning this carbon monoxide into carbon dioxide before it reaches the atmosphere. However, catalytic converters do not always work efficiently. One thing you should never do is close a garage door with the car engine running.

Figure 7.20 This child is very close to car exhausts. Burning petrol in car engines produces highly toxic carbon monoxide

Asthma and atmospheric pollution

Figure 7.21 One in five children in the UK now suffers from asthma

Asthma affects about 5 million people in the UK. An asthma attack is caused when the airways in the lungs become inflamed and narrow, making it difficult to breathe. The inflammation is caused by the body's own immune system and is an allergic reaction to something that is breathed in. Asthma sufferers are usually sensitive to more than one substance in the atmosphere that can trigger an attack.

But how do people become asthmatic in the first place? The answer is that scientists are not sure. There seem to be many causes. A person may have over-sensitive airways. There may be a family history of asthma, suggesting a genetic link, or it may be an allergy to a particular compound that triggers asthma. One theory suggests that a viral infection when someone is very young can make the airways very sensitive. Another theory links asthma to centrally heated homes and dust mites. What we do know is that asthma is far more common in the developed world, and some people believe that one major cause is atmospheric pollution by vehicles.

We know that sulphur dioxide, an atmospheric pollutant that causes acid rain, can trigger an attack in someone who is already an asthma sufferer. We also know that people suffering from lung diseases have real problems when the air pollution levels are high. What scientists have not been able to prove is that atmospheric pollution *causes* asthma. There are scientific studies that suggest a link but there is still not enough evidence for scientists to agree. One problem is deciding how much exposure someone has to atmospheric pollution. A person who stays indoors all day, even if they live in the middle of a city, is exposed to a very different level of atmospheric pollution than someone who cycles to work.

Sulphur is found in fossil fuels. When a fossil fuel is burnt, sulphur dioxide forms. It is a major cause of **acid rain**. Acid rain is responsible for killing vegetation and damaging stonework.

As part of this course you will be asked to interpret and evaluate data about respiratory disease and air pollution. It may be that you will consider a graph similar to the one below to identify a trend, such as the increase in asthma due to a particular atmospheric pollutant. Always remember that just because a particular piece of evidence *suggests* a link, there may be other factors that are responsible.

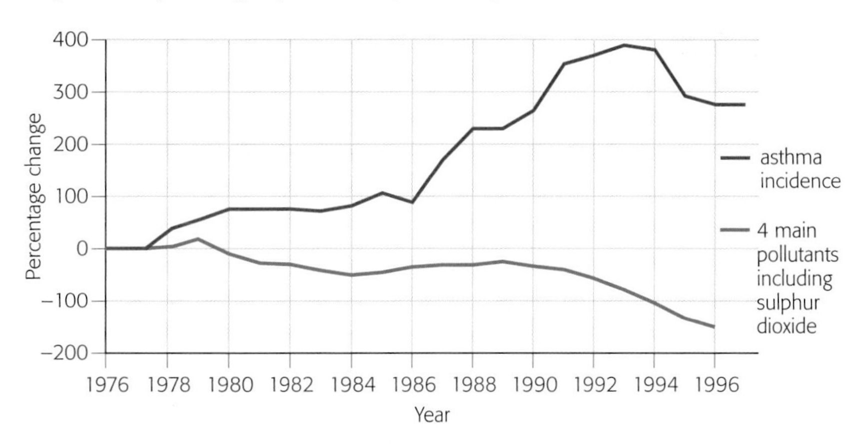

Figure 7.22 From this graph of the percentage change in incidence of asthma and the concentration of atmospheric pollutants, do you think there is evidence for a link?

Summary

- Crude oil is a complex mixture of compounds, nearly all of them hydrocarbons.
- Crude oil is separated at the refinery into fractions in a fractionating column.
- A fraction is a mixture of compounds with boiling points that lie in a particular range of temperatures.
- The different fractions have different names and different uses:
  - gases – bottled gas
  - petrol – car fuel
  - naphtha – making chemicals
  - kerosene – aeroplane fuel
  - diesel oil – lorry and bus fuel
  - fuel oil – fuel for power stations and ships
  - bitumen – road surfaces.
- H The higher up the fractionating column a fraction condenses off, the lower its boiling point range, the smaller its molecules, the runnier the liquid and the easier it is to ignite.
- Incomplete combustion occurs when there is not enough oxygen present to burn all the fuel to carbon dioxide. When carbon-containing fuels combust incompletely they produce carbon monoxide and carbon.
- Carbon monoxide is highly toxic and lowers the ability of the blood to carry oxygen round the body. This can be fatal.
- H Asthma is a respiratory disease with many causes. There is not enough evidence at the moment to link the start of the disease to atmospheric pollution.

Questions

7.15 Crude oil is a mixture of compounds. Which two elements do almost all of its molecules contain?

7.16 Before crude oil enters the fractionating column, what must happen to it?

7.17 In a household the first to show the symptoms of carbon monoxide poisoning are very small children. Why do you think this is?

H

7.18 Look at the graph in Figure 7.22.

a What is the trend in the incidence of asthma from 1976 to 1994?

b The graph of percentage change in the four atmospheric pollutants suggests there is no link between them and the incidence of asthma. Explain why.

c Imagine you are a health campaigner who believes there is a definite link between sulphur dioxide in the atmosphere and asthma. What argument could you put forward to question the interpretation given in part b?

E Obtaining useful substances from air and salt

Liquid nitrogen is wonderful stuff. Nitrogen has a boiling point of −196 °C, which means the liquid is very cold. It is also very unreactive and so it can be put to an amazing number of uses. The food industry uses it to freeze all manner of foods, from pies and dairy products to meat and strawberries. As it is so cold it freezes food instantly. It also has the added benefit that it removes oxygen from around the food, meaning bacteria can't grow.

Hospitals use liquid nitrogen to transport organs for transplants around the country, and to preserve human eggs and sperm. The oil industry uses it to build up pressure in oil wells and so pump up more crude oil. Some farmers even pour liquid nitrogen onto very muddy patches of land to freeze the mud, so that tractors can go over it without sinking.

Figure 7.23 Strawberries frozen using liquid nitrogen

● Fractional distillation of liquid air

We have already looked at the fractional distillation of crude oil. It is possible to fractionally distil air in much the same way. Unlike crude oil, which was heated to form a gas before it went into the fractionating column, air must first be cooled down until it is a liquid before it can be sent into the column. Before the air gets really cold, water and carbon dioxide are removed as solids. The air is then cooled down to –200 °C to liquefy it.

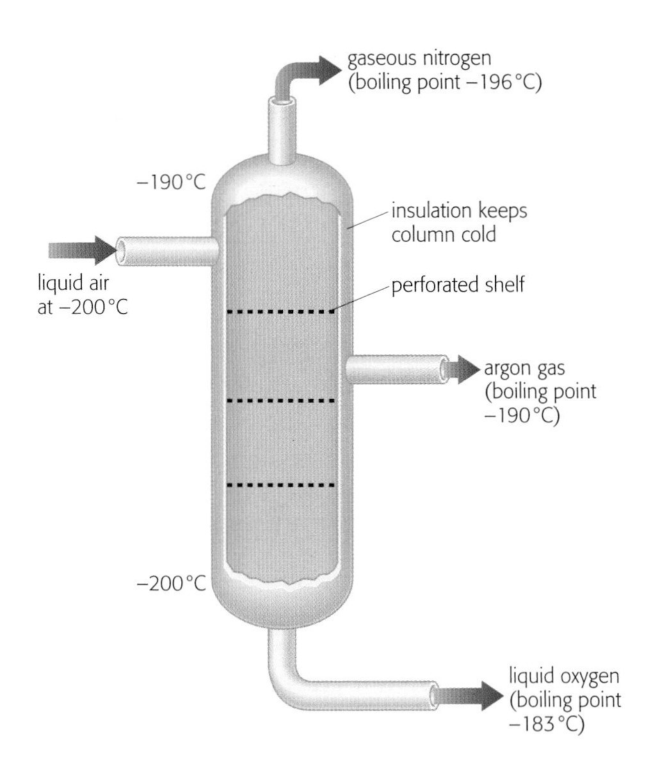

The liquified air enters near the top of the fractionating column, which is at –200 °C at the bottom and –190 °C at the top. Oxygen stays as a liquid, because its boiling point is –183 °C. It falls to the bottom of the column and is collected as liquid oxygen. Nitrogen travels up to the top of the tower and comes off as a gas. The other major fraction of liquid air is argon, which is tapped off in the middle of the column, because its boiling point is in between that of oxygen and nitrogen.

Figure 7.24 The fractional distillation of liquid air

Extracting useful chemicals from salt

Salt – sodium chloride – is found dissolved in the sea or buried underground as rock salt. In the UK we obtain most of our salt by mining rock salt. There are huge deposits of it in Cheshire. The sodium chloride was deposited millions of years ago when the sea that covered Britain evaporated.

Figures 7.25 and 7.26 show two ways of mining rock salt.

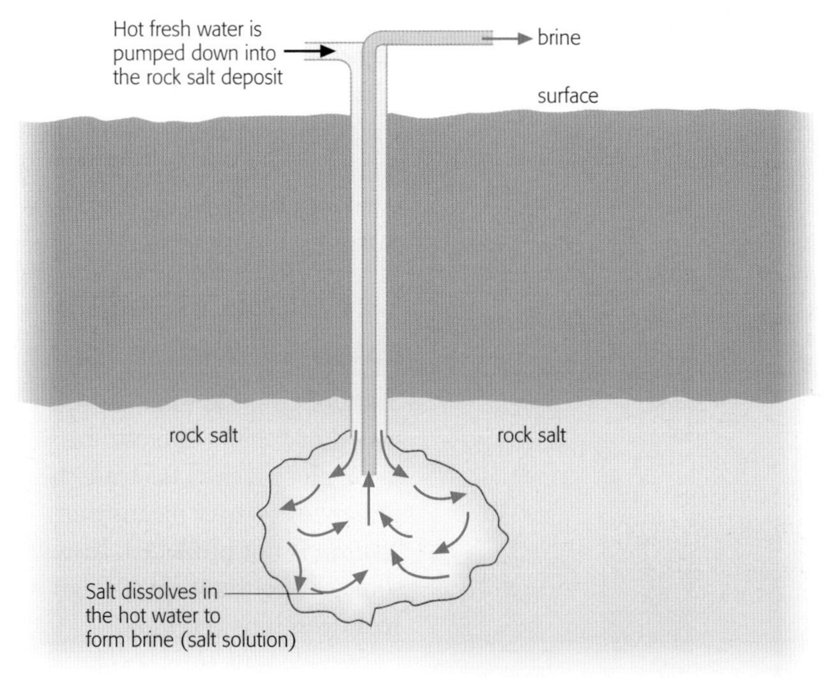

Figure 7.25 Hot water is pumped down into the rock salt. This dissolves the salt to form a solution called brine. The brine is then pumped up to the surface where it is stored in brine lakes until it is needed

Figure 7.26 This mine at Winsford in Cheshire produces 100 000 tonnes of salt each week. The mining equipment needs to be large because the rock salt layer laid down millions of years ago is very thick

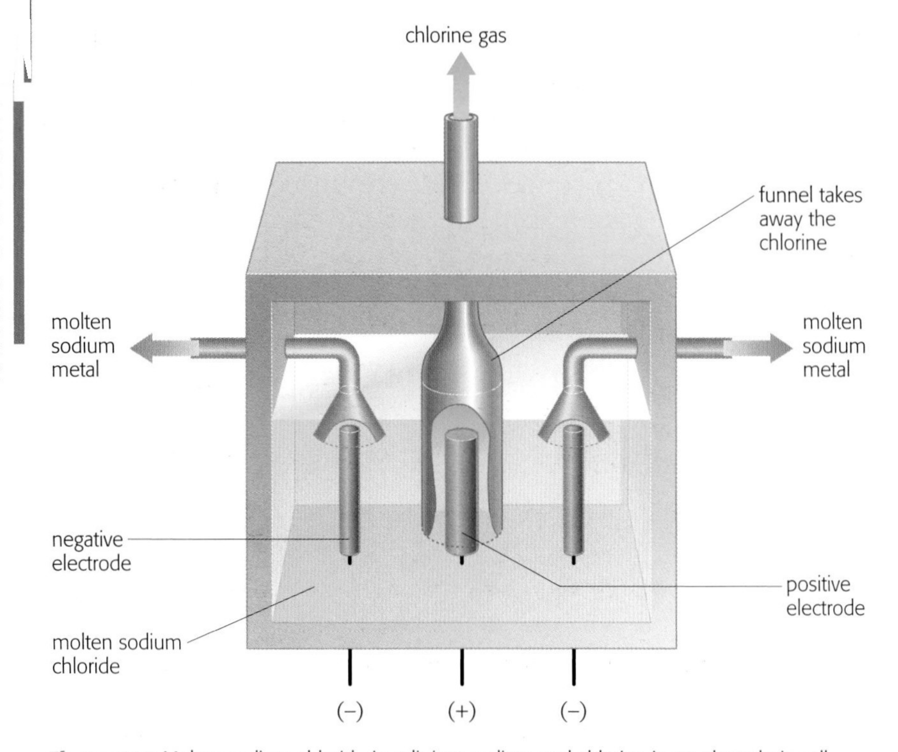

Figure 7.27 Molten sodium chloride is split into sodium and chlorine in an electrolysis cell

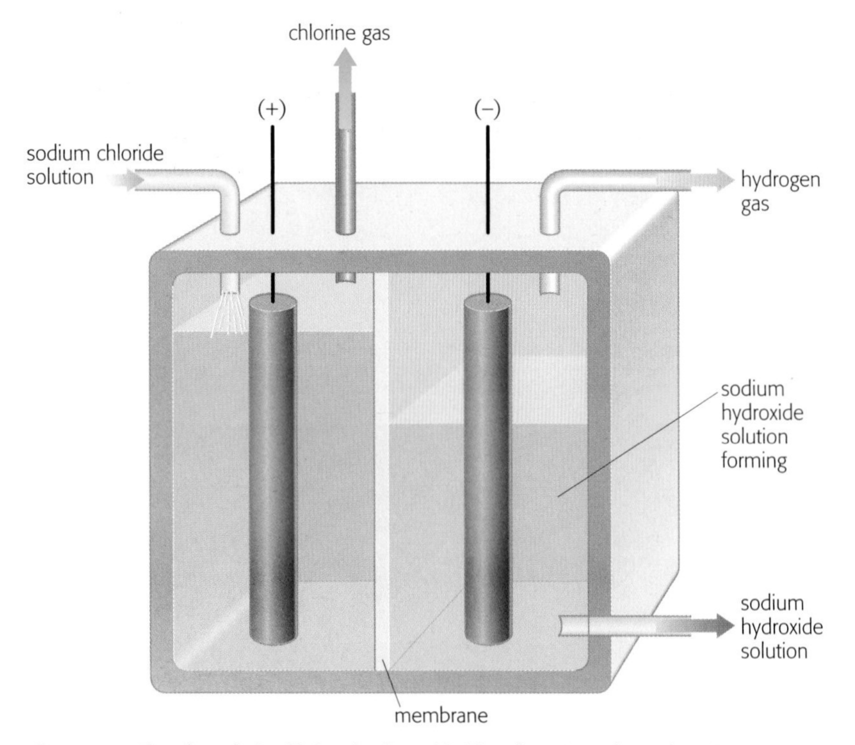

Figure 7.28 The electrolysis of brine (sodium chloride solution) produces three important chemicals – chlorine, hydrogen and sodium hydroxide solution

Electrolysis of molten sodium chloride

In Chapter 6 (page 111) we saw that reactive metals need to be extracted by **electrolysis**. This is the splitting of a substance using electricity.

Solid sodium chloride contains particles of sodium and chlorine called **ions** bonded together. Sodium ions are positive ions with the symbol Na^+, and chloride ions are negative ions with the symbol Cl^-. When the sodium chloride is melted, an electric current can split it up and separate the negative ions and the positive ions. Sodium ions, because they are positive, are attracted to the negative **electrode**. Sodium metal forms here and floats to the surface of the molten sodium chloride. See Figure 7.27. At the positive electrode, chloride ions turn into chlorine gas.

Electrolysis of sodium chloride solution (brine)

Sodium chloride solution – salt water, or brine – can also be split using electricity. This time the presence of water means that sodium is not produced. Instead, some of the water splits to give hydrogen ions (H^+) and hydroxide ions (OH^-). The hydrogen ions are attracted to the negative electrode and become hydrogen gas. The chloride ions (Cl^-) go to the positive electrode and become chlorine gas. See Figure 7.28. This leaves sodium ions (Na^+) and hydroxide ions (OH^-) still in the solution. The sodium hydroxide solution formed is a very important and useful chemical.

How we use the chemicals we obtain from salt

Table 7.2 Useful substances obtained from sodium chloride

Substance	Uses
Sodium	Tiny amounts of sodium vapour are used in yellow street lamps. It is also used to conduct heat away from nuclear reactors
Chlorine	This is added to water supplies and swimming pools to kill bacteria. It is also used as a bleach, for example in paper making
Hydrogen	This is used in the manufacture of margarine, to make vegetable oils solid
Sodium hydroxide	This is used to make soaps and detergents

Summary

- **Nitrogen, oxygen and argon are obtained by the fractional distillation of liquid air.**
- **Sodium chloride is found underground as rock salt and also dissolved in seawater.**
- **Molten sodium chloride can be electrolysed to produce sodium and chlorine.**
- **Sodium chloride solution, also called brine, can be electrolysed to produce hydrogen, chlorine and sodium hydroxide.**

Questions

7.19 Give two reasons why liquid nitrogen is used to freeze foods.

7.20 Why do you think the '*chloralkali* industry' is the name given to the industrial electrolysis of brine?

7.21 Sodium hydroxide is used to make some oven cleaners. Find out why people are advised to wear rubber gloves when using oven cleaners made from sodium hydroxide. [Hint: You can find information about sodium hydroxide in Chapter 6, page 123.]

8 Designer products

A Materials fit for purpose

Nearly all of us either participate in a sport or enjoy watching others do it. Your interest may be in traditional sports such as football, rugby, tennis or badminton, or it may be in extreme sports like skateboarding, rock climbing or mountain biking. Whatever your sport, the materials used in the clothing and equipment are extremely important – at the highest level they may make the difference between winning and losing.

Take Wimbledon, for example. Until the 1960s all tennis racquets were made out of wood, then some players began to experiment with steel and aluminium. Although these were lighter and moved through the air more quickly, they lacked the flexibility of wood and so were less accurate. The advent of racquets made from carbon fibre finally meant the end for wooden racquets and these were last seen at Wimbledon in 1987. Carbon fibre now crops up in many items of sports equipment, from Formula 1 racing cars to skateboards and ice hockey sticks, because it is light and very strong.

Figure 8.1 This skateboard is made from carbon fibre

Sports clothing has not escaped the advance of new materials. One of the first such advances was the invention of something that you probably take for granted – elastic. Before this, shorts were held up by cord tied around the waist and this did not stretch with the body. Modern sportswear may now be entirely made with a fabric that includes elasticated material such as Lycra. This makes the clothes cling to the body and so improves aerodynamics.

Figure 8.2 Wayne Rooney's shirt is made from 'sport wool'

Professional footballers' shirts are made from 'sport wool', which is a mixture of natural wool and polyester. The wool wicks away (soaks up and removes) sweat from the skin, while the polyester is an ideal material on which to print colourful designs and logos. Even the humble sock does not escape improvements in fabric material and design. Sports socks are now 'high wicking', which means they remove sweat from the skin quickly. They may also be 'anti-friction' – the socks are made from two layers, one of a material that grips the shoe and the other a material that grips the skin. Friction (rubbing) occurs between these two layers and not at the skin's surface, making the socks much more comfortable for runners.

Lycra®

Figure 8.3 These are some of the first bathing costumes, made of wool

In the box on page 152 we saw that the properties of a material are really important. Makers of clothing choose materials with properties that are going to support the way the clothes are to be used. Consider the history of swimwear. One of the most efficient ways to move through the water is simply in your skin, naked, because most swimwear slows you down slightly. This slowing down is called drag and it occurs because there is friction between you and the water.

The first bathing costumes, in the mid-19th century, were made of wool. Wool gets heavy when its wet, so it wasn't very comfortable and it didn't look good when you came out of the water! It also created a lot of drag in the water, so you couldn't swim very fast.

Natural silk is made by silk worms and it takes a lot of silk worms to make enough silk for a bathing costume, which is why they were so expensive.

In the Olympics of 1924, the elite swimmers wore silk swimwear, which was lighter, created less drag and also felt comfortable to wear. The problem was that swimwear made from silk was incredibly expensive.

Today's swimwear technology has moved a long way from wool and silk. The material of choice since the 1980s has been a mix of nylon and the elasticated material **Lycra**, which keeps the material close to the body, like a second skin.

On labels you will see the word Lycra printed with a little ® after it, as in the heading of this section. The ® in Lycra® tells you that Lycra is a legally registered trademark (TM) or brand name. Only DuPont manufacture Lycra. Its more general name is elastane, or spandex in the USA.

Lycra is an artificial fibre, invented by Joseph Shivers while working for the company DuPont in the 1950s. It's an amazing material and can be stretched to 600 times its length and return to its original shape.

Before Lycra was invented, thin threads of rubber were used for elasticated waistbands and also for stretchy undergarments like girdles that keep a body's shape. However, rubber reacts with sweat, body oils and detergents, and it slowly loses its elasticity. Lycra does not react with these chemicals and can be washed without losing its elastic properties. This makes it ideal for clothing. It is also much lighter than rubber. Lycra is never used on its own but always mixed in with other fibres. In swimsuits it may be up to 20% of the fibre used, the rest being nylon. The figure-hugging qualities of this fabric make it also ideal for cycling shorts.

Figure 8.4 Olympic swimmers may shave 100th of a second off their time because of the properties of the materials in their swimsuits. This can make the difference between a gold or silver medal

Apart from sports, Lycra has revolutionised the entire clothing industry by helping all sorts of garments retain their shape. Not much of it is needed. If 2% to 5% Lycra is blended with any other textile material, from polyester to wool, it will improve clothing fit, make the garment last longer and lessen unwanted creases. It can even be inserted into leather to make shoes more comfortable.

TM means the name is a trademark – a trade name that has been established by use.

Thinsulate™

By blowing air into melted plastic, fibres can be made to come out of tiny holes in nozzles. The fibres are made even thinner by blasting more air at them. These fibres, called microfibres, are about 1000 times thinner than your hair. They are then splattered onto a screen in a tangle, and this is the basis of **Thinsulate**, one of the most effective thermal insulating materials known. This process was invented in the 1960s by the 3M company.

You will find Thinsulate inside many garments where its thermal insulation property is a must. These include ski jackets, gloves and even shoes. Deep-sea divers use it in their dry suits (Figure 8.6).

What makes it such a superb insulator is the fineness of its fibres. Air is trapped in between them and it is the air that provides the insulation because it is a very poor conductor of heat. What makes it so desirable in clothing is its lightness and, as the name suggests, it is so efficient as an insulator that you only need a thin layer. This means that ski jackets can be less bulky than those with natural padding, allowing the skier to move more freely and more comfortably. Its remarkable properties do not stop there because it does not absorb water well, so it works even when damp, and it keeps its properties even after being washed several times.

Figure 8.5 A close-up view of Thinsulate

Figure 8.6 A dry suit is exactly what it says. It is a totally waterproof garment. Thinsulate inside the dry suit traps a layer of air next to the body, keeping the diver warm in cold seas

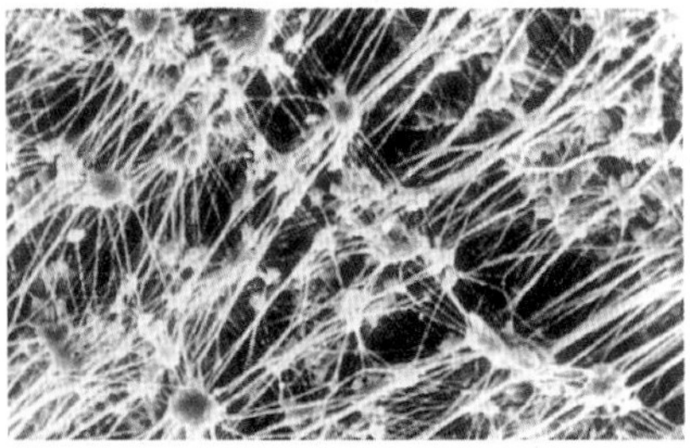

Figure 8.7 The microscopic structure of the Gore-tex membrane. One pore is about 0.0002 mm in diameter

Gore-tex®

You can now see that the properties of Lycra and Thinsulate determine how they are used. **Gore-tex** is another example of a material that has properties that make it fit for a purpose. Gore-tex is a membrane that is waterproof but still allows water molecules from sweat to escape. So how does it do this?

The answer lies in the structure of the membrane. To make the fabric, a piece of Teflon is stretched at high temperature into a very thin sheet called the Gore-tex membrane. (The Gore family invented

You can read about the discovery of **Teflon** on page 159.

the process.) When it is pulled so thinly, the membrane has many billions of tiny holes in each square centimetre (Figure 8.7). These holes, or pores, are 20 000 times smaller than a water droplet, so liquid water just cannot get through. However, water vapour can escape because the water molecules are widely separated in the vapour, and the Gore-tex pores are 700 times bigger than a water molecule.

Now think of how this can be used by rock climbers or walkers. A jacket made of nylon, or other material, can be lined with Gore-tex and no rainwater can get in through the membrane. However, even on a cold, windy day we perspire (sweat) when we are doing physical activity. The pores in the membrane allow water molecules to the outside and so the skin stays comfortably dry. We say that the Gore-tex is a **breathable** fabric because it will allow water vapour to pass out through the membrane. Fabrics that are not breathable leave the skin feeling cold, clammy and uncomfortable.

Figure 8.8 The Gore-tex lining of this cagoule allows perspiration to pass out of the garment but no rainwater can get in

Smart materials

Smart materials are materials that can change their properties in response to an external stimulus such as a change in temperature or the flow of an electric current. Smart materials are now big business because their properties can improve the functioning of clothing or equipment. They are particularly useful in extreme sports, such as skiing and bob sleighing. (Extreme sports are those that give the sports person an exhilarating feeling of excitement, often due to the danger involved and the skill required to stay safe.)

Stimulus means a change that makes something (or someone) behave differently.

Phase change materials

Phase change materials (PCMs) were originally designed to protect astronauts from sudden changes of temperature. They are smart materials because they change from a solid to a liquid and back again according to the outside temperature and the temperature of the body. The same properties are now being used in specialist sports clothing.

Let's imagine that you are a skier. When you are skiing you will get hot in your ski jacket. But the fibres of your ski jacket are coated with encapsulated microgranules of phase change materials. When you get hot these absorb some of your heat energy and melt, and this cools you down. At the bottom of the ski slope you decide to wait and take the ski lift back up again. You start cooling down. The microgranules also start to cool down and change back to a solid. As they change their phase from liquid to solid they release heat, and this helps you to stay warm.

At the moment very few garments are coated with phase change materials, but it is likely that these materials will play an increasingly important part in the sports clothing of the future.

Remember that a **smart material** can change its properties in response to an external stimulus – in this case a change of temperature.

Thermochromic and photochromic materials

If you think about the *thermo* in thermochromic, it tells you that this smart material will react to the external stimulus of heat (thermal energy). *Chromic* comes from the Greek word *chroma*, which means colour. So clothing such as T-shirts made from **thermochromic** material will change colour with changes in temperature.

Now if you look at the word **photochromic** you can probably work out that photochromic dyes will change colour in response to sunlight. These dyes can be again be used for fun T-shirts, but also for military clothing – the camouflage patterns change when sunlight strikes.

Carbon fibre sports equipment

Nowadays all top tennis players use racquets made of carbon fibre, because it combines lightness, strength and flexibility, making the racquets perform better than those made from other materials such as wood, plastic or aluminium. So what is carbon fibre?

The components of a composite material are chosen so that their properties enhance the use of the material.

Carbon fibre is the name given to a composite material of plastic reinforced by very thin fibres of carbon. **Composite materials** are materials made up of more than one component. Concrete is a composite because it is made up of the components cement, sand and gravel. Reinforced concrete is another composite that has steel rods in the middle of it, giving added strength. Carbon fibre composites are made in a similar way, but instead of concrete and steel rods, it is plastic reinforced by carbon fibres. Carbon fibres on their own are brittle but when they placed inside plastics they form a composite material that is flexible and very strong.

Wool, cotton and artificial plastics like polyester, Lycra and Teflon all contain thousands of carbon atoms arranged in long chains in their molecules. If tiny threads of these molecular chains are heated very slowly in ovens, what is left is nearly all carbon, in tiny filaments or threads. In the 1960s, scientists in Farnborough perfected this way of making carbon fibres and discovered the best material to heat in the first place. The fibres they made were five times stronger than the same mass of steel, and when they incorporated them inside plastic the material they produced is what we now know as **carbon fibre**.

Figure 8.9 The properties of carbon fibre – lightness, strength and flexibility – make it ideal for a tennis racquet

Carbon fibre is expensive to produce because it is difficult to ensure that the liquid plastic flows evenly around the carbon fibres. It is a process that is often done by hand, and that makes it expensive.

In the Barcelona Olympics in 1992 a British cyclist, Chris Boardman, won the gold medal and set a new world record for the 4 km Pursuit race. He owed his success partly to his carbon fibre bike. The carbon fibre frame was very light, strong, and had been moulded to be as streamlined as possible, so that the air would not create drag as the bike was being ridden at speed. In the Atlanta Olympics of 1996, every cyclist competing had one of these so-called 'super bikes'. Even the wheel rims were made of carbon fibre instead of steel. However, such was the expense of producing the aerodynamic frame using carbon fibre that these super bikes were thought to give wealthier countries an unfair advantage. Consequently they have been banned from future Olympics – at least for the time being.

Figure 8.10 Chris Boardman on his carbon fibre bike

Carbon fibre technology is still advancing. Carbon fibres have recently been developed that are hollow inside. This makes them lighter and stronger when in carbon fibre composite material – see page 161.

As we saw on page 152, another piece of equipment that has benefited from carbon fibre is the skateboard. To perform an 'Ollie', a skateboarder jumps and the skateboard stays in contact with the feet. At the moment the world record for the highest Ollie is 112 cm, set in 2004. Carbon fibre has transformed the wooden skateboard of the 1950s to a very high-tech product. What makes the new boards so superior is the properties of the carbon fibre.

Summary

- Materials for sports equipment and clothing are chosen because of their properties.
- Lycra is a material that is very stretchy. It is used in combination with other materials to improve the properties of textiles. It does not react with water, body chemicals or detergents.
- Thinsulate contains tiny fibres called microfibres. These trap air to make a thin but very effective insulating layer.
- Gore-tex is a thin layer of Teflon that has billions of tiny pores in it. It is waterproof but allows water vapour to escape. This property is called breathability.
- Smart materials can change their properties in response to an external stimulus.
- Phase change materials are smart materials that maintain the temperature around the body in sudden changes of external temperature.
- Thermochromic materials change colour with changes in heat; photochromic materials change colour in sunlight.
- Composite materials are materials made up from more than one substance.
- Carbon fibre is the name given to composite material that contains fibres of carbon inside plastic. It is very strong and very light.

Questions

8.1 **a** What properties of Thinsulate make it a useful material in ski jackets?
b Explain how Thinsulate gets its insulating properties.

8.2 What advantages does Lycra have over elastic for the waistbands of shorts?

8.3 Gore-tex is a waterproof, breathable fabric.
a What is meant by *breathable*?
b Explain how the structure of Gore-tex allows it to have the properties of breathability and waterproofness.

8.4 What is meant by the term *smart material*?

8.5 Give three uses of carbon fibre in sports equipment. For each use explain how the properties of carbon fibre make it a good choice of material.

B Discovering new materials

Post-it notes have come to be regarded as essential office stationery and they are often just as important in the home and at school. They can mark the pages of books or they can be used to leave messages. What is so good about them is that they stick to a surface but can be easily removed without leaving a mark or damaging the surface. This important property is due to the layer of adhesive at the top of one side of the paper.

It might surprise you to know that *Post-it* notes haven't been around very long, although their story begins in 1968. A senior research chemist, Spencer Silver, working for the company 3M, invented an adhesive (glue) that was like no other adhesive that had been invented before. When looked at under a microscope it didn't form a smooth film but looked like a cobbled street of shiny pebbles. The pebbles were 'microspheres' that sparkled in the light. These microspheres formed a single layer when coated onto surfaces and the layer was bumpy, so the adhesion (stickiness) wasn't great. For 5 years people at 3M tried to think about how this adhesive could be used. At one time they thought about coating noticeboards with it so that notices and reminders could be stuck to them. There was a major problem with this, as all the dust and dirt around in the air stuck to the board, making it look dirty and spoiling the adhesion.

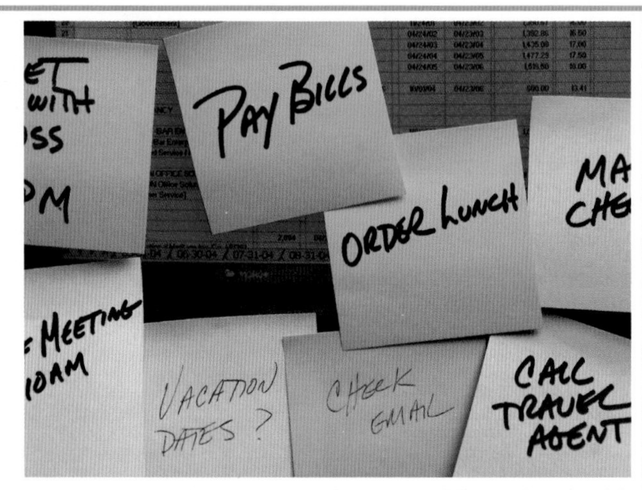

Figure 8.11 *Post-it* notes seem to be an essential part of modern life

In 1974 Art Fry, another chemist at 3M, was singing in his church choir. He would mark the pages of his music books with pieces of paper but these would often fall out. It was then he realised how Silver's adhesive could be used and the idea of the *Post-it* note was born. It took another 5 years for the concept to become a product ready for mass production. In 1980 the *Post-it* note went on sale and the rest, as they say, is history.

Post-it is another registered trademark, so you will see it as *Post-it*® on packaging.

Finding uses for new materials

The story of the adhesive used on *Post-it* notes shows you that sometimes scientists create new materials but the applications (uses) of the material don't become obvious till later. The same can be said of many of the materials we have already discussed. When the Gore family stretched Teflon into a thin sheet, nobody realised that there would be billions of tiny pores that would give it the property of breathability, and make it so useful as a waterproof lining. Indeed, Teflon itself was once a novel material with no obvious use.

Teflon®

The discovery of Teflon is an example of serendipity in science. Serendipity means *lucky accident*, but a famous scientist called Louis Pasteur once said that 'Chance favours the prepared mind'. This was certainly true in 1938 when Roy Plunkett discovered Teflon. He was working with a gas called tetrafluoroethene. One day he could not get any of this gas from one of the steel gas cylinders that was supposed to be full of it. Instead of jumping to the conclusion that the gas must have escaped, he weighed the cylinder and found that it had the same weight as when it was full. He then decided to saw the cylinder in half and examine its contents. What he found was a white powder, now called **p**oly(**t**etra**f**luoro**e**thene) or PTFE for short. It is better known by the brand name **Teflon**.

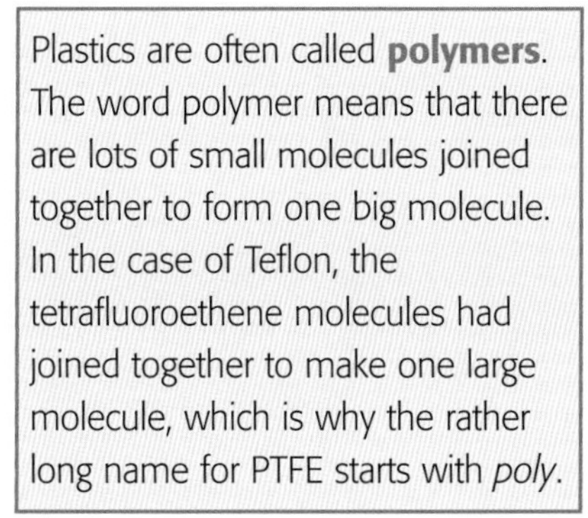

Plastics are often called **polymers**. The word polymer means that there are lots of small molecules joined together to form one big molecule. In the case of Teflon, the tetrafluoroethene molecules had joined together to make one large molecule, which is why the rather long name for PTFE starts with *poly*.

The properties of Teflon are quite amazing. It has a virtually friction-free surface (in fact it is considered to be the most slippery substance ever invented), it is resistant to heat and chemicals, it is water repellant, and it does not conduct electricity at all – it is a perfect electrical insulator. Its first use was for machine parts, where a low-friction surface was needed. In the 1960s a new application was discovered, as a coating for 'non-stick' pans. It is also used in plumbers' tape (PTFE tape), which makes screw fittings joining water pipes watertight, and electrical insulation tape. Its slipperiness even makes it useful for coating human joint replacements.

Figure 8.12 A familiar use of Teflon is as a coating on non-stick pans

Kevlar®

Stephanie Kwolek, a chemist working for DuPont, invented this plastic. It is in the same family of plastics as nylon but it has one major structural advantage. Its very long molecules are almost straight so the molecules can line up very close beside each other. This makes Kevlar very strong. A 7 cm diameter steel cable has a breaking strain of about 40 tonnes. An identical Kevlar cable has the same breaking strain but is five times lighter.

So you can see that one of the properties of Kevlar is its strength and another is its lightness. Also, it does not melt and it's not flammable. It can be made into fibres and this is key to an important use of Kevlar – as bulletproof vests. A web of Kevlar fibres is strong enough to stop a bullet. The Kevlar fibres absorb the energy of the bullet and spread this to other fibres in the web.

Figure 8.13 This police officer is wearing a bulletproof vest made from Kevlar

Nanoparticles

Nano comes from the Greek word for dwarf. A nanometre is a unit with the symbol nm. A nanometre is 0.000 000 001 metre or 10^{-9} m.

The 13 nm gold nanoparticle contains about 10 000 gold atoms.

A new scientific revolution in materials technology is happening at this very moment. It is set to transform our lives in a similar way to the discovery of plastics in the last century. It therefore seems quite surprising that these new materials are based on very tiny particles called **nanoparticles**. A nanometre (nm) is a millionth of a millimetre. This is about 80 000 times smaller than the thickness of one hair on your head. Nanoparticles are particles that range in size from 100 nm down to about 1 nm. Particles of material this size can show completely different properties than normal-sized lumps. Take gold, for example. Everyone knows that gold is a yellow metal, but a nanoparticle of gold actually looks red. We can't see one nanoparticle of gold because it is only 13 nm wide, but Figure 8.14 shows you what a solution of gold nanoparticles looks like.

Figure 8.14 A solution of gold nanoparticles in water appears red

You can't make a gold nanoparticle by grinding down gold – it's just too small to produce this way. Chemists 'grow' nanoparticles of gold using a chemical reaction, stopping the reaction when the particles reach the required size.

Gold can have quite different properties when it is in the form of artificially made nanoparticles. The ability to manipulate atoms and molecules is producing some astounding possibilities. Look back at Figure 5.24, page 99. It shows an atomic artwork using 27 carbon monoxide molecules. This ability is also helping to produce completely new materials.

Nanotechnology now

Nanotechnology is the application of the properties of nanoparticles to making new materials and finding new uses for nanoparticles. It may involve creating structures by actually building them up molecule by molecule.

You may think you've had no contact with nanoparticles yet but it's likely that you are wrong. If you have used a sunscreen lotion recently, you have probably put nanoparticles on your skin. These nanoparticles are of either zinc oxide or titanium oxide and they are able to absorb and reflect the Sun's harmful ultraviolet rays. They are transparent to visible light. Sunscreens used to contain much larger particles of these compounds that were not transparent, and this meant they left your skin feeling sticky and looking white. The advent of nanoparticles has made sunscreen lotion invisible on the skin, and it doesn't have that sticky feel any more.

Figure 8.15 Invisible sun protection, thanks to nanoparticles

Pilkington, a UK manufacturing company, now make window glass that cleans itself. The technology involves the use of a thin film of titanium dioxide, which is only a few nanometres thick. The film of titanium dioxide on the surface of the glass reacts with organic dirt particles when exposed to sunlight. The dirt is broken down and when it rains, the surface, which is hydrophilic (water loving), washes away the dirt.

Figure 8.16 Self-cleaning windows use nanotechnology

The clothing industry has also been quick to exploit some of the remarkable properties of nanoparticles. Stain-resistant fabrics have a coating of nanoparticles on their surface. Any liquid spilt on them tends to form beads that do not soak into the fabric, so they can be simply wiped away. These fabrics are not only useful for ties, but are used for upholstery, for example aircraft seats.

Silver nanoparticles have even found their way into our socks. As nanoparticles, silver does not allow microbes like bacteria and fungus to grow, so this stops socks becoming smelly and the skin feeling itchy. A more important use for this anti-microbial property is in bandages that prevent microbes from infecting the wound.

Nanotubes and nanocomposites

Figure 8.17 Nanotubes of carbon

Nanotubes were first discovered in 1991 in Japan. They are ultra-thin sheets of carbon rolled up to form hollow cylinders that are only a few nanometres in diameter. These nanotubes have some fascinating properties. For a start they are very strong and can be used to make **nanocomposites**. We have already looked at composites when we considered carbon fibre (pages 156–157).

Carbon nanotubes have a breaking strain 100 times greater than steel of the same mass. When encased in a plastic we have a new carbon fibre composite that is even stronger, and is already finding a use in making much stronger car bumpers.

Silicon dioxide nanoparticles can form another nanocomposite that strengthens conventional carbon fibre. They fill in the tiny holes in the material, which weaken the structure, making a nanocomposite that is stronger than the original carbon fibre. The carbon fibre tennis racquet is already benefiting from this nanotechnology.

More uses of nanoparticles

Some other potential uses of nanoparticles in the near future are:

- drug delivery via nanotubes
- production of tiny wires and conducting nanocomposites
- hydrogen fuel storage in nanotubes
- more efficient fuel cells
- better adhesives that require less volatile solvents, which makes them better for the environment
- smaller computers with much bigger memories
- paints that weigh less and use less solvent
- stronger, lighter composite materials
- fabrics that can monitor your heart and blood pressure.

In the title of this section we use the term 'nanotechnologies' since, as we have seen already, there is more than one type of nanotechnology – ranging from the invention of new nanocomposite materials to the use of nanotubes for delivering drugs.

The uncertainties of nanotechnologies

Up to now we have considered the wonderful benefits that are already being exploited as a result of nanotechnologies. However, we have seen that the properties of nanoparticles can be very different from the properties of much larger particles. Nanotechnologists are also able to manipulate atoms and molecules to build materials with properties that we are only just beginning to understand. Some scientists and members of the public point out that there could be potential harm from nanotechnologies, and that all production of nanoparticles should cease until we know more.

In 2003 Prince Charles expressed the fears of some scientists and environmentalists, saying that we need to properly investigate the risks associated with nanotechnologies and the nanoparticles they produce. Figure 8.18 shows some of the media headlines following the Prince's intervention.

Figure 8.18 Media coverage of worries about nanotechnologies

Serious Health Risks from Nanoscience

Life On Earth To Be Destroyed By Nanobots

GREY GOO TO TAKE OVER THE WORLD

Nano Nightmare Predicted

There are two issues to consider here:

1 What *are* the risks of nanotechnologies, now and in the future?
2 How does the media present these risks?

We'll examine these two issues in turn.

Risks of nanotechnologies

The Royal Society, the UK's most prestigious scientific body, was asked by the UK Government to investigate nanotechnologies and report on the risks. In 2004 a report was produced. Here are some of the findings:

- Nanoparticles are so small that they have different properties than larger particles of the same material. This means that nanoparticles should be treated as new chemicals and identified as such on product labels. It also means that proper safety testing should be carried out each time a nanoparticle is used in a product.
- There is no evidence that nanoparticles used in sunscreens and cosmetics pass beyond the dead skin cells that form the first layer of human skin. This means they do not invade the body. However, if the skin is damaged or broken by cuts or eczema they could enter the body and their effects on internal body cells are not known. We do know that nanoparticles can be small enough to pass into cells through cell membranes.
- Some scientists are using nanoparticles, such as nanotubes, to allow the targeted delivery of drugs to various organs in the body. More research is required to check that these nanoparticles are not affecting tissue cells far away from the intended target.

Larger animals in the food chain eat large numbers of smaller animals. This means that if the smaller animals have nanoparticles in their bodies, these could become more concentrated in the larger animals. See Chapter 1 for information about food chains.

- Nothing is known about the effect of nanoparticles on species other than humans, or whether nanoparticles can accumulate in the food chain.
- We do not know how nanoparticles react with air, soil and water. Until more is known, the recommendation is that the release of nanoparticles into the environment is avoided as much as possible.

What the 2004 report tells us is that there are some risks associated with nanotechnologies, but that these risks are difficult to assess as too little is known about this new science. Does this mean that the general public should panic about nanoparticles? Almost certainly not, as we have been living with nanoparticles in the air ever since we discovered how to make fire. Nanoparticles are present in the air in our homes, as cookers often release them.

What do we know about the effect of airborne particles on our health? There is some evidence that some particles released by diesel engines (called particulates) may have a damaging effect on some people's health, but the concentration of diesel particles in the air is much greater than that predicted for nanoparticles.

It is important to remember that many nanoparticles are in the structure of composites, and are not free to get into the environment.

Presentation of risks by the media

As you can see from Figure 8.18, some of the media headlines do not give a balanced picture. The term 'grey goo' actually comes from a science fiction book published more than a decade before nanotechnology became a reality. This book, *Engines of Creation* by K.E. Drexler, imagined nanomachines that were able to reproduce themselves and spread over the Earth, creating a grey sludge. This is not how nanoparticles function, and it doesn't fit at all with our current understanding of nanoscience. But does this mean that this will never happen? One of the problems science is often faced with is, how do you prove a negative? For example, how can anyone say for sure at this point that tiny, self-reproducing machines that have the potential to take over the world could *not* be made? We can only gather evidence and say that the risk is so small as to be virtually non-existent.

However, the media is right to point out that there are risks associated with nanoparticles, and that at the moment we don't know enough to say confidently that there is no potential harm to human health or the environment. The nub of the issue is that members of the general public need to be drawn into a debate about this technology based on fact, rather than by often misleading headlines and scare stories.

Summary

- **Scientists sometimes create new materials with novel properties, and the applications only become obvious later. Good examples of this are the adhesive on *Post-it* notes, and Teflon.**
- **Teflon is a plastic that has a virtually friction-free surface, is heat resistant and unreactive, and an electrical insulator.**

- Kevlar is a plastic that is very strong when made into fibres. It is very light, very heat resistant and does not melt, and is non-flammable. Some of these properties make it an ideal material for bulletproof vests.
- Nanoparticles have a size between about 1 nm and 100 nm (nm stands for nanometre, which is a millionth of a millimetre).
- Nanoparticles often show different properties than the same materials on a larger scale.
- Sunscreen lotions may contain nanoparticles of titanium oxide or zinc oxide. These nanoparticles are transparent; when they are present as larger particles they are opaque and white.
- Self-cleaning windows, anti-microbial wound dressings, stain-resistant fabrics and transport mechanisms for drugs are some of the current uses of nanoparticles.
- Nanoparticles can be combined with other materials to form nanocomposites, such as a new form of carbon fibre that uses carbon nanotubes.
- The use of nanotechnologies in the future has the potential to change our lives in ways that are difficult to imagine at the moment.

H

- There are uncertainties and risks associated with the use of nanotechnologies.
- The portrayal of nanotechnologies in the media can be unbalanced and alarmist.

Questions

8.6 Why is the adhesive on *Post-it* notes fit for the purpose for which it is used?

8.7 Teflon was discovered by accident. Which properties of Teflon make it a good material to use:

- **a** for non-stick pans
- **b** for artificial human joints?

8.8 Kevlar is used in bulletproof vests because of its properties.

- **a** What are the properties of Kevlar that make it a very good material for this use?
- **b** Think about the properties of Kevlar and suggest other uses for this material. Use the internet to find out if you are right.

H

8.9 Design a list of properties for the products below. As you compile your list, consider how the product will be used.

- **a** a ski glove
- **b** a pole vaulter's pole
- **c** nail varnish

8.10 Nanotechnologies are just beginning to take off, with many new uses just around the corner.

- **a** Explain what is meant by *nanoparticle* and *nanocomposite*.
- **b** Why is titanium oxide used in sunscreens? What property of titanium oxide occurs at the nanoparticle size, which is different for larger particles?

H

8.11 Why do you think media coverage sometimes emphasises the negative aspects of a science story?

C Alcoholic drinks

However unlikely this may seem, scientists in Italy are developing a pill that contains the contents of a glass of red wine but without the alcohol, or the water. They are doing this because recent studies have suggested that a glass of red wine a day may prevent some forms of cancer and help protect people from heart disease. Some hospitals in the UK recommend to patients recovering from heart attacks that they drink red wine.

Some of the chemicals in red wine are antioxidants (see page 113) and help to reduce the build-up of cholesterol. Other chemicals are even now being discovered in red wine that appear to prolong the life of some animals, and may be further developed to prolong the life of humans.

The process that the Italian researchers are using to produce their pill is freeze-drying. This is the same method that coffee makers use to produce instant coffee granules. It is a process that does not damage the molecules of the coffee or wine. Under the right conditions, ice can change straight into a gas without going through the liquid water stage. You might see this on a very cold day when frozen puddles can shrink but remain frozen. Wine can be frozen in containers and the pressure lowered by pumping some of the air out. The water vaporises away from the wine and is removed, which is why the process is called freeze-drying. It also removes the alcohol. The remaining ingredients are compacted and made into tablets.

Figure 8.19 The contents of this glass of red wine may soon be available in a pill, but without the alcohol

Some food nutritionists have said that there do appear to be benefits from taking the red wine pill. It could be particularly beneficial for those who do not drink alcohol, for religious or other reasons.

Sugars are carbohydrate molecules. These contain carbon, hydrogen and oxygen atoms, and can usually be written as the general formula $C_x(H_2O)_y$. Notice that the *hydrate* part of the name carbo*hydrate* is due to the ratio of hydrogen and oxygen atoms, which is the same as in water.

Making beer and wine

Beer and wine are called alcoholic drinks because they contain the alcohol **ethanol**. Ethanol is a compound of carbon, hydrogen and oxygen, with the formula C_2H_5OH.

Alcoholic drinks have been made for thousands of years using a microbe called yeast. Recipes for brewing beer written by the Babylonians have been found that date back to 4300BC.

Yeast is a single-celled organism that feeds on **sugars** in the absence of oxygen. As well as producing the energy it needs to survive and grow, it produces two waste products – ethanol and carbon dioxide. The process by which yeast converts sugars to alcohol is known as **fermentation**. Different sugars are present in fruit; the yeast first converts these to **glucose** and then to alcohol:

$$\text{glucose} \xrightarrow{\text{yeast}} \text{ethanol} + \text{carbon dioxide}$$

This fermentation reaction occurs in the absence of oxygen. Winemakers prevent oxygen from entering their brews by using air locks (see Figure 8.20). The air lock also prevents bacteria from entering and oxidising the wine to vinegar. Carbon dioxide can escape as bubbles of gas through the water.

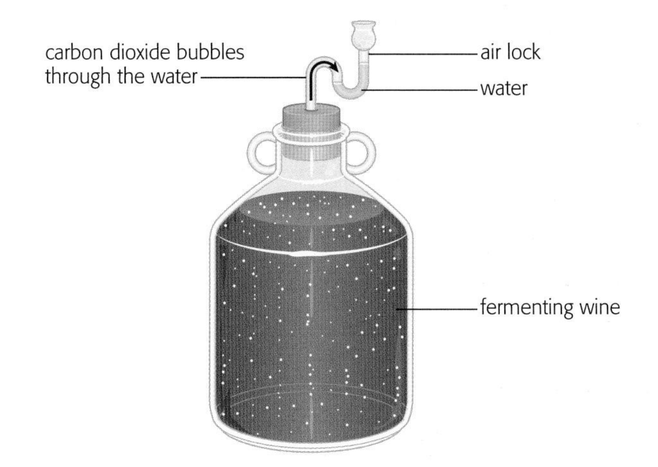

Figure 8.20 The fermenting wine is protected by an air lock. The carbon dioxide produced can escape through the water but oxygen and bacteria cannot enter

Wine is made from grapes. The fruit is crushed and yeast is added to the mixture. The yeast feeds on the sugars in the grape flesh, and ferments them to ethanol and carbon dioxide. When the concentration of alcohol in the fermenting wine reaches 15%, the yeast dies because alcohol at this concentration kills it. It is not possible to produce wine with an alcohol content of more than 15%. To make drinks with greater alcohol content, more ethanol is added to produce 'fortified' wines such as port and sherry, or the wine is distilled to make spirits such as brandy.

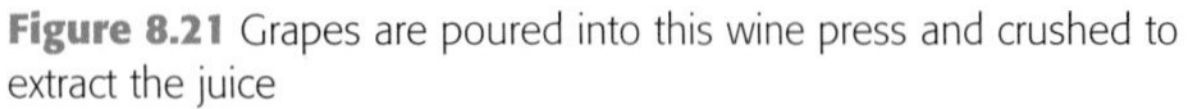

Figure 8.21 Grapes are poured into this wine press and crushed to extract the juice

Figure 8.22 This fermenting brew will produce bitter, a type of beer

Beer is the name given to lager, bitter, stout and ale. It is brewed from mixtures of hops, barley, water and yeast. The barley provides the sugars; hops are often added for flavour and also to prevent the growth of certain bacteria. Different strains of yeast produce different tasting beers. Bitter, for example, is usually made using a yeast that floats on the top of the liquid, whereas lager is produced from a yeast that ferments at the bottom of the brew.

Drinking alcohol

Alcohol is a legal drug that has an immediate effect on someone's mood. It makes people feel relaxed and happy in social situations and provides relief from stress. It is rather odd, then, that it is classed as a depressant drug. It is called this because it depresses, or slows down, the function of the brain and nervous system. The first effect of drinking alcohol is to slow down the part of the brain that causes inhibitions. In greater quantities, alcohol depresses the part of the brain responsible for coordination, and this is when speech becomes slurred and people have difficulty walking in a straight line. If very large quantities of alcohol are drunk then there will be loss of consciousness. The last part of the brain to be depressed is the part that controls breathing and blood pressure. If too much alcohol is consumed then death will ensue.

Benefits and risks

In 1997 the World Health Organisation concluded that one glass of wine every other day reduced the risk of heart disease. Drinking in moderation is also thought to prevent high blood pressure in young women and the onset of Alzheimer's disease in older people. However, there are real risks attached to drinking too much alcohol, either in one session or regularly.

The UK Government has produced guidelines for men and women based on **units** of alcohol. One unit of alcohol is 10 cm^3 or 8 grams of alcohol. The guidance used to relate to the number of units drunk in one week, until doctors realised that people were saving their units until a Friday night and then binge drinking, which is actually quite dangerous. The current guidelines are shown in Table 8.1. They are different for men and women because, on average, men have a greater volume of water in their bodies than women in which to distribute the alcohol.

Table 8.1 Guidelines on intake of alcohol

	Men	Women
Safe (as recommended by UK Government)	3–4 units per day with regular alcohol-free days	2–3 units per day with regular alcohol-free days
Hazardous	21–50 units per week	14–35 units per week
Dangerous	>50 units per week	>35 units per week

Figure 8.23 All of these alcoholic drinks contain one unit of alcohol

As you can see from Figure 8.23, lagers of different strengths have different amounts of alcohol in them. This makes it difficult for people to work out how many units they are drinking. Similarly, a bottle of alcopop usually contains about 1 unit of alcohol, but this can vary. You can read the volume of alcohol as a percentage on the label.

So what are the risks of regularly drinking above the Government's guidelines? They can be many, including cancer of the mouth and throat, cancers of the intestinal tract, breast cancer, ulcers, liver disease, weakening of the muscles, as well as social and psychological problems. Heavy drinkers are reckoned to lose 12 years from their life span. Alcohol is a poison. The liver removes it at the rate of 1 unit per hour, but long-term exposure to excess quantities of alcohol can cause the cells of the liver to become overactive and die. This is called cirrhosis of the liver and the damage is irreversible.

Some facts about alcohol consumption

- Only 20% of the alcohol drunk enters the bloodstream through the stomach. The rest enters from the small intestine.
- The valve between the stomach and the small intestine stays closed when the stomach is full of food. Alcohol is absorbed into the bloodstream much more slowly from the stomach than from the small intestine, allowing the liver to break it down before it builds up too much.
- Drinking on an empty stomach allows the alcohol to flow to the small intestine and it is absorbed more quickly there. The liver does not have time to break it down fast enough, so levels in the bloodstream rise quickly.
- Fizzy drinks like champagne contain carbon dioxide bubbles. These are released in the warmth of the stomach and tend to keep the stomach valve open. This means that the alcohol goes to the small intestine where it is absorbed more quickly.
- Alcohol is a diuretic. This means that it removes water from your body. In fact people urinate more water than the volume of alcoholic drink they consume, leaving them dehydrated.
- The headache symptom of a hangover is a result of the pain sensors on the outside of the brain reacting to the brain shrinking away from the skull as it loses water.

Alcohol – the social issues

- Out of an adult population in the UK of 43 million people, 36 million people drink regularly and 7.3 million drink above the Government's guidelines, with 4 million being classed as heavy drinkers and a further 800 000 as problem drinkers. Another 400 000 are dependent on alcohol, i.e. they are alcoholics.
- Half a million hospital admissions each year are alcohol related. About one in three patients at Accident and Emergency has been drinking before going in.
- 80% of fire deaths are due to alcohol consumption.
- 33 000 deaths each year in the UK are thought to be due to alcohol.
- 50% of murders can be blamed on the drinking of too much alcohol.
- 33% of divorces are thought to have alcohol as one of the contributing factors.

- 8 million working days are lost each year due to alcohol-related illnesses, which costs society £1600 million per year.
- Drinkers are more likely to have casual sex, leading to unwanted pregnancies and sexually transmitted diseases.

Summary

- Yeast is a single-celled organism that ferments sugar to produce alcohol.
- The alcohol in alcoholic drinks is called ethanol, C_2H_5OH.
- Fermentation is the process of converting sugars to ethanol and carbon dioxide.
- Wine is made from crushed grapes and yeast.
- Beer is produced from various mixtures of barley, hops, water and yeast.
- Alcohol is a mood-altering drug. Small amounts make people relax and lose their inhibitions.
- Too much alcohol in the blood causes a lack of coordination and slurred speech. Very high levels of alcohol cause unconsciousness and even death.
- There may be harmful consequences of drinking alcohol in excess of the UK Government guidelines, such as an increased risk of some forms of cancer and liver disease.
- There are many social issues relating to alcohol, ranging from alcohol-related deaths to unwanted pregnancies and sexually transmitted diseases.

Questions

8.12 What is meant by *fermentation*?

8.13 **a** What is the chemical name of the alcohol in alcoholic drinks?

b The formula of this alcohol is C_2H_5OH. How many atoms of carbon, hydrogen and oxygen are present in one molecule of this alcohol?

8.14 **a** Draw and label apparatus you could use in the laboratory to collect the gas coming off from the fermentation of a glucose solution. [Hint: You may need to look back to page 119.]

b What test could be used to show the identity of the gas collected?

8.15 Consider the effects of alcohol on the brain and nervous system. Why is it unsafe for people to drive when they have been drinking alcohol?

8.16 A bottle of Vladivar™ Veba™ is a sweet and sour cherry drink containing vodka. The label tells you that there is 4.1% alcohol by volume. The bottle contains 275 cm^3 of the flavoured drink. How many alcohol units are in this bottle? [Hint: You will need to look back in this section to find the volume of 1 unit of alcohol.]

D Designer products and foodstuffs

It is easy to tell if a banana is ripe because it turns from green to yellow, but it is not so easy to do this with a pear because it looks just about the same whether it is ripe or unripe. Shoppers usually test the ripeness of a pear by giving it a squeeze. If it feels very firm it is not yet ripe, while if it is soft it is ripe. The problem with giving a pear a squeeze is that the tissue becomes bruised and then develops a brown patch. If you are a greengrocer this is the last thing you want to happen because the pear will become unsaleable. This is where scientists in New Zealand come in. They have invented intelligent packaging called *Ripesense* that can actually tell you when a pear is ripe.

When fruit is ripening it gives off an aroma (smell) and this is the basis for the intelligent packaging. Chemicals in a label react with the aroma compounds and change colour from red, which means unripe, through orangey-red, which means half ripe, to yellow, which means the pears are fully ripe. This allows the customer to choose the stage of ripeness required. If the fruit is to be eaten straight away then the label will need to be yellow.

Figure 8.24 *Ripesense* packaging changes colour to show the customer how ripe the pears inside are

Intelligent packaging

The label shown in Figure 8.24 that can tell you the ripeness of a pear is one example of **intelligent packaging**. Intelligent packaging responds to changes in conditions in some way, usually changes that occur on the inside of the package.

Removing oxygen

Many foods need to be kept away from oxygen if they are to be kept fresh and not deteriorate and 'go off'. Oxygen reacts with fats and oils and turns them rancid. This means that foods such as meat, dairy products and fish develop an unpleasant smell and taste. Pizza bases and other baked products will turn mouldy in the presence of oxygen, because oxygen allows moulds to grow. The nutritional value of food can also be lowered when it reacts with oxygen.

Chilled packaged foods need to be kept fresh for as long as possible so that their shelf life (their 'display until' date) is extended. These foods are packed in an atmosphere that excludes oxygen, otherwise they will change colour, lose their flavour, lose their nutritional value, become rancid or grow mould, depending on the foods inside the package. They may be packed in an inert atmosphere, or they may be vacuum packed, so that oxygen is removed from around them. The problem remains that most transparent film packaging is made of some type of plastic, and all plastics will allow oxygen through, some faster than others. Intelligent packaging will actually remove oxygen that enters the packaging.

The process of removing oxygen in intelligent packaging is called **oxygen scavenging**. Some simple oxygen scavengers are based on iron, or iron compounds, which react with the oxygen and become oxidised.

You can find out more about oxidation by looking at Chapter 6, page 109.

These are usually present in powdered form, in sealed sachets inside the food packet – see Figure 8.25. Other methods incorporate an oxygen-scavenging compound as part of the plastic packaging.

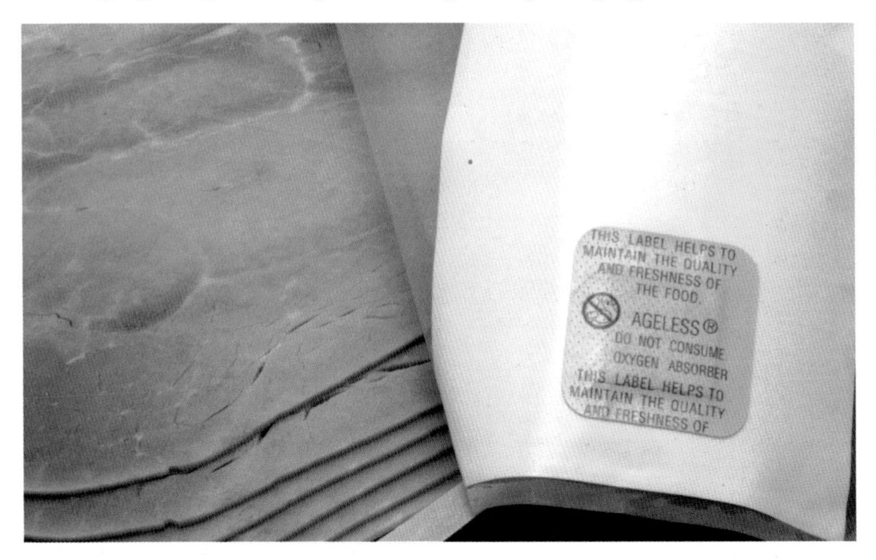

Figure 8.25 The label is actually a sachet containing an iron compound. This reacts with oxygen as it enters through the plastic film, keeping the meat fresh for longer

Another method relies on a chemical reaction we met in Chapter 6. A mixture of nitrogen with about 5% hydrogen can be used as an oxygen-free atmosphere in bottles of wine, beer, fruit juices and sauces. The hydrogen reacts with oxygen as it enters the bottle cap. Normally hydrogen will not react with oxygen at room temperature, but a **catalyst** of palladium metal will allow this reaction to happen:

Catalysts speed up chemical reactions. The reason hydrogen does not appear to react with oxygen at room temperature is because the reaction is very slow. Palladium speeds the reaction up.

$$\text{hydrogen} + \text{oxygen} \xrightarrow{\text{palladium catalyst}} \text{water}$$

$$2H_2 + O_2 \xrightarrow{\text{palladium catalyst}} 2H_2O$$

The palladium is incorporated into the bottle cap as a fine powder on a foam wad, surrounded by a gas-permeable membrane. **Gas-permeable** means that gases can move through. Oxygen then reacts with the hydrogen before reaching the contents of the bottle.

Removing water

Water allows microbes such as mould and bacteria to grow. These food-spoiling microorganisms particularly affect fruit, vegetables and baked products like breads. The water is often produced after the product is packed. For example, vegetables and fruit go on respiring in the pack and this produces water vapour, which if not removed will cause the food to spoil. The water can be removed by intelligent packaging, keeping food fresh for longer.

Mayonnaise – a designer product

Mayonnaise is a thick, creamy sauce made from vegetable oil, vinegar or lemon juice, and egg yolk. The story of its invention goes back to 1756 when the British were fighting the French at Port Mahon on the island of Minorca. The victors on this occasion were the French, and the chef of the victorious commander was ordered to prepare a special meal. The sauce for the meal was to have been made from cream and eggs but the chef had run out of cream, so he added olive oil – and 'Mahonnaise', named after Port Mahon, was the result.

Figure 8.26 The oil is floating on top of the vinegar (which is almost all water)

Normally oil and water do not mix (see Figure 8.26) unless you give them a good shake. When shaken, one liquid forms tiny droplets that spread through the other liquid. This mixture of tiny drops of one liquid dispersed in another liquid is called an **emulsion**. But it does not last long and very soon the two layers re-form.

The secret of mayonnaise – a mixture of oil and vinegar (which is mostly water) – lies in the addition of egg yolk, which keeps the two liquids dispersed almost indefinitely.

Egg yolk contains an **emulsifier**. This is a long molecule that will keep oil and water mixed in an emulsion. The emulsifier molecule has two parts – a water-loving (**hydrophilic**) part and a water-hating (**hydrophobic**) but oil-loving part (Figure 8.27).

The emulsifier molecule in egg yolk is lecithin.

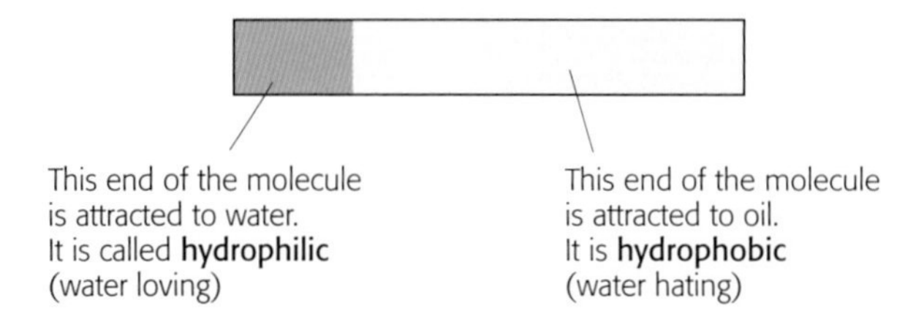

Figure 8.27 An emulsifier molecule

In mayonnaise, when the oil is broken up into droplets these are surrounded by the oil-loving parts of the emulsifier molecules. As you can see in Figure 8.28, the hydrophilic parts stick out into the water and keep the droplet suspended because they are attracted to the water. The emulsion in mayonnaise is called an oil-in-water emulsion because the oil droplets are spread out in the vinegar.

Another oil-in-water emulsion is milk. The emulsifier is a protein. If milk is left to stand, some of the oil, in the form of cream, will separate out of the emulsion and rise to the top.

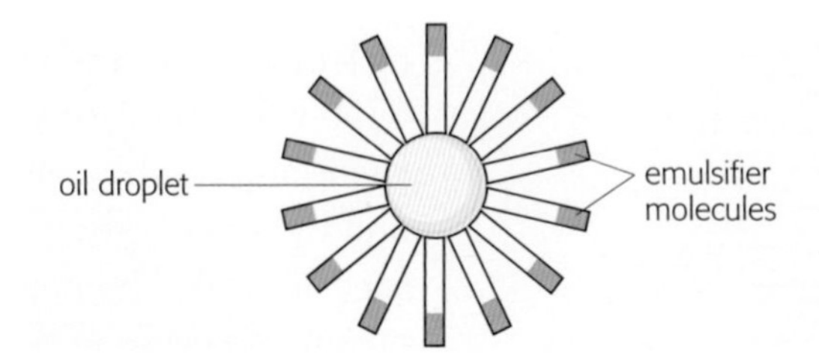

Figure 8.28 Emulsifier molecules surround the oil droplet and keep it dispersed in the water

Summary

- Intelligent packaging reacts to changes in conditions inside the package, to keep food fresh for longer.
- Intelligent packaging can remove oxygen or water.
- Oxygen inside food packages causes:
  - fats and oils in dairy products and meat to go rancid
  - the nutritional value of some foods to be reduced
  - the colour of some foods to change
  - moulds to grow on baked products, fresh fruit and vegetables.
- The main problem with water inside food packaging is that it encourages food-spoiling microorganisms to grow.
- Emulsions are made up of two liquids that do not normally mix. The droplets of one liquid spread through the other.
- Emulsifiers keep the droplets of one liquid spread throughout (dispersed in) another liquid.
- Egg yolk contains an emulsifier which keeps vegetable oil droplets in mayonnaise suspended in the watery vinegar.
- Emulsifier molecules have a water loving (hydrophilic) part and an oil-loving, water-hating (hydrophobic) part.

Questions

8.17 When tomatoes ripen they produce a gas called ethene. They can be stored in packaging that absorbs this gas and prevents them from ripening.

a What is the general name given to the type of packaging that absorbs the ethene gas?

b Why is the packaging given this name?

8.18 Explain why oxygen and water need to be removed from the inside of the packaging of many foods.

8.19 Butter is a water-in-oil emulsion.

a What does the term *emulsion* mean?

b What does *water-in-oil* tell you about the emulsion?

c Draw an emulsifier molecule and label its two parts.

d Draw emulsifier molecules surrounding the droplets in butter. Label the droplets as either oil or water.

9 Producing and measuring electricity

A Battery power

Mobile phones are useful. The latest phones have many interesting features – they store photos and messages, remind you of appointments, allow you to watch streaming movies and listen to music from the internet. All of this takes energy, and that energy has to come in a portable form. The answer, of course, is a battery.

Each new phone comes with advice on how to recharge its battery. Should you recharge every night, or wait until the battery indicator is low? If you don't follow the manufacturer's advice, you could end up with a phone that doesn't work when you want it to.

Figure 9.1 Mobile phones – freedom or frustration?

Battery choice

When designers invent new phones, they have to be sure that they have made the best choice of battery. They want it to be:

- rechargeable
- small and lightweight
- a good energy store.

A phone whose battery runs down quickly is almost worse than useless.

There are many other portable devices which make use of batteries as their energy store: laptop computers, cameras, MP3 players, calculators and watches. You almost certainly have something that you use regularly which depends on battery power.

Getting more from your battery

There are several different types of battery that are used in mobile phones. It helps if you know which type you have, so that you can learn how to make the most of it and avoid finding yourself with a flat battery.

PhoneBatteries 4U

for bargains, for all phones, for you!

	Type	Rating	
	NiMH	4.8 V 800 mAh	▶ Buy Now
	NiMH	3.6 V 600 mAh	▶ Buy Now
	Li-ion	3.6 V 600 mAh	▶ Buy Now
	Li-poly	3.6 V 1800 mAh	▶ Buy Now
	Li-poly	6.0 V 3000 mAh	▶ Buy Now

Figure 9.2 Which one to choose?

- **Nickel–cadmium (NiCd):** This type of battery is the most tricky to look after. The problem is that you need to let it run down fully before you recharge it, and that can be awkward. If you recharge while the battery is still half-charged, it may never work below this level again. Also, NiCd batteries have cadmium in them, a very poisonous metal which must be recycled carefully – don't dump it!
- **Nickel–metal hydride (NiMH):** This type is lighter than a NiCd battery, and doesn't contain cadmium. The disadvantage is that it may not last as long – they tend to fail after a few hundred recharging cycles. If you keep getting 'battery low' messages while you are making a call but the phone is all right on standby, you may have a NiMH battery that needs replacing.
- **Lithium ion (Li-ion):** This type of battery is even smaller and lighter, and you can top it up whenever you want to. But these are the most expensive, so are only fitted in expensive phones.
- **Lithium polymer (Li-poly):** These can be easily squeezed into a new shape to fit into an awkwardly shaped battery compartment.

Phone battery voltages are all multiples of 1.2 V, and this tells you something about how a battery is constructed – it is made of two or more single cells, each providing 1.2 V. The cells are connected together in series (end-to-end) so that their voltages add up.

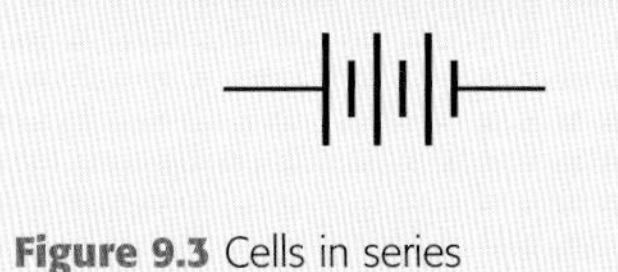

Figure 9.3 Cells in series

If you need to buy a new mobile phone battery, you will have to make sure you buy the right one. There are hundreds of different ones, and each is suitable for only a few designs of phone. You can learn a lot from reading the detailed information given for each one.

1 You can find out the type of battery (NiMH, Li-ion, etc.).
2 You can discover the **voltage**. Most provide a voltage of 3.6 V or 4.8 V. Some give 2.4 V or 6.0 V.
3 Finally, you will find that some batteries can go for longer before they need recharging. Find the figure with units Ah (amp-hours) or mAh (milliamp-hours). This tells you the **capacity** of the battery. The greater the capacity, the longer the phone battery will last.

Figure 9.4 The bigger the battery, the greater its capacity and the longer it will last. The size D battery has 20 times the capacity of the AAA one

Ordinary batteries (the sort you might use in a torch or an MP3 player) come in different sizes – AAA, AA, C and D. Each of these provides the same voltage (1.5 V for alkaline batteries) but the bigger the battery, the greater its capacity and the greater the time for which it can supply a particular current. A battery of capacity 4.0 Ah can provide a current of 1 A for 4 h, or 0.5 A for 8 h, and so on.

Worked example

For how long can a battery of capacity 1.5 Ah supply a current of 0.3 A?

Time = capacity / current = 1.5 Ah / 0.3 A = 5 h

Testing batteries

You can test the output of a battery using a resistor, a voltmeter and an ammeter.

1. Connect up a battery with a 10 Ω resistor and an ammeter in series (making a simple loop circuit). A current will flow through the resistor and be measured by the ammeter.
2. Connect a voltmeter in parallel with the battery. This will indicate how its voltage changes.
3. Record current and voltage at intervals of time. How do these readings change?

Alternatively, you can use voltage and current sensors and a datalogger to record the values over a period of time, and display the results using a computer.

It isn't safe to connect an ammeter directly to a battery – you must include a resistor in the circuit.

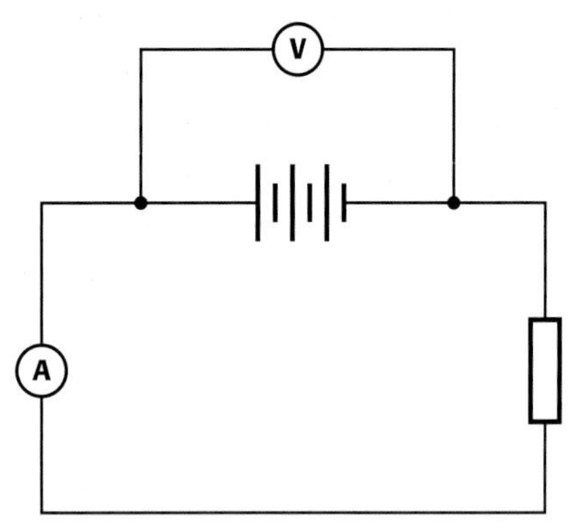

Figure 9.5 Testing the output of a battery

Running down

A mobile phone has to keep in touch with the local network, in case there is a message or a call for you. It sends out a signal – a short burst of microwaves – every few seconds, so that the nearest phone mast knows where your phone is. This doesn't take much energy, but it does mean that the battery is providing a small current, and eventually the battery will run down. If you make a call, the current from the battery will be much greater and the battery indicator will go down more quickly.

What's happening when a battery runs down? A battery contains chemicals which react together to provide the voltage between the battery terminals. This voltage pushes the current round the circuit. Eventually all of the chemicals have reacted and the battery is discharged. At this point, the voltage drops quite rapidly to zero.

Today's phones are cleverly designed to use as little current as possible. They have a circuit inside which monitors what's going on – are you making a call, typing in an SMS message, or just on standby? Then it activates only those parts of the phone's circuitry that are needed. It even works out how close you are to the nearest mast, so that it transmits a signal that is only just strong enough to be received by the mast. This clever technology helps make the most of the available battery power.

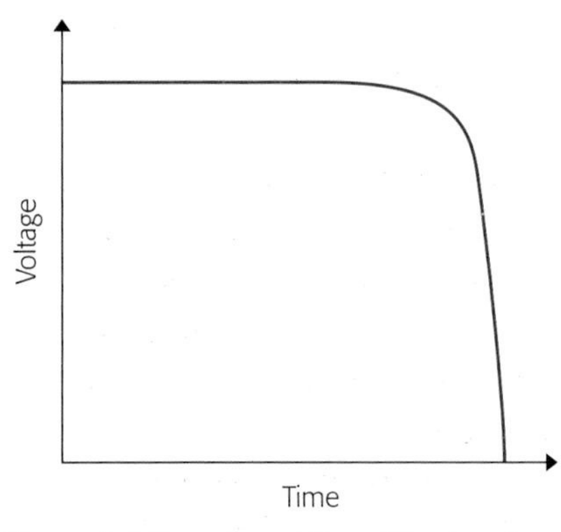

Figure 9.6 The voltage falls rapidly when the battery runs down

Charging up

When you recharge a battery, the charger provides a voltage which is a bit bigger than the battery voltage. It pushes current the other way round the circuit; the chemical reaction inside the battery goes into reverse, and the battery is recharged.

A voltmeter, connected in parallel with the battery, can show how the voltage of the battery recovers. It goes up rapidly at first, then gradually approaches its maximum value. The charger should be left on for a good while because the individual cells inside the battery may be slightly different, and one may become fully charged before the others. The charger detects when the battery reaches the fully charged voltage, and switches off.

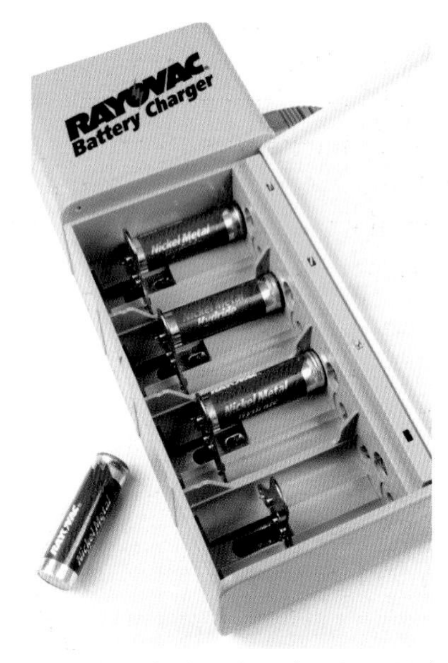

Figure 9.7 Batteries being recharged in a charger

A battery may become warm as it is recharged. This is a sign that some of the energy you are putting in is being wasted – you get less out of the recharged battery than you put in.

Figure 9.8 A solar-powered phone box in France

Solar cells

Batteries are useful because they are a convenient, portable supply of electricity. However, you need to be near to a mains electricity supply to recharge them.

In out-of-the-way places, such as mountainous or desert regions, roadside phones use **solar cells** (Figure 9.8). A solar cell uses the energy of sunlight to produce electricity. It doesn't store energy; usually it charges up an ordinary chemical battery, which then provides the current to work the phone. Solar cells are finding many other uses – for example, to power refrigerators in which medications are stored in clinics in rural Africa.

Mains or battery?

Batteries and solar cells have two terminals, labelled positive (+) and negative (−). We picture electric current flowing from the positive terminal, around the circuit, and back to the negative terminal. If you connect an ammeter in series in the circuit, the current must flow through it and it registers the amount of current flowing. The current flows in one direction only, and is known as **direct current (d.c.)**.

In fact, if you had a super-high-powered microscope and could see what was going on inside the wires, you would find that the current is a flow of tiny particles called **electrons**. These are much smaller even than atoms, and they are negatively charged. They actually flow out of the negative terminal of the supply, round the circuit, and in to the positive terminal. (The electrons flow this way because their charge is negative, so they are attracted by the positive terminal – opposite charges attract. When scientists first decided which terminal to call positive, they had never heard of electrons and they didn't know which way the charge flowed.)

Strictly speaking, we should say that an electric current is a flow of electric charge. Since electrons are charged particles, a flow of electrons is an example of an electric current.

The mains electricity supply is different. It has live and neutral connections, instead of positive and negative. The current flows back and forth in the circuit, first out of the live connection and then out of the neutral connection. It repeats this back-and-forth cycle 50 times a second, so we say that the mains frequency is 50 Hz (hertz). Current which flows back and forth like this is called **alternating current (a.c.)**.

Figure 9.9 Current flow and electron flow in d.c. and a.c. circuits

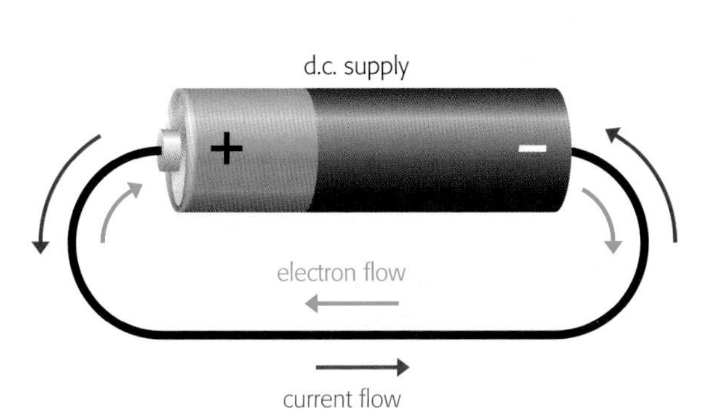

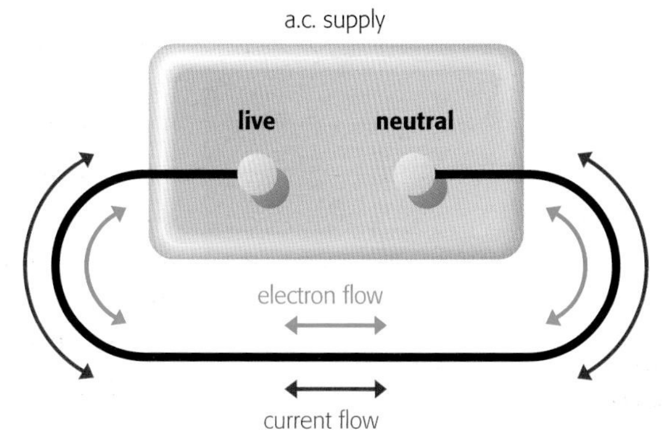

Figure 9.10 A battery recycling bin – a rare sight in the UK

Counting the costs

Mains electricity is cheap. It costs perhaps 8p to run a 1 kilowatt heater for an hour. If you used AA batteries to run the heater, you would get through something like 2000 in an hour to produce the same amount of heat energy! That would be expensive.

So mains electricity is good for supplying a lot of energy quickly and cheaply. That's why we use it for things like heaters, washing machines, TV sets and lighting. But it's not portable in the way that batteries are – you need to have a continuous circuit between your appliances and the power station.

How do solar cells compare? These have no running costs – the energy of sunlight is free – however, they are expensive to buy in the first place. And it takes a large area of cells – several square metres – to collect enough sunlight to provide the electricity for a household.

So for certain uses, batteries seem the best option. But we have to think about the environmental costs of using them. They often contain hazardous chemicals and so must be disposed of carefully. In the UK, most of us throw used batteries in the dustbin with other domestic waste. In some countries, any shop that sells batteries has to provide a recycling bin so that used batteries can be returned to the manufacturer for safe recycling.

Summary

- **An ammeter, connected in series, measures the current flowing through a component.**
- **A voltmeter, connected in parallel, measures the voltage across a component.**
- **Batteries supply direct current (d.c.). The mains supplies alternating current (a.c.).**
- **The capacity of a battery is given in amp-hours.**
- **Electric current is a flow of negatively charged electrons.**

Figure 9.11 A few pulls on the cord is enough to charge up this phone for a 5-minute call

Questions

9.1 A car battery is labelled '12 V, 100 Ah'. It is made up of 6 individual cells.

- **a** What voltage is provided by each cell?
- **b** How are they connected together?
- **c** It takes a current of 8 A to work the car's headlights and sidelights. If the driver leaves the lights on after parking, how long will it be before the battery is flat?

9.2 Figure 9.11 shows Christopher Jung with his wind-up phone, which he invented when a student in Berlin.

- **a** What advantages does this phone have over other mobile phones?
- **b** Where might it be useful?
- **c** What other devices might benefit from a system like this?

9.3 A mobile phone designer must choose which type of battery to fit to a new model of phone. Make a list of all the different factors which should be taken into account when choosing a battery.

9.4 Explain how you would estimate the capacity (in amp-hours) of a small battery. Support your explanation with a circuit diagram.

9.5 Table 9.1 shows the battery capacities recommended for different appliances. Suggest why these different appliances need different sizes of battery.

Table 9.1 Recommended battery capacities

Appliance	Battery capacity (Ah)
mobile phone	0.5
laptop computer	4.0
disabled person's scooter	35

B Sensing change

Property owners get nervous. The bigger their house and the more possessions they keep in it, the more likely they are to fit a security system.

- Windows and doors are fitted with switches that detect if they are opened when they should not be.
- Movement detectors respond to anyone moving around inside the house, or in the grounds.
- Temperature sensors detect when something warmer than its surroundings is nearby – perhaps a person or a car.

These detectors may simply switch on a light, or they may sound an alarm or even contact the local police. All of this happens automatically, making use of electrical components that act as sensors.

Figure 9.12 Street lighting is controlled by electrical sensors

As people become more concerned about security, local authorities fit more street lighting. People feel safer when the streets are lit up at night. Again, these lights come on automatically, as night falls.

However, not everyone agrees that extra lighting is the best policy. After all, burglars need to find their way around, too, and it may actually help them if there is good lighting in the area where they want to operate.

Light-dependent resistors

Some street lights use a timer which switches them on at night and off again at dawn. This works well, but the timer needs to be reset as the seasons change. It is better to use a device which detects when it gets dark and switches the light on automatically. Such a device is a **light-dependent resistor (LDR)**. You may have used LDRs in Technology classes.

Figure 9.13 This street lamp has an LDR fitted on top

The circuit in Figure 9.14 on the next page shows how to investigate an LDR. It is connected to a battery which makes a current flow through it. The ammeter shows the current. If you move your hand or a piece of card across the LDR, the current drops. This tells us that with less light falling on the LDR its **resistance** has increased, so that it is harder for the battery to push current around the circuit. Move your hand away again and the current reading on the ammeter goes back up again – the resistance of the LDR has decreased.

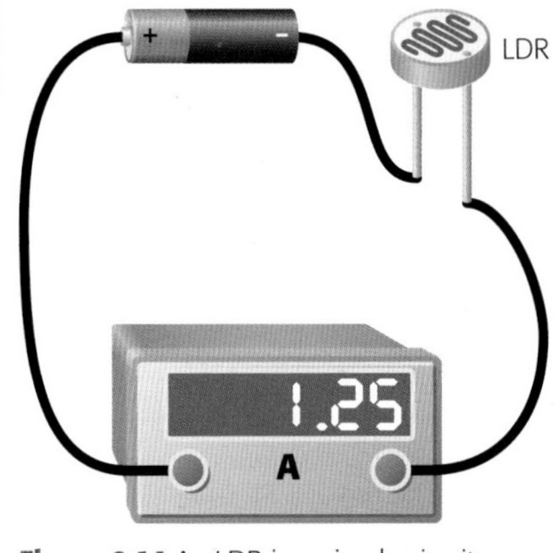

Figure 9.14 An LDR in a simple circuit

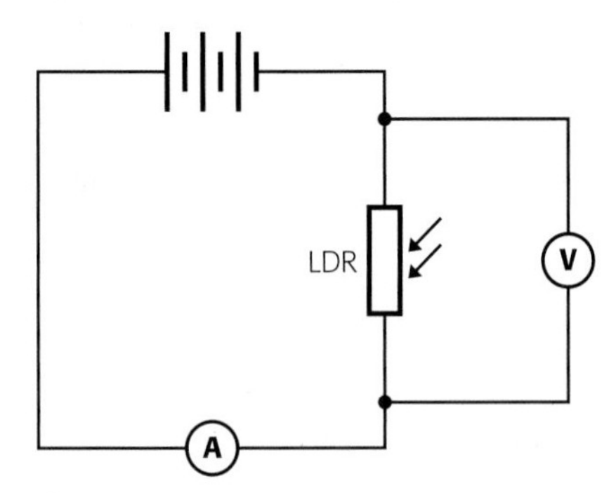

Figure 9.15 Circuit for measuring the resistance of an LDR. Note the symbol for the LDR

In some books you will see a slightly different symbol for an LDR – a circle is drawn around the resistor.

It may help you to remember that I stands for current by thinking of *Intensity*.

How does it work? The current in the circuit is a flow of electrons. The electrons find it hard to flow through the LDR. However, when light falls on the LDR, this gives the electrons more energy and they find it easier to flow. We see a bigger current flowing, and we say that the resistance of the LDR is less.

The greater the resistance in a circuit, the smaller the current that flows (for a given voltage)

Measuring resistance

To measure the resistance of the LDR, you need to know two things: the current flowing through it and the voltage across it. This means that, as well as an ammeter, you need to include a voltmeter in the circuit. You should remember that:

- an ammeter is connected in series in a circuit (so that the current flows *through* it)
- a voltmeter is connected in parallel with a component (so that it measures the voltage *across* the component).

The circuit diagram in Figure 9.15 shows how to do this.

Calculating resistance

Once you have a measurement of the current flowing through the LDR and of the voltage across it, you can work out its resistance.

$$\text{resistance} = \frac{\text{voltage}}{\text{current}}$$

You have to use the correct units: current in amps (A) and voltage in volts (V). Then the answer comes out in ohms (Ω). (This is the Greek letter omega.)

You may see this equation written in symbols:

$$R = \frac{V}{I}$$

Note that the letter I is used to stand for current.

Worked example

An LDR is connected to a 5.0 V supply. A current of 0.01 A flows through it. What is its resistance?

Here are the quantities we know:

- voltage = 5.0 V
- current = 0.01 A

Now we can work out the value of the LDR's resistance by substituting these in the formula for resistance:

resistance = voltage / current = 5.0 V / 0.01 A = 500 Ω

So the LDR's resistance is 500 Ω.

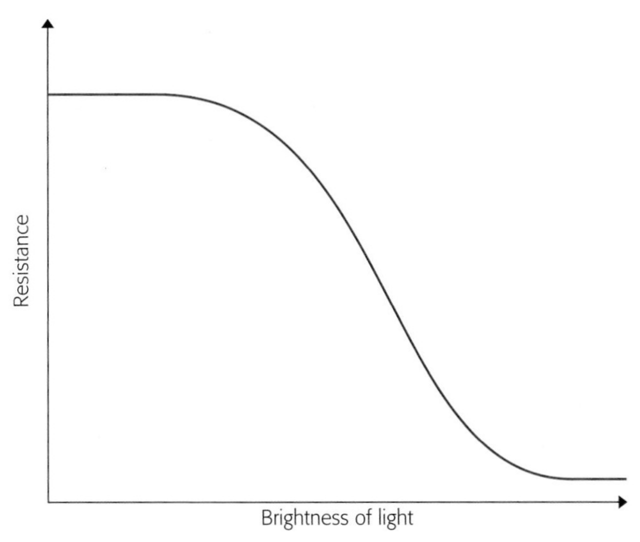

Figure 9.16 How the resistance of an LDR depends on the brightness of light falling on it

Sensing light

The resistance of an LDR goes *down* as the brightness of the light goes *up*. You can see this represented in the graph of resistance against brightness (Figure 9.16). Note that the graph is not a straight line; it's more like a downward step.

When the LDR is in darkness, its resistance is high, perhaps a million ohms. In bright light, it is a few hundred ohms.

LDRs are used in street lights, as we have seen. They are also used as the light meter in some cameras. There are many other places where it is useful to detect varying light levels – for example, in barcode readers and in computer scanners. These, however, usually use light-sensitive diodes rather than LDRs.

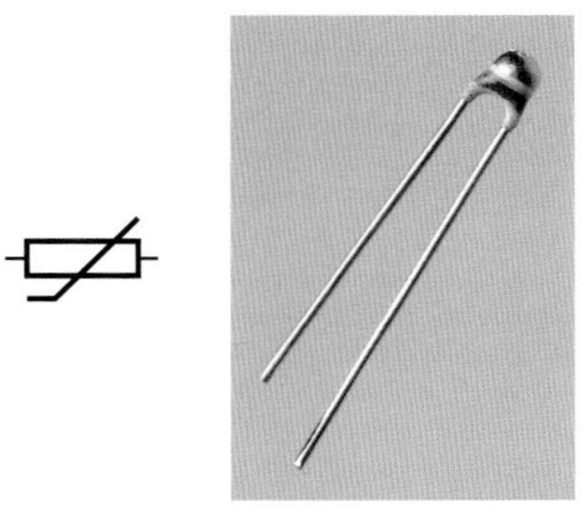

Figure 9.17 The symbol for a thermistor, and an example of one type of thermistor

Thermistors

It can also be useful to detect changing temperatures. For this we need another kind of resistor – a **thermistor** (short for thermal resistor). These have a lot of uses, for example:

- in fire alarms – to detect the rising temperature of a house on fire
- in fridges and freezers
- in fish tanks

– in fact anywhere where you need an automatic system to warn of a change in temperature and possibly turn on a heater or cooler.

The graph in Figure 9.18 shows that the resistance of a thermistor *decreases* suddenly as the temperature *increases*. The electrons that are trying to move through the thermistor have more energy at higher temperatures and so can move more freely.

There is a second type of thermistor whose graph goes the other way – its resistance increases as the temperature goes up.

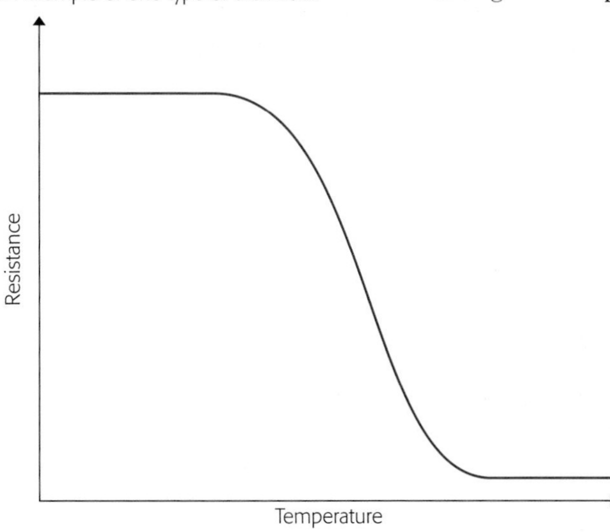

Figure 9.18 How the resistance of a thermistor changes as the temperature changes

Summary

- The resistance of a component is given by:

  resistance = voltage / current $R = V / I$

- The resistance of a light-dependent resistor (LDR) decreases when light shines on it.
- The resistance of a thermistor changes (usually decreases) as its temperature increases.

Questions

9.6 List some uses of light-dependent resistors (LDRs) and thermistors.

9.7 There is a type of thermistor whose resistance increases as it gets hotter. Sketch a possible graph of resistance (*y*-axis) against temperature (*x*-axis) to represent this. [Hint: Figure 9.18 may help you.]

9.8 **a** Draw the circuit you would set up to measure the resistance of a thermistor.

b Suggest how you could set about finding how its resistance changed with temperature.

9.9 Copy and complete the table. Be careful not to mix up the symbols for quantities and the symbols for their units.

Quantity	Symbol for quantity	Unit	Symbol for unit
resistance	R	ohm	
voltage			
current			

9.10 A thermistor is connected to a 10 V power supply. At a particular temperature, a current of 0.20 A flows through it. Calculate its resistance at this temperature.

C Resistors and lamps

Georg Ohm was a German scientist and teacher. His father took a great interest in maths and science, and forced his children to study these subjects. When Georg went to university, however, he didn't attend to his studies, but spent his time dancing, skating and playing billiards. Father Ohm made him leave university and get a job. Georg became a teacher and in time took up the private study of mathematics.

In the 1820s, when he was in his thirties, he had a teaching job in Berlin which allowed him to carry out experiments on electricity in the well equipped physics laboratory. He wanted to know what happened when he connected a piece of wire to a battery – if the voltage across the wire was increased, how did the current flowing through the wire change?

He managed to show that doubling the voltage made twice the current flow. In other words, current is proportional to voltage. This is what is now known as Ohm's law.

Figure 9.19 Georg Ohm (1787–1854)

It is easy to check Ohm's law in a school lab. You need a variable power supply or batteries, a length of wire, an ammeter to measure the current and a voltmeter to measure the voltage. Increase the voltage, and watch the current increase. Easy. However, Georg Ohm managed this at a time when there were no ammeters or voltmeters – they hadn't been invented. So how did he do it?

To start with, he used some old-fashioned batteries called voltaic piles. By connecting these together, he could increase the voltage across the wire in steps.

But how did he go about measuring the current? Here, he had to use the fact that a current in a wire produces a magnetic field (think of electromagnets). Place a compass near a wire with a current and the needle will turn round. The bigger the current, the greater the force on the compass needle.

So Georg connected his voltaic piles to a metal wire, placed a compass next to it, and measured the deflection of the compass needle. In this way, he was able to show that the greater the voltage, the greater the current.

At first, other scientists dismissed this work, saying that it was really just some experimental results, and what was needed was a theory of *why* current is proportional to voltage. Fortunately for Georg, there was a new theory around at that time, of the way heat conducts through solids. He was able to adapt this theory to explain his electricity results.

Eventually Ohm became a professor of physics.

Checking up on Professor Ohm

Figure 9.20 shows how to check Ohm's law today. This is the same circuit as is used to measure the resistance of a component, except that there is a variable power supply in place of a fixed voltage. A length of wire is used as the object under investigation.

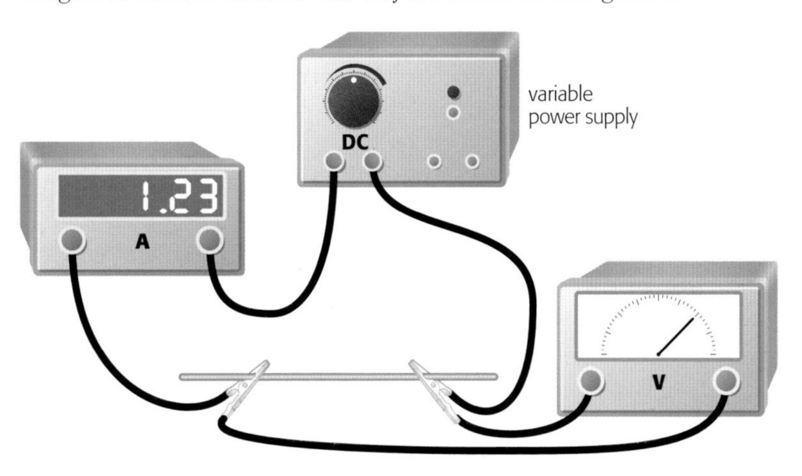

Figure 9.20 Verifying Ohm's law

Recall that you need one measurement of voltage and current to calculate the resistance of a component, using:

$$\text{resistance} = \frac{\text{voltage}}{\text{current}} \qquad R = \frac{V}{I}$$

However, it is usually better to make several measurements of a quantity, and find an average. This is for two reasons:

- You may have made a slip while making one measurement; this will stand out if you make several measurements.
- Each measurement is likely to be slightly imprecise. If you read an ammeter, is the current 0.15 A or 0.16 A? Is the voltage exactly 4.00 V, or might it be 3.99 V or 4.01 V? Making several measurements can help to reduce these effects and so increase the **precision** of your answer.

So here is what you do when you have connected up the circuit:

1. Switch on the power supply, and set it to give a voltage of, say, 2.0 V. (Read this from the voltmeter, not from the dial on the supply.)
2. Read the value of the current from the ammeter. (Check that the reading is steady.)
3. Repeat this, increasing the voltage in steps of, say, 2.0 V. Record the results in a table.
4. Plot a graph of current against voltage.

Voltage (V)	Current (A)
2.0	0.03
4.1	0.06
6.0	0.10
8.0	0.13
10.1	0.16

Table 9.2 Experimental results

Table 9.2 and Figure 9.21 show some sample results.

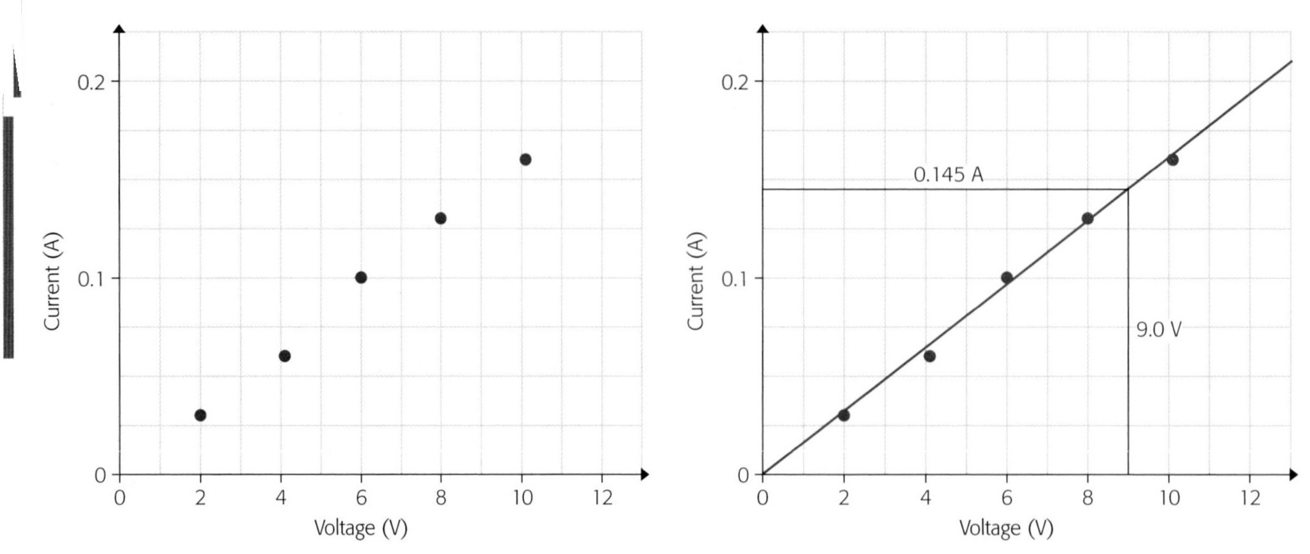

Figure 9.21 Experimental results plotted, and the best straight line through the points

When you have marked the points on the graph, look at them carefully. Do they appear to fall on a straight line? In this case, it looks as though they do (although they don't form a perfect straight line – we shouldn't expect them to). Use your judgement to draw the best straight line.

The line should ideally go through the origin of the graph, because we know that when the voltage is zero, the current is also zero. However, don't *force* the line to go through the origin, because there may be some unknown reason why it doesn't – this might be something to investigate later.

Interpreting the graph

The graph of current against voltage for a length of wire is a straight line. This shows that current is proportional to voltage, at least for the wire you are investigating. Ohm was right!

From the graph, we can find the resistance of the wire. Take a point on the line, near the top end. Read off the values of current and voltage. Calculate the resistance. For our example results:

resistance = voltage / current = 9.0 V / 0.145 A = 62 Ω

Why is this better than using single values of current and voltage? By using the line, we are using *all* of the data we have collected. Look what happens if we just use one point. In Table 9.3 you can see the data from Table 9.2 again, but the table is extended to show the values of resistance calculated from the experimental values.

Table 9.3 Experimental results and calculated resistance

Voltage (V)	Current (A)	Resistance (Ω)
2.0	0.03	66.7
4.1	0.06	68.3
6.0	0.10	60.0
8.0	0.13	61.5
10.1	0.16	63.1

You can see that each row shows a slightly different value for the resistance. We could find the average of these, but using the straight line is better.

What the graph tells us

The graph in Figure 9.21 shows that the current which flows through a wire is proportional to the voltage across the ends of the wire. Ohm found this to be true for many types of wire.

Another way to say this is that the resistance of the wire doesn't change, even if the current through the wire changes.

If you connect the wire into the circuit the other way round, the voltage will be reversed (negative) and the current will flow in the opposite direction (also negative). Combining forward and reverse results gives a graph like the one shown in Figure 9.22. It is a straight line passing through the origin of the graph.

We find graphs similar to this for many types of wire, and also for electronic components called **resistors**. These are used in circuits to control the amount of current that flows. Resistors are very ordinary components; their circuit symbol is simply a rectangular box.

Current

Negative voltage

Voltage

Negative current

Figure 9.22 Current–voltage characteristic for a wire

The meaning of resistance

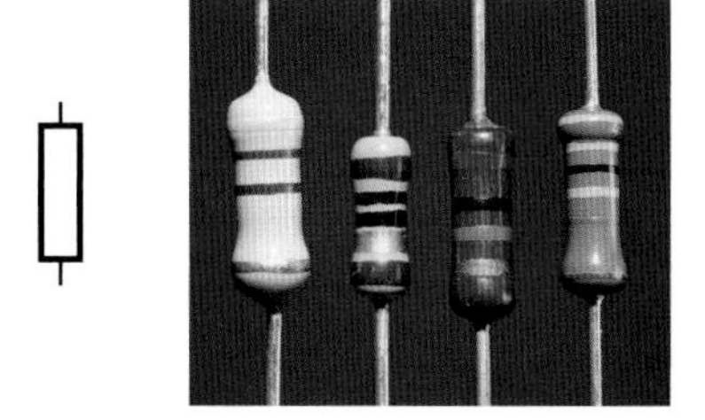

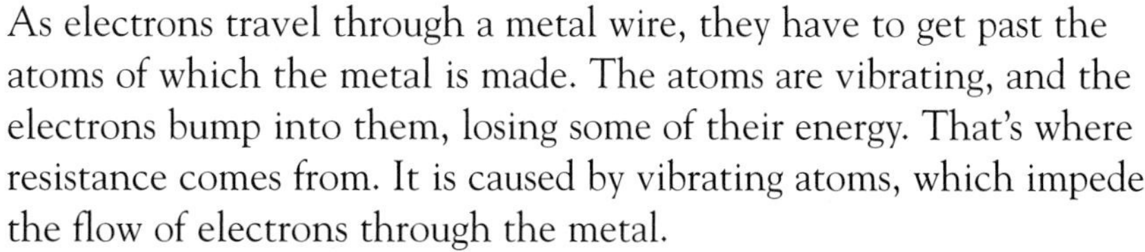

Figure 9.23 The symbol for a resistor, and some examples of resistors

As electrons travel through a metal wire, they have to get past the atoms of which the metal is made. The atoms are vibrating, and the electrons bump into them, losing some of their energy. That's where resistance comes from. It is caused by vibrating atoms, which impede the flow of electrons through the metal.

The equation for resistance is $R = V / I$. We could write this in units:

$$\text{ohms} = \frac{\text{volts}}{\text{amps}}$$

You can think of resistance as telling us how many volts it takes to make a current of 1 amp flow through a component.

- It takes 10 V to make 1 A flow through a 10 Ω resistor.
- It takes 100 V to make 1 A flow through a 100 Ω resistor.

It's the voltage that gives the electrons the push they need to flow through the metal.

The filament lamp

There is a different result if we repeat the Ohm experiment with a filament lamp in place of the wire. A filament lamp contains a curled-up length of thin wire that is designed to get hot and glow when a big current flows through it. This time, the current–voltage graph is curved (Figure 9.24).

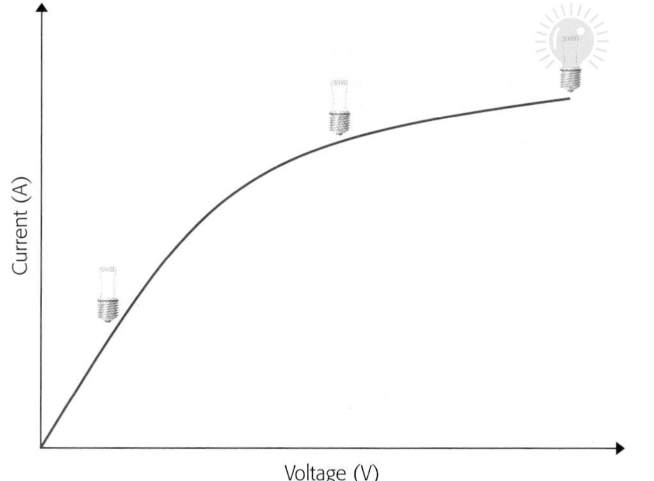

Figure 9.24 Current–voltage characteristic for a filament lamp

Can we think what is going on in the filament to produce this curve?

- First thought: As the voltage is increased, the current also increases, but more and more slowly. Less current is getting through than we might have expected. The current is not proportional to the voltage.
- Second thought: What is happening to the resistance of the filament lamp? It is getting harder for the current to flow through the lamp as the voltage is increased, so its resistance must be increasing.
- Third thought: What is going on in the metal? The wire filament is getting hot – very hot. Its atoms are vibrating more and more, so much that the filament glows. This makes it increasingly hard for the electrons to whiz through the metal without bumping into the atoms and losing their energy.

It is the electrons' lost energy that makes the filament glow.

So a filament lamp is different from a resistor. It doesn't obey Ohm's law. Its resistance increases as the voltage across it increases (because it gets hot).

Using ICT

Instead of an ammeter and a voltmeter, you can use a current sensor and a voltage sensor to measure the current and voltage in a resistance experiment. A datalogger will store the results and a computer will show the graph. This can be a better way of doing the experiment – you simply turn the dial on the power supply to increase the voltage, and the datalogger records the current and voltage. It records many more data points than the average human being can manage (or can be bothered to do). The disadvantage is that you have to be sure to increase the voltage *slowly*, so that the resistor or filament lamp has time to adjust. (The lamp must be glowing steadily when you record the values.)

Another thing that you have to think about, in a datalogging experiment as in all other experiments, is how accurate are your measuring instruments? Suppose you have a cell that supplies exactly 1.50 V. A voltmeter connected to it will not necessarily give this exact reading. We say that the meter is **inaccurate**. A digital meter may look very accurate because it says 1.500 V, but you cannot be sure that this is exactly correct. Another, similar meter might read 1.503 V or 1.498 V. It depends how well they were manufactured and tested before they were sold, or whether they have deteriorated in use. You can be even less certain with sensors and dataloggers, because the data remains hidden in the computer until you decide to look at it. It is worth comparing the results from a datalogging experiment with one that uses voltmeters and ammeters.

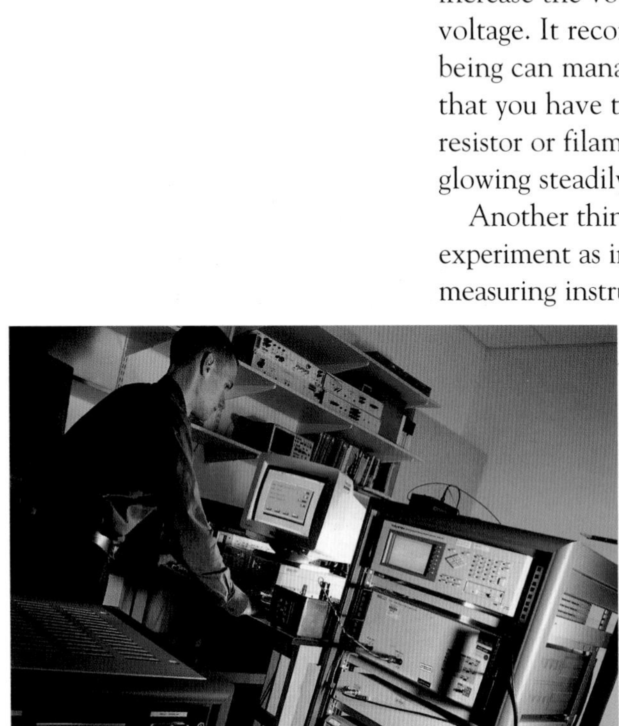

Figure 9.25 This scientist is testing equipment to make sure that the readings it gives are accurate

Further calculations

The equation for resistance can be rearranged in different ways.

$$\text{resistance} = \frac{\text{voltage}}{\text{current}} \qquad R = \frac{V}{I}$$

$$\text{current} = \frac{\text{voltage}}{\text{resistance}} \qquad I = \frac{V}{R}$$

$$\text{voltage} = \text{current} \times \text{resistance} \qquad V = IR$$

Worked example 1

What current will flow through a 5 Ω resistor when there is a voltage of 15 V across it?

We are trying to find the current, so we use:

current = voltage / resistance

Substituting in the values gives:

current = 15 V / 5 Ω = 3 A

Worked example 2

What voltage is needed to make a current of 2 A flow through a 20 Ω resistor?

We are trying to find the voltage, so we use:

voltage = current × resistance

Substituting in the values gives:

voltage = 2 A × 20 Ω = 40 V

Summary

- **For some components (e.g. resistors), the resistance doesn't change as the current flowing changes.**
- **For other components (e.g. filament lamps), the resistance increases as the current increases, because the temperature rises.**

Questions

9.11 Which components are being described here?
  a Resistance decreases when light shines on it.
  b Resistance increases as it gets hotter; gives out light.
  c Resistance decreases over a narrow range of temperature.

9.12 Look at the equation: $V = I \times R$.
  a What quantity does each letter stand for?
  b What is the unit of each quantity?

9.13 What voltage is needed to make a current of 2 A flow through a 20 Ω resistor?

9.14 What is the resistance of a lamp if a voltage of 12 V makes a current of 3 A flow through it?

9.15 The table alongside shows how the current through a lamp changes as the voltage across it is increased.
  a Draw a circuit diagram to show the circuit you would use to obtain these results. Label the components with their names.
  b Copy the table and complete the third column to show the resistance of the lamp at each voltage.
  c Draw a graph of current against voltage.

Voltage (V)	Current (A)	Resistance (Ω)
2.0	0.50	
4.0	0.92	
6.0	1.30	
8.0	1.60	
10.0	1.84	
12.0	2.00	

D Electricity and life

Luigi Galvani was an Italian scientist who worked in the late 18th century. At the time, little was understood about electricity. People performed many experiments with static electricity, and they were able to give each other strong electric shocks. But they didn't really understand what was going on – they didn't know about electric charge.

Galvani was one of many people at the time who thought that electricity was vital to life. He imagined that there was electricity in our bodies that made us alive, so he set out to perform electrical experiments on biological materials.

Figure 9.26 shows one of the ways he did this. In this famous experiment, he used frogs' legs. He touched a foot and the spine with two metal rods, one silver and the other brass, and then brought the metal rods together. The leg twitched as the muscles were affected by electricity. Galvani believed that it was the legs themselves that generated the electricity, and this he called 'animal electricity'.

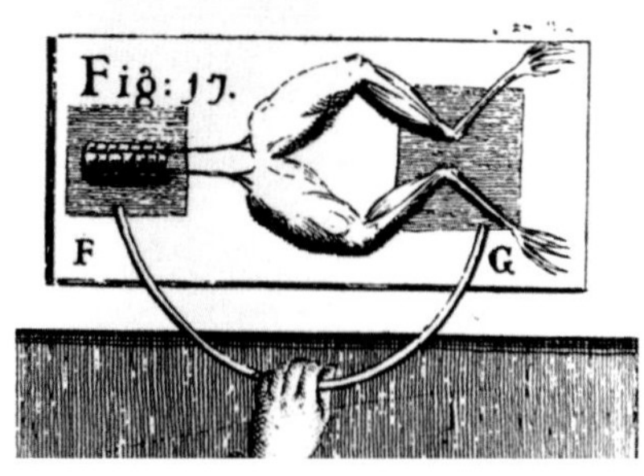

Figure 9.26 Galvani's experiment on frogs' legs. The silver rod is on the left, the brass on the right

We now know that Galvani was wrong. An alternative idea was proposed by Alessandro Volta, the Italian scientist after whom the 'volt' is named. He realised that it was the two different metals which generated the electricity, not the frog's legs. He published his ideas in 1800, and at the same time showed how to make a type of battery that could be used to supply electricity for many different experiments. His 'pile' consisted of alternating copper and zinc discs, with discs of paper soaked in acid in between (Figure 9.27).

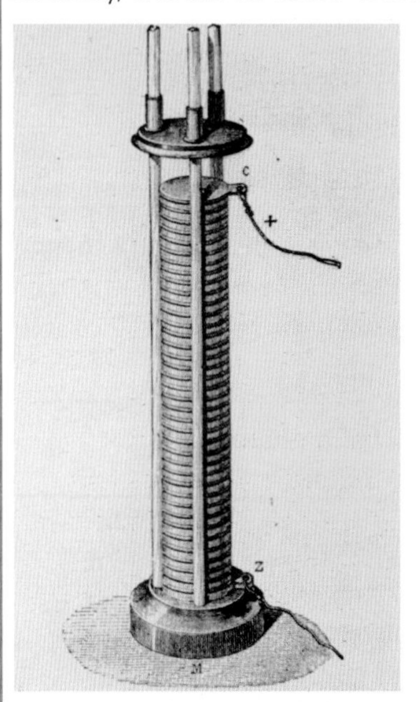

Figure 9.27 Volta's electric pile, the first electrical battery

The work of Galvani and others led many people to believe that it was electricity that made living matter different from dead matter. The writer Mary Shelley knew about these ideas when in 1818 she wrote her novel *Frankenstein*, in which a scientist constructs a body from parts of corpses and brings it to life using electricity.

Although it is true that electricity plays an important role in the human body, experiments like those of Dr Frankenstein are much more likely to lead to the death of the experimenter than to the resurrection of a dead body.

● Electricity inside you

There are electric currents at work in our bodies. An electrical signal travels along a nerve (see page 40) – it is the quickest way to send a message from the central nervous system to other organs in the body. Electrical signals can also help doctors to know what's going on inside the body.

Think of an electrocardiogram, which is a way of monitoring the electrical activity of the heart (Figure 9.28). The patient has several electrodes connected to the upper body. The electrodes pick up the changing voltage at different points in the body, and these are displayed as a chart. A trained nurse or doctor can tell if the heart is functioning satisfactorily or not.

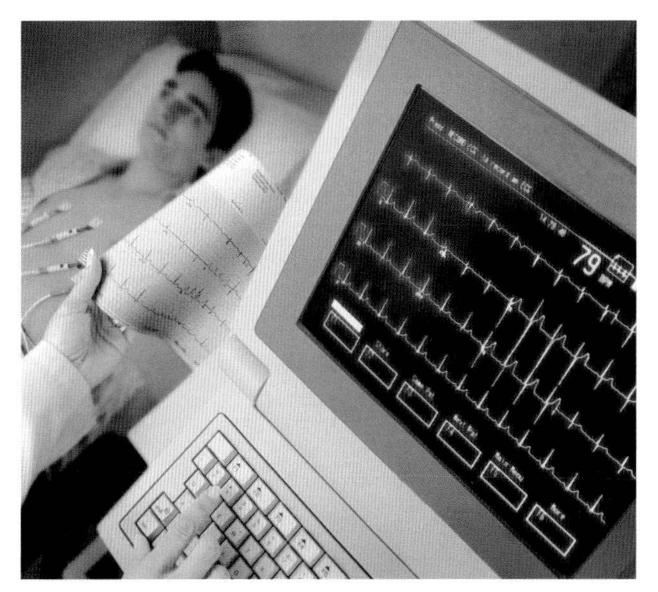

Figure 9.28 An electrocardiogram charts the electrical activity of the patient's heart

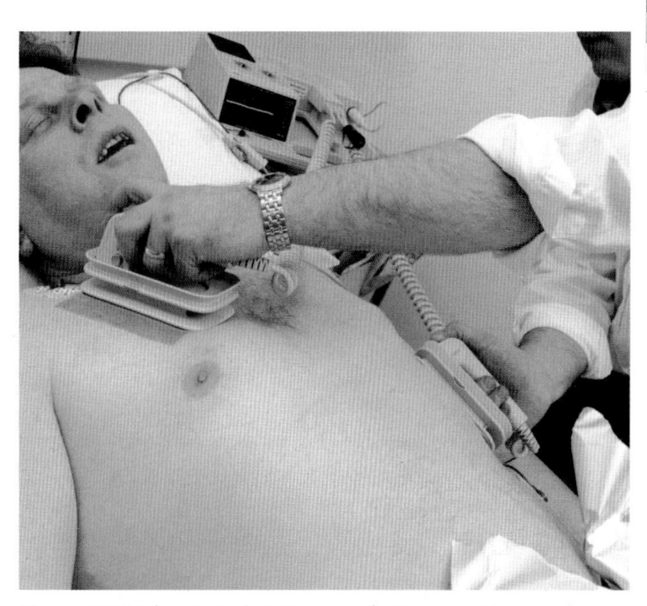

Figure 9.29 A heart attack victim – an electric current is passed through the heart to try to jolt it into normal activity

The heart normally works in a steady, rhythmical way. A heart attack can destroy this rhythm, and that's where electricity can help again. Look at Figure 9.29. You may have seen this done in a TV medical drama; metal plates are placed on the patient's chest, everyone stands clear, and the patient is given an electric shock. The idea is that this should jolt the heart so that it restarts with a regular rhythm once more. It usually works on the TV but, in real life, only a small proportion of patients is successfully revived in this way.

Scientific ideas of electricity and the body have developed a great deal since the days of Galvani and Volta, and they are still developing. For example, the brain is an area of great interest because its working depends on electrical signals. Scientists and doctors are trying to find a way of curing some medical conditions (such as Parkinson's disease and epilepsy) using electricity. Small electrodes may be implanted in the brain and fed with tiny electric currents. This approach is still at a rather crude stage, because we don't know enough about the way the brain works. But in science it is often necessary to start in this basic way and then hopefully more subtle techniques can be developed as our understanding grows.

Changing times

H

Our growing scientific understanding of electricity has brought great changes to our lives. In the 20th century electrical technology developed very rapidly. As a result of scientists studying phenomena, the new understanding of what is going on can allow technologists to come up with new inventions. For example, research into superconductivity has led to the development of very powerful superconducting electromagnets.

Table 9.4 shows just some of the developments in our understanding of electricity and the applications related to them. In the next section we will look at one of these applications – how we can generate electricity using a moving magnet.

Scientific phenomenon	What it is	Technological applications
Heating effect of current	A current loses energy as it passes through a resistance, and generates heat	Heating, lighting
Electric motor effect	A coil carrying an electric current in a magnetic field is made to move	Transport, pumps, car starter motors, fans, CD players, loudspeakers, many other devices
Electromagnetic induction	A current flows when a conductor is moved in a magnetic field	Generators, dynamos, microphones, telephones
Radio waves	A rapidly varying current in an aerial produces electromagnetic waves	Radio and TV broadcasting, radar, mobile phones
Superconductivity	Some substances have no electrical resistance at all at low temperatures	Powerful electromagnets, MRI scanners, Maglev trains
Semiconductor behaviour	Some substances are half way between conductors and insulators	Transistors, computer chips, microprocessors, lasers

Table 9.4 Electrical technology

Figure 9.30 This Maglev (magnetically levitated) train is supported above its track by powerful electromagnets. Some Maglevs use superconducting electromagnets. It is roughly 100 years since superconductivity was discovered, but it is only now, with sophisticated computing power available, that Maglev trains can be designed and built

The dynamo

We began this chapter with a close look at batteries and how they are used to supply electricity. Most of the electricity that we use, however – mains electricity – is produced in a different way, using generators in power stations. If you have a bike with a dynamo, you have a miniature version of a generator.

The two important parts of a dynamo are:

- a magnet, which turns as the bicycle wheel turns
- a coil of wire, which is connected to the bicycle's lights.

As the magnet turns round, it makes an electric current flow in the coil and round the circuit to the lights. The faster it turns, the bigger the current. When the magnet stops turning, the current stops – this is the disadvantage.

Bicycle dynamos are now quite rare. Most cyclists use battery-powered lights these days.

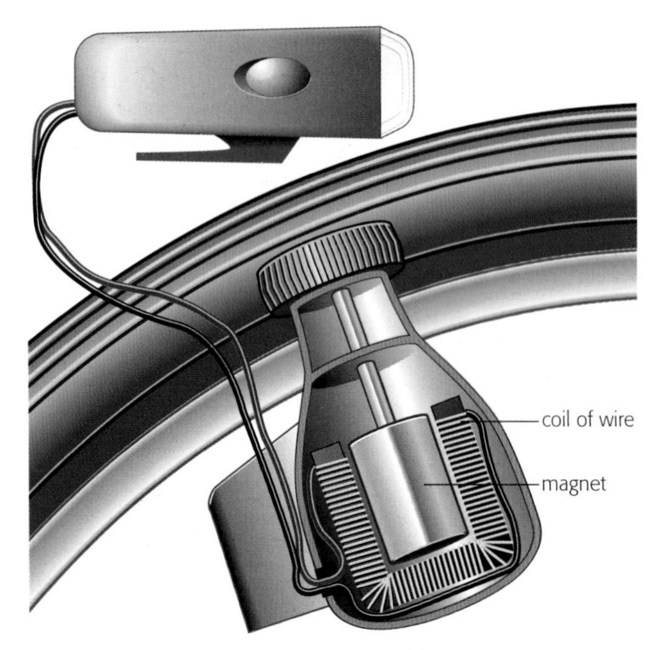

Figure 9.31 A bicycle dynamo

How a dynamo works

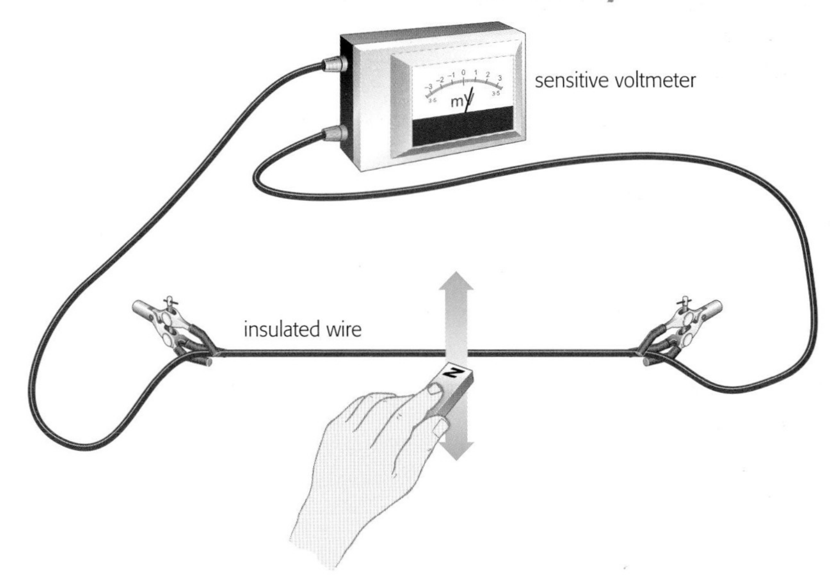

Figure 9.32 Inducing a voltage in a wire

To understand how a dynamo works, we need to start with something a bit simpler: a wire connected to a voltmeter, and a magnet. Look at Figure 9.32. If the magnet is moved up and down next to the wire, the needle on the voltmeter moves back and forth. We say that there is an **induced voltage** in the wire.

Observing the effect in more detail:

1 Move the magnet upwards: the needle moves to the right – a positive voltage has been induced.
2 Move the magnet downwards: the needle moves to the left – now a negative voltage has been induced.
3 Hold the magnet still: the needle returns to zero – no voltage is induced.

This last observation shows that *the magnet must be moving* to induce a voltage. Here are some further observations:

4 Move the magnet faster: the induced voltage is greater.
5 Turn the magnet round and use the other pole: the induced voltage is reversed.
6 Move the wire instead of the magnet: the meter shows that a voltage is induced.

It was Michael Faraday who first discovered how to induce a voltage in a wire using a moving magnet. He realised that it was better to use a coil of wire than a single strand – each turn of wire is affected by the magnet, and the induced voltage is much bigger. (That's why you have to use a sensitive voltmeter when you only have one strand of wire.)

Figure 9.33 Inducing a voltage in a coil

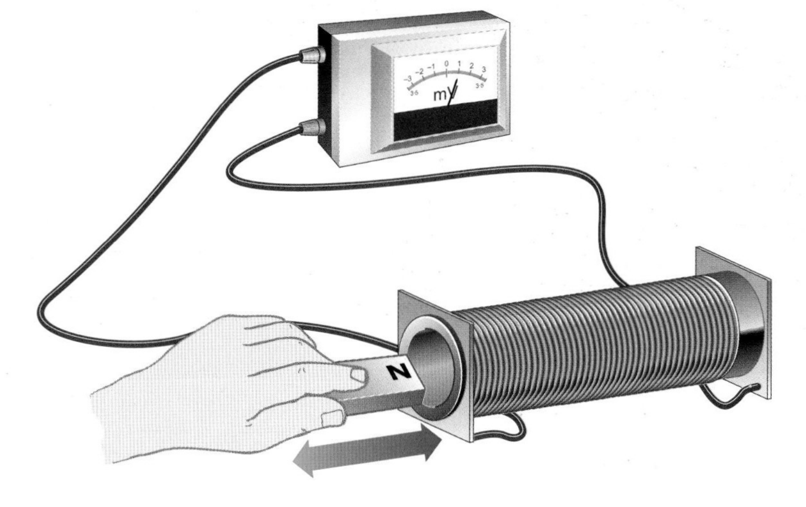

Here is how we think of the process of inducing a voltage in a wire. Picture the magnetic lines of force coming out of the magnet's pole. When the magnet is moved past the wire, the lines of force are cut by the wire. It is this cutting that makes a current flow in the wire. The faster the lines of force are cut, the bigger the induced voltage.

Figure 9.34 Michael Faraday (1791–1867) gave popular scientific lectures on electricity and magnetism. His audience here includes many young people – this was the origin of the Royal Institution Christmas Lectures

Here are the rules Faraday discovered about the factors which affect the size and direction of the induced voltage.

To increase the induced voltage:

- move the magnet faster
- use a stronger magnet
- use a coil of wire with more turns.

To reverse the direction of the induced voltage:

- move the magnet in the opposite direction
- use the opposite pole of the magnet.

Michael Faraday was a scientist. He was fascinated by the phenomena that he was studying, and he thought of simple ways to investigate them. At the same time, he was aware that his discoveries would soon have technological applications, although he couldn't have imagined the part that induced voltages would play in our lives. Today, large generators in power stations use the principles discovered by Faraday to generate the electricity used in homes, offices, schools, hospitals, shops and factories. A single generator can produce enough electricity to power thousands of homes.

Summary

- **Our scientific understanding of electricity has made a great many applications possible: in health, the home, travel, communications, and so on.**
- **In particular, a magnet spinning inside a coil of wire forms a dynamo, which can be used to generate electricity.**
- **The voltage induced in a generator can be increased by increasing the speed of rotation, the strength of the magnetic field, and the number of turns on the coil.**
- **The induced voltage can be reversed by reversing the magnetic field or the direction of motion.**

Questions

9.16 A dynamo can be used to generate electricity.

  a Which part of the dynamo spins around?

  b Which part of the dynamo is fixed?

  c In which part does the electric current flow?

9.17 A student is investigating what happens when a bar magnet is moved into a coil of wire. The ends of the wire are connected to a sensitive voltmeter. When she pushes the north pole of the magnet into the coil, the needle of the meter shows that a positive voltage has been induced.

  a Suggest two ways in which the student could produce a bigger deflection of the needle.

  b Suggest two ways in which the student could make the needle deflect in the opposite direction.

9.18 A car has a battery to power its lights and the starter motor. It also has a generator. What is the purpose of this?

You're in charge

A Electricity and the modern world

Large areas of the country woke up to find their electricity supply had failed this morning. The recent cold snap resulted in a sudden surge in demand which the distribution system could not cope with. Traffic lights stopped working, causing traffic chaos during the morning run to work. Hospitals switched to emergency generators to keep their operating theatres at work.

A spokesman for PowerNorth told us, 'A few power lines are down. Engineers are working on this as we speak. We hope to have the system up and running again by this evening.'

Meanwhile, over 100 passengers remain stranded aboard an underground train.

Figure 10.1

We rely on electricity in many ways. Think about the electrical systems in a typical home:

- lighting
- heating, hot water
- telephone, computer, internet
- TV, radio, other entertainment systems
- fridge, freezer
- cooker, microwave oven

and so on. Even a gas-fired central heating system relies on an electric pump.

Without a reliable electricity supply, our lives would be very different. Just a century ago, most of these desirable electrical systems were not available to most people in the UK – and they are still unavailable to many people around the world.

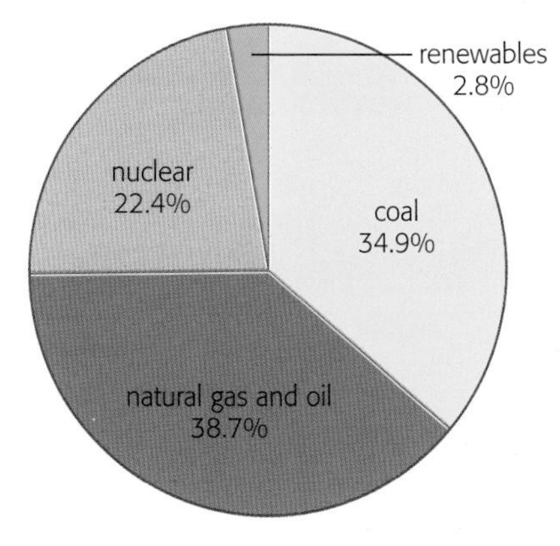

Figure 10.2 UK electricity production in 2003

Generating power

Where does our electricity come from? The simple answer is, from power stations. However, there are a great variety of these, using different energy sources and different technologies. Figure 10.2 shows the proportions of our electricity supply that come from different types of power station.

You can see that most of our electricity is generated in power stations that burn fossil fuels – mostly natural gas and coal. Roughly one-fifth is generated in nuclear power stations. That leaves just a small fraction that is generated from **renewable sources** such as wind and hydro power.

The UK Government has been encouraging the electricity supply companies to make more use of renewable energy resources, and the number of wind turbines has been increasing rapidly in recent years. There are several reasons for this push for renewables:

- One day, fossil fuels will run out – in particular, gas and oil reserves are limited.
- Burning fossil fuels produces harmful waste gases. Carbon dioxide adds to the greenhouse effect, and other waste gases cause acid rain.
- Nuclear power stations produce highly radioactive waste.

Figure 10.3 Our old power stations are harming the environment

Nuclear fuels as well as fossil fuels are non-renewable, so one day we will be forced to find alternatives. Because of the damage these fuels do to the environment, the sooner we switch to alternatives, the better. The question is, can renewables do the job?

Evaluating renewable sources

The main ways of generating electricity from renewable energy sources in the UK are:

- **Wind power:** The UK is an island nation, and it's quite windy. Winds blow off the Atlantic for most of the time, so that is quite a good resource. A wind generator consists of two main parts: the turbine, which is a set of blades that are made to turn by the wind, and the generator itself, which is turned by the turbine so that it generates electricity.
- **Solar power:** Some buildings are fitted with solar cells, usually on the roof. These generate electricity so long as there is light falling on them. The brighter the light, the greater the power they generate. At night they produce nothing, so it is usual to connect them to batteries which store energy during the day for use at night.
- **Hydroelectricity:** Many rivers in mountainous areas have been dammed. Water from the reservoir behind the dam flows past turbines inside the dam, and these turn a generator.
- **Tidal power:** As the sea's tides ebb and flow, large currents of water flow back and forth. Underwater turbines spin as the current flows past, and this turns a generator. This technology is made use of in Scandinavia, but it is still at the experimental stage in the UK.

- **Wave power:** The seas around the UK are wavy, as well as windy. These surface waves make a good energy resource. A wave generator is usually in several sections; as the waves pass by, the generator flexes up and down. This motion of the different sections is used to generate electricity.

Figure 10.4 Wind turbines (left) and a wave generator (right)

One of the attractions of these renewable energy resources is that they don't require any fuel. They are 'free' energy resources – but there are many other costs to be considered when deciding which, if any, to use. For example, is the technology reliable? Fossil fuels have served us well for a century – could we switch to renewables and still have a reliable electricity supply?

To **evaluate** the different possibilities, we can divide up the different factors that need to be considered under four headings – technical, economic, environmental and social.

Technical considerations

Suppose we decide to make more use of tidal power. The machines for doing this, like underwater wind turbines, have only been around for a few years. They will be improved over the next decades, but at present this is a young technology. So we always need to think, is the technology reliable? Is it sufficiently developed to make the best use of the resource?

Another technical question – is there enough of the resource? In the UK, most rivers that could be dammed without interfering with shipping are in the Highlands and are already the sites of hydroelectric schemes. These provide about 10% of Scotland's electricity, but we can't expect to get much more from this source.

Similarly, the wind is a rather spread-out resource. It takes hundreds of wind turbines covering many square kilometres to replace a single coal- or gas-fired power station. And what if the wind doesn't blow? We expect a continuous supply of electricity, so fossil-fuel power stations must be kept running, ready to supply electricity as a back-up to wind farms.

Economic considerations

Although there are no fuel costs, there are other financial costs to think about. There are the costs of building the power station in the first place, of maintaining the equipment, and of dismantling (decommissioning) it at the end of its life. And there's the cost of distributing the electricity from the site of generation.

Environmental considerations

We have already mentioned the pollution which can arise from burning fossil fuels and from using nuclear fuels. Renewables don't create these same problems, but we should remember that there are waste materials produced when, for example, solar cells are manufactured, and when they are disposed of.

Electricity generation can have other impacts on the environment. Power stations, even wind turbines, are thought of by some people as ugly and noisy. They may spoil an attractive area of countryside, they may disturb the natural habitats of wildlife, or they may simply take up a lot of space.

Social considerations

Some power stations can be a security risk. Nuclear stations, especially, could be a terrorist target. The materials they use and the waste they produce might also be a problem – they might be used in bombs, for example.

Employment is something else to think about. If we changed to an electricity supply system based on renewables, would this lead to more jobs, or more unemployment?

Table 10.1 A table is a good way of evaluating two or more alternatives

	Wind power	Solar cells
Technical considerations	Turbines catch only a fraction of the available energy of the wind. Wind doesn't blow all the time, so a back-up supply is needed.	Solar cells catch only a fraction of the available energy of sunlight. The Sun doesn't shine all the time, so battery back-ups are needed.
Economic considerations	High initial cost; low running costs; no fuel costs.	Very high initial cost; low running costs; no fuel costs.
Environmental considerations	Some people object to wind farms in open countryside. They occupy a lot of space, and some say they are noisy.	A large area of cells is needed (several m^2 per person). There is no noise. Hazardous wastes are produced during manufacture.
Social considerations	No security problems. Jobs made in manufacturing turbines.	No security problems. Jobs made in manufacturing solar cells.

● Counting the costs

It's quite easy to see the advantages of renewable energy resources for generating electricity. However, these have to be balanced against any negative considerations.

To make a fair comparison, you really need to be able to work out the **lifetime costs** of each method of generating electricity – in money terms, this is original cost, running costs including fuel, and decommissioning costs. You also need to know how much electricity will be produced for this amount of money. Then you can work out the cost of each unit of electricity.

The chart in Figure 10.5 shows the outcome of a study comparing the cost in money terms of generating electricity from some of the possible sources. It is much harder to include the environmental and social costs.

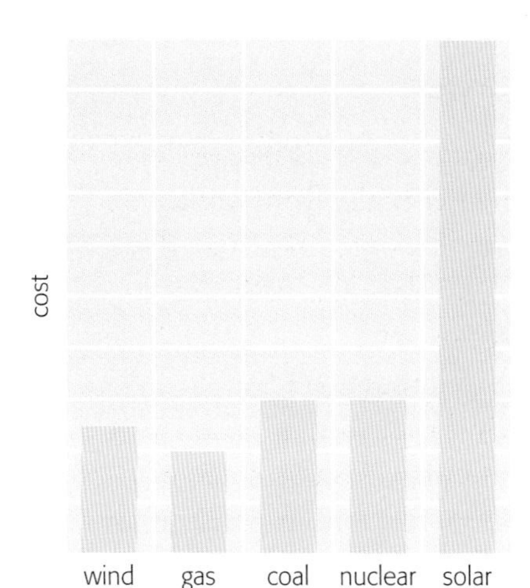

Figure 10.5 The relative cost of generating electricity from different sources

You can see that natural gas is the winner – electricity produced from gas is the cheapest. So why are new wind farms being built?

It is Government policy to generate more electricity from renewable resources, with a target of about 20% of all electricity coming from renewables by the year 2020. This is mainly to meet targets to reduce greenhouse emissions. By encouraging wind farms, the idea is that more money will be invested in developing better, more efficient wind turbines that will produce more electricity for the same money. Then the price of electricity from wind power will fall. It also means we will be better prepared for the day when fossil fuels become scarcer and more expensive.

It is also clear from the bar chart that solar cells will have to become a lot cheaper if they are to be a viable alternative for large-scale renewable energy.

Choosing technology

In an industrialised country like the UK, there is a national electricity supply system. A large number of power stations feed electricity into the Grid, which is the network of cables that criss-crosses the country. In the UK, these cables are carried across the countryside on long lines of pylons – over 22 000 of them. In towns, the cables are usually buried underground.

Figure 10.6 Cables and pylons in the UK countryside

Pylons and cables are ugly and can spoil rural views. Would it not be better to put all the cables underground? There is a choice to be made between overground and buried cables.

How can we decide? It seems an attractive idea to bury cables – the obvious advantage is that there is less damage to the landscape. Another advantage is that the cables are less likely to be damaged in a storm. However, there are disadvantages, too.

Figure 10.7 In many US towns, cables are slung between poles in the street rather than being buried underground as in UK towns

H

Figure 10.8 At the end of a line of pylons, the power cables disappear underground on the way to their final destination

- It costs perhaps 40 times as much to bury the cables.
- If anything goes wrong, it is difficult to find and repair the fault. Consumers would be without power for much longer, and there might be severe disruption as cables were dug up and repaired.

So a balance must be struck. How much do we value an attractive landscape? Should we be prepared to pay to protect that landscape, both in financial terms, and in terms of possible disruption if faults arise?

We often have to make choices like this between two competing technologies. Sometimes we can decide simply by comparing financial costs; more often, we have to balance these costs against other considerations, such as environmental concerns.

Summary

- **In order to evaluate new technologies (such as renewable energy resources), we need to take account of technical, economic, environmental and social considerations.**
- H **When considering whether or not to implement a new technology, we need to be able to weigh up its benefits and drawbacks, including costs.**

Questions

10.1 **a** Name four renewable energy resources from which electricity can be generated.

b Name four non-renewable energy resources from which electricity can be generated.

10.2 Give as many reasons as you can why it is desirable to switch to electricity generation from renewable resources.

H **10.3** Tidal power is a renewable energy resource from which electricity could be generated in the future. Under the headings *Technical*, *Economic*, *Environmental* and *Social*, list aspects of tidal power that should be taken into consideration when deciding whether it should play a part in our future electricity supply.

You could give your answer in the form of questions, for example: 'Environmental: Will the turbines affect fish and other sea creatures?'

B Electric motors

Peugeot is a car manufacturer with a large factory in Coventry. They are looking to the future by designing electric cars. The cars have a large battery under the bonnet; this powers electric motors which turn the wheels. The idea is that the use of battery-powered cars, particularly in towns and cities, will reduce the amount of polluting gases in the air. Many people develop respiratory problems because they are exposed to the exhaust fumes of petrol and diesel cars.

In an experiment in Lambeth, South London, owners of Peugeot electric cars have recharging points outside their homes. Provided they don't travel too far from home, they can recharge whenever they like.

Figure 10.9 Recharging an electric car at a special roadside power point in South London

Of course, the electricity used to charge the car's batteries must come from somewhere. If it is generated in a fossil-fuel power station, then waste gases are produced there. However, if we can eventually replace fossil fuels with renewable energy resources, we will end up with a much greener transport system.

Electric transport isn't completely new – there have been electric trams for over 100 years.

Motors everywhere

We use electricity because it makes our lives easier. Think about washing clothes: you could do this by hand, but it's much easier to use a washing machine. An electric motor does the hard work for you. It turns the drum back and forth to wash the clothes, and then it spins the clothes to remove most of the water.

Electric motors have so many uses, that there are far more motors than people in the UK! Figure 10.10 shows some familiar places where motors are used.

Figure 10.10 Motors have many uses

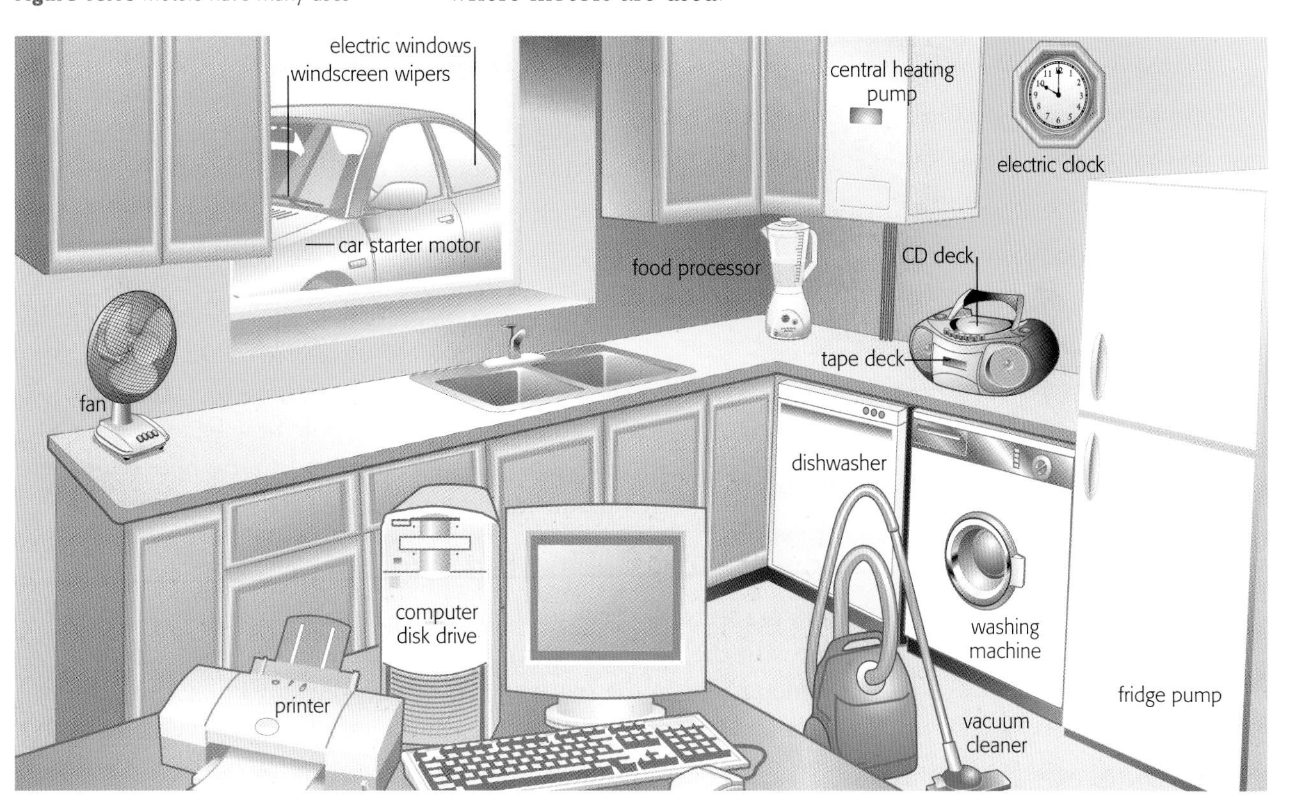

Inside a motor

Figure 10.11 A cutaway view of an electric motor

The photo shows the inside of an electric motor. It's rather complicated, but you can see the most important parts:

- At the centre there is a shiny silver **coil** of wire – this acts as an electromagnet.
- Around the coil are some **magnets** (coloured blue).
- At front right there are **connections** for the electric current to flow in and out of the motor.

When a current flows through the coil of the motor, it turns. You can see the motor's axle sticking out on the left – that's the part that spins round. You have probably used small electric motors similar to this in Technology lessons.

How it works – explanation 1

> Recall that an electromagnet is a coil of wire with an electric current flowing through it. One end of the coil acts as a north magnetic pole, the other as a south magnetic pole.

There are several ways of thinking about how an electric motor works – we'll look at two. First, we will use ideas about electromagnets.

Figure 10.12 shows a very simple electric motor. In the centre is the coil of wire. There are two magnets on opposite sides of the coil. (You may have the opportunity to make a model motor like this.)

How does the electric current get into and out of the coil? It does this through the **commutator**. This you can see is in two sections, with the two ends of the coil connected to them. Two **'brushes'** (really springy pieces of metal) press on to the commutator sections. Current from a battery flows in through one brush, around the coil, and out through the other brush.

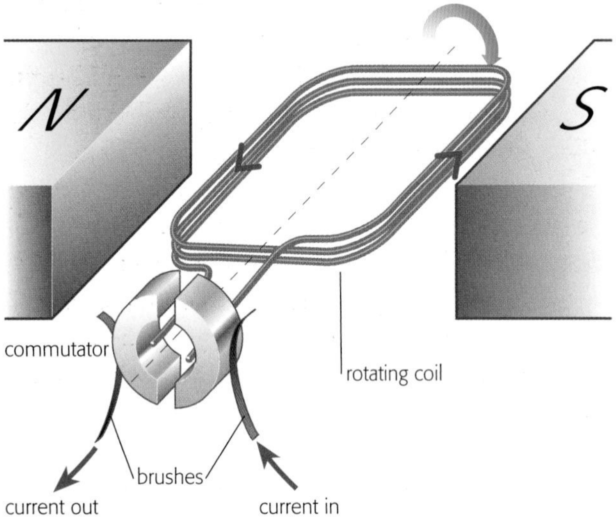

N
S
commutator
rotating coil
brushes
current out
current in

Figure 10.12 A simplified electric motor

Why does the coil spin round? In the position shown, the coil is an electromagnet with its north magnetic pole (N) on top. This is attracted round towards the S pole of the field magnet, and repelled by the N pole of the other field magnet. So the coil starts to turn.

Now, here's the clever part. The commutator rotates with the coil. It is designed so that, as the coil turns, its connections to the battery are swapped over. With every half-turn, the current is made to flow the opposite way round the coil, so that its magnetic poles are reversed. The side of the coil that was the N pole now becomes the S pole. This is attracted round to the N pole of the field magnet, so the coil keeps spinning round.

How it works – explanation 2

Here is a second way of explaining how an electric motor works. We start with the idea that, when an electric current flows, it creates a magnetic field in the space around it. (You already know this – think about an electromagnet; it's just a current flowing in a wire, producing a magnetic field. The wire is coiled up to concentrate the field.)

Figure 10.13 shows what happens when a current in a wire is placed between two magnets. The wire is pushed out of the magnetic field. The reason is that the magnetic field produced by the current is repelled by the magnetic field of the magnets.

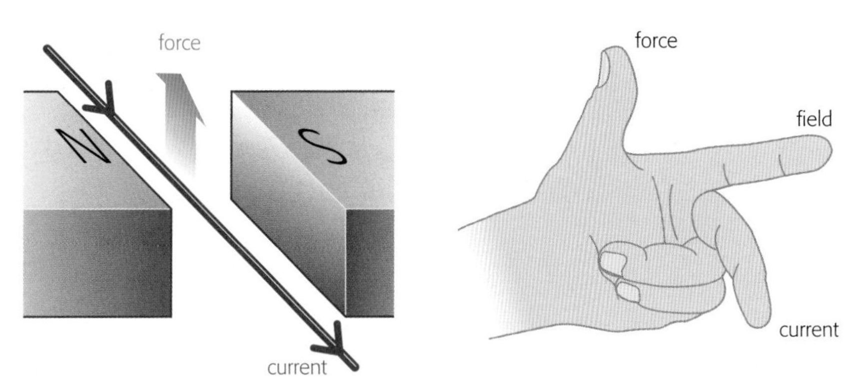

Figure 10.13 Fleming's left hand rule for the force on a wire in a magnetic field

There are three things here whose directions are important:

- the current
- the magnetic field of the magnets
- the force on the current.

From the diagram, you can see that these are all at right angles to each other. An electrical engineer called Ambrose Fleming devised a rule, **Fleming's left hand rule**, which shows the relative directions of these. Arrange the thumb and first two fingers of your left hand at right angles to each other. Then they show the relative directions of the force, field and current, as shown in Figure 10.13.

Now we can apply this rule to the electric motor. We will think about the coil as a simple rectangle, to clarify things. Look at the two long, straight sides of the coil in Figure 10.14. In one, the current is flowing towards the left; Fleming's left hand rule tells us that it will feel a force upwards. In the other side of the coil, the current is flowing towards the right and it will feel a force downwards. These two forces cause the coil to turn.

As in the first explanation, the commutator reverses the direction of the current in the coil every half-turn, so the coil keeps on turning in the same direction.

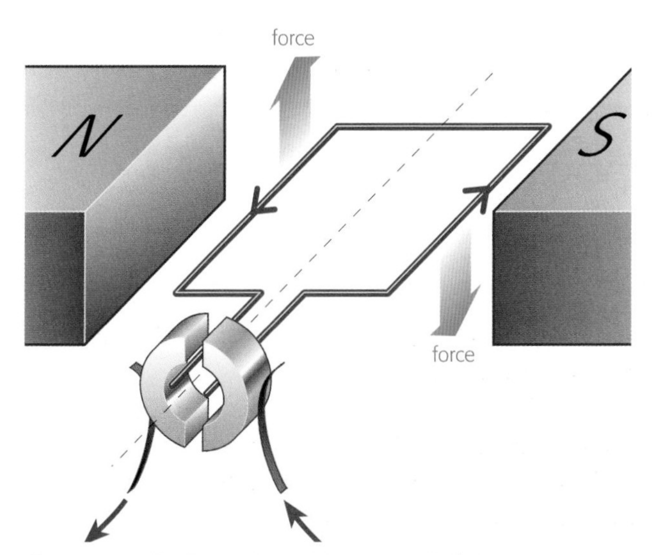

Figure 10.14 The forces on a coil in a magnetic field

Summary

- In an electric motor, a current flows in a coil of wire.
- A magnetic field produces forces which cause the coil to turn.
- The commutator reverses the direction of the current every half turn, so that the coil keeps turning in the same direction.

Questions

10.4 Name these four parts of an electric motor:

a a device to reverse the direction of the current automatically

b provide a magnetic field

c two devices which lead the current into the rotating part

d acts as an electromagnet

10.5 Use Fleming's left hand rule to decide the direction of the force on the current-carrying wires in diagrams **a** and **b** in Figure 10.15.

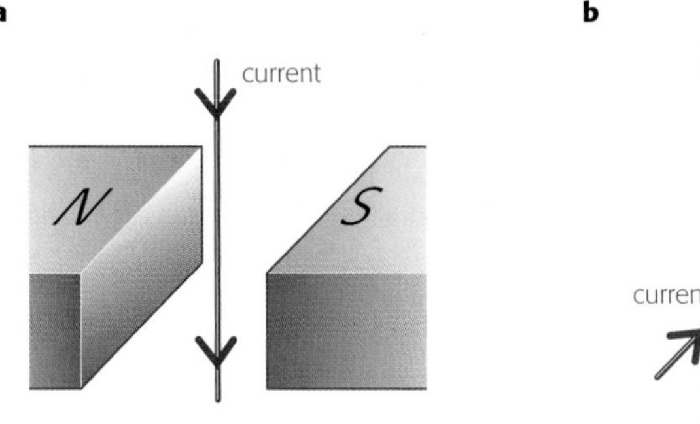

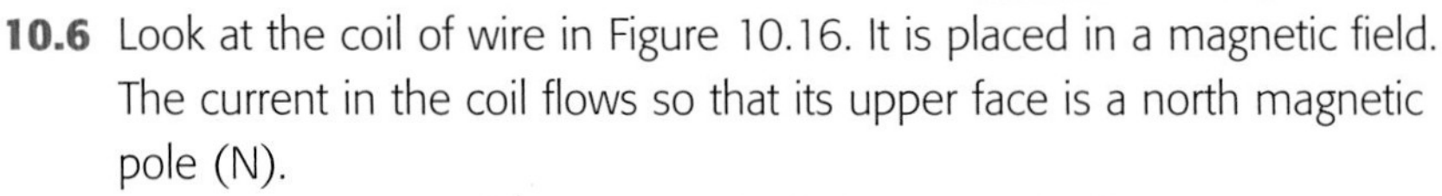

Figure 10.15

10.6 Look at the coil of wire in Figure 10.16. It is placed in a magnetic field. The current in the coil flows so that its upper face is a north magnetic pole (N).

a Which way will the coil turn, clockwise or anticlockwise? Make a sketch to show your answer.

b Add force arrows to the coil in your sketch to show the forces that cause it to turn.

c Explain what would happen if the current flowed through the coil in the opposite direction.

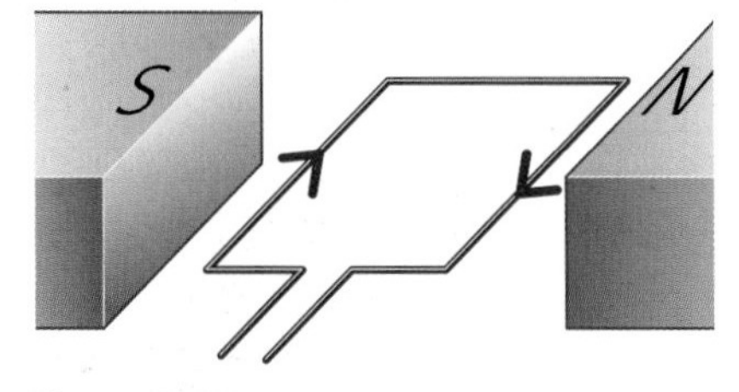

Figure 10.16

C Paying for power

Every year, about 1500 households in the UK have their electricity supply disconnected. This usually happens because the homeowner has failed to pay the bill.

There are certain categories of people who cannot be disconnected, however – the elderly and the disabled, families with young children, and people on benefits. This is an agreement between the electricity companies, following distressing cases where vulnerable pensioners were disconnected and who then lived for months without electricity. But people who regularly fail to pay bills may have their meter replaced by a pre-payment meter. They have a card or 'chip-holder', which they must pay to 'charge up' at a shop or post office.

Electricity prices depend on the price of oil and gas. These started rising in 2003–4 and rose further in 2005. With increasing demand for oil, it is likely that electricity will become yet more expensive, so it is worth knowing which appliances in your home use most electricity, and thinking about how to reduce your consumption by wasting less.

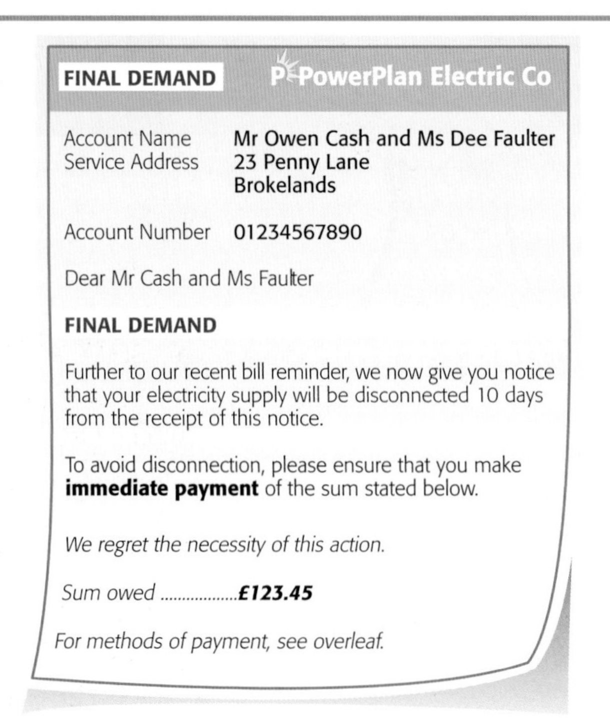

FINAL DEMAND PowerPlan Electric Co

Account Name Mr Owen Cash and Ms Dee Faulter
Service Address 23 Penny Lane
Brokelands

Account Number 01234567890

Dear Mr Cash and Ms Faulter

FINAL DEMAND

Further to our recent bill reminder, we now give you notice that your electricity supply will be disconnected 10 days from the receipt of this notice.

To avoid disconnection, please ensure that you make **immediate payment** of the sum stated below.

We regret the necessity of this action.

Sum owed ***£123.45***

For methods of payment, see overleaf.

Figure 10.17 A final warning

Figure 10.18 A rating label from an electric fan heater

Read the label

Most electrical goods have a label that gives important technical information. This includes the correct voltage of the supply, whether it should be direct current or alternating current (and the frequency of the alternating supply), and so on.

To find out about the energy demands of the appliance, look for its **power rating**. This is given in watts (W) or kilowatts (kW). This tells you how fast the current delivers energy to the appliance.

1 watt = 1 W = 1 joule per second
1 kilowatt = 1 kW = 1000 W

In other words, the **power** of an appliance is the rate at which electrical energy is transferred to it. The appliance then changes this energy into other forms.

electrical power = rate of transfer of electrical energy

$$\text{power} = \frac{\text{energy transferred}}{\text{time taken}}$$

For example, a 60 W light bulb uses 60 J of electrical energy each second.

Appliances that produce a lot of heat are the ones which tend to have the highest power rating. See Figure 10.19.

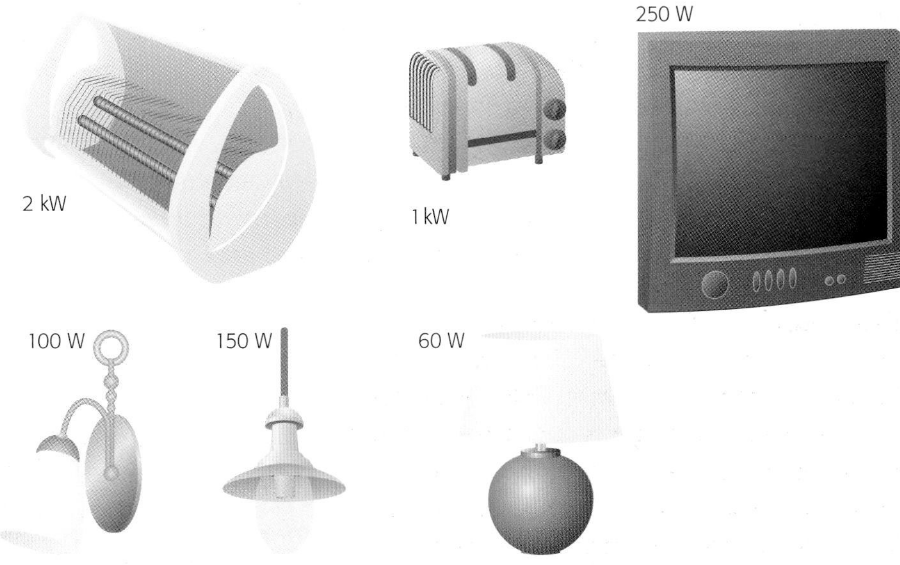

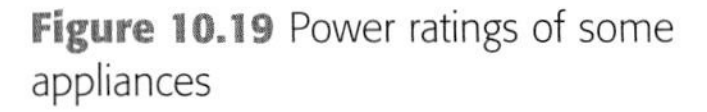

Figure 10.19 Power ratings of some appliances

Current, voltage and power

The power of an appliance depends on two quantities:

- the current flowing through it (in amps, A)
- the voltage pushing the current through (in volts, V).

Both of these combine to give the value of the power:

power = current × voltage

Why are both current and voltage involved? You should recall that current tells you how much *charge* flows each second, and *voltage* tells you how much *energy* is carried by each unit of charge. Together, they tell you how much energy is carried each second, i.e. the power.

Worked example 1

What is the electrical power of a lamp if a current of 0.25 A flows through when connected to a 240 V mains supply?

Here, we are given the current and the voltage. We can simply multiply them together to find the power.

Power = current × voltage = 0.25 A × 240 V = 60 W

So the power of the lamp is 60 W.

H

Worked example 2

What current flows through a 120 W television set when connected to a 240 V mains supply?

Here we need to find the current, so we must rearrange the equation for power:

power = current × voltage

current = power / voltage

So current = 120 W / 240 V = 0.5 A

A current of 0.5 A flows through the television.

Small units

For some components, such as the electronic components in a computer, the voltage involved may be very small, smaller than a volt. In this case, the voltage may be measured in millivolts (mV).

1 millivolt = 0.001 V   1000 mV = 1 V

Similarly, very small currents are measured in milliamps (mA).

1 milliamp = 0.001 A   1000 mA = 1 A

Figure 10.20 You probably have a meter like this at home

● Read the meter

When we pay for electricity, we are paying for the energy it brings us. Somewhere, coal is burning and steam is turning a turbine, or the wind is making a wind turbine spin. Almost instantly, that energy is available to us. As we use the energy of electricity, a meter records how much we use.

The meter is usually positioned close to the 'fuse box' (see page 213), where the electricity comes in to the building. Alternatively, it may be in a white box on the outer wall of the house or flat, where it is easier to read. In older meters, you can see a metal wheel slowly turning. The more appliances you have switched on, the faster it turns – it is really a kind of electric motor connected up to a series of dials, which record your how much electricity you use.

A meter doesn't record your energy consumption in joules, because you probably use several million joules every hour, on average! It uses a more convenient unit, the **kilowatt-hour**.

1 kilowatt-hour (1 kWh) is the energy used when you run a 1 kW appliance for 1 hour

So if you run a 2 kW appliance for 5 h, you use 10 kWh, because 2 kW × 5 h = 10 kWh.

We can write a general equation:

energy used (kWh) = power (kW) × time (h)

1 kilowatt-hour is sometimes known as a **unit** of electricity.

Note this is the same as the equation we had earlier:
power = energy / time
so energy = power × time

Paying the price

We pay for the number of units (kWh) of electricity we use, as recorded by the electricity meter. At regular intervals, the meter is read and the number of kWh used is calculated by comparing with the previous reading. The electricity company sets the cost of each kWh, from which the cost of the electricity used is calculated:

cost = number of kWh used × cost per kWh

Worked example 3

In the course of a week, a 2 kW heater is used for 60 hours. If each kWh of electricity costs 8p, what is the total cost of this?

Cost = number of kWh used × cost per kWh

= power (kW) × time (h) × cost per kWh

= 2 kW × 60 h × 8p = £9.60

Summary

- **Electrical power = rate of transfer of electrical energy = energy transferred / time**
- **Electrical power = current × voltage**
- **Power is measured in watts (W) or kilowatts (kW).**
- **Electrical energy is measured in joules (J) or kilowatt-hours (kWh).**
- **Cost of electricity = power (kW) × time (h) × cost per kWh**

Questions

10.7 A heater is labelled '1.5 kW'.

a What quantity does this indicate?

b How much energy does the heater transfer in 1 s? (Give your answer in joules, J.)

c How much energy does it transfer in 1 h? (Give your answer in kWh.)

10.8 An electric motor has a current of 3 A flowing through it when it runs off a 100 V supply. What is the electrical power of the motor?

10.9 A rented holiday cottage has an electricity meter which the owner reads at the beginning and end of each stay. Figure 10.21 shows the readings on the meter before and after a week-long stay.

a How many kWh of electricity have been used?

b If the cottage owner charges 15p per kWh, what will the bill be?

c If the owner pays the electricity company 8p per unit, how much profit has he made?

Before:

0006736

After:

0006866

Figure 10.21

10.10 A 4 kW heater is used for 2 hours.

a How many kilowatt-hours of energy does it use?

b If each kilowatt-hour costs 8p, what is the cost of using the heater?

10.11 A current of 0.40 A flows through a lamp when it is running from the 240 V mains supply. What is the electrical power of the lamp?

10.12 An electric kettle rated 2 kW runs from a 240 V mains supply.

a How much energy does the kettle transfer in each second?

b What current flows through the kettle element when it is switched on?

10.13 A student leaves a 150 W lamp on for 10 hours while he is out of his room. How many kWh of electricity has he wasted?

10.14 The table gives information on four different electrical appliances. Copy and complete the table.

Appliance	**Operating voltage (V)**	**Operating current (A)**	**Power (W)**
cooker	240		7000
motor	200	3.5	
car headlamp		3.0	36
torch	4.5		6.0

D Energy efficiency

There is a lot of debate about four-wheel-drive cars, which have become fashionable over the last few years. They tend to be big, and they are often criticised for consuming too much fuel. Cars that might seem more at home driving around in rough conditions on a country estate are now a common sight on the school run in busy cities. In America, these cars are often described as 'gas guzzlers'.

All new models of car are put through a standard test procedure to determine their fuel efficiency. The idea is to determine how much fuel they use when driving in typical urban and motorway conditions. The results are published so that car buyers can compare different makes and models. The figures are given using two different units:

Figure 10.22 A 'gas guzzler'

- miles per gallon (mpg) – how many miles the car will travel on one gallon of fuel
- litres per 100 km (l/100km) – how many litres of fuel are consumed for every 100 km travelled.

Motor engineers have designed increasingly efficient car engines, which will give more miles (or kilometres) for a given quantity of fuel. Other design features also contribute – a more streamlined shape is more fuel-efficient. Four-wheel-drives tend to be heavy and not very streamlined, and these factors contribute to their wastefulness. Surprisingly, though, test figures show that many are more efficient than some more conventional cars – so they cannot all be condemned out of hand.

Making more of energy

For lots of reasons, it makes sense to get as much as possible out of the energy we consume. We want a car to take us as far as possible, using as little fuel as possible – so we should choose one with the best **fuel efficiency**. What about light bulbs?

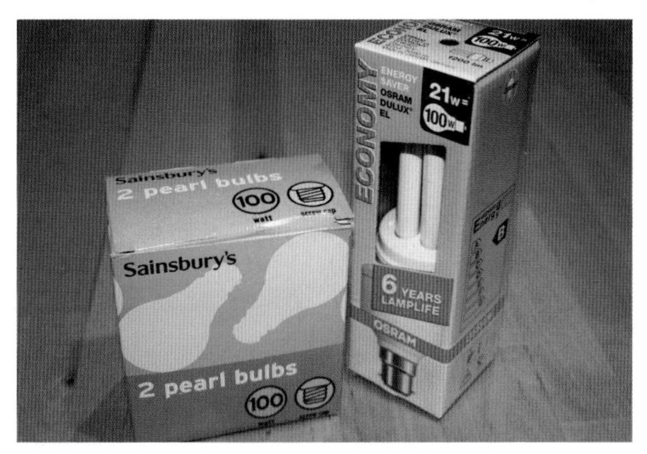

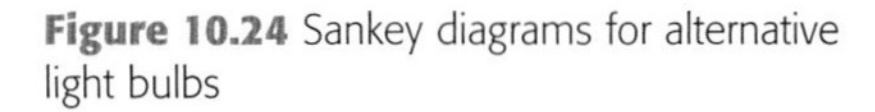

Figure 10.23 We can choose energy-efficient light bulbs

The important thing is to be able to compare one bulb with another. One of the types of bulb in the photograph is an ordinary filament bulb; the other is an **energy-efficient** lamp – a type of compact fluorescent tube. Why do we say this second type is more efficient?

We supply a bulb with electrical energy, and we want it to give us light energy. Unfortunately, most of the energy we supply to a light bulb is transformed to heat energy. (That's why light bulbs become too hot to touch.) So the electrical energy is transformed like this:

electrical energy → light energy (useful) + heat energy (wasted)

Sankey diagrams (arrow diagrams) show that the energy-efficient lamp gives a lot more light for 100 J of electrical energy input (Figure 10.24).

Figure 10.24 Sankey diagrams for alternative light bulbs

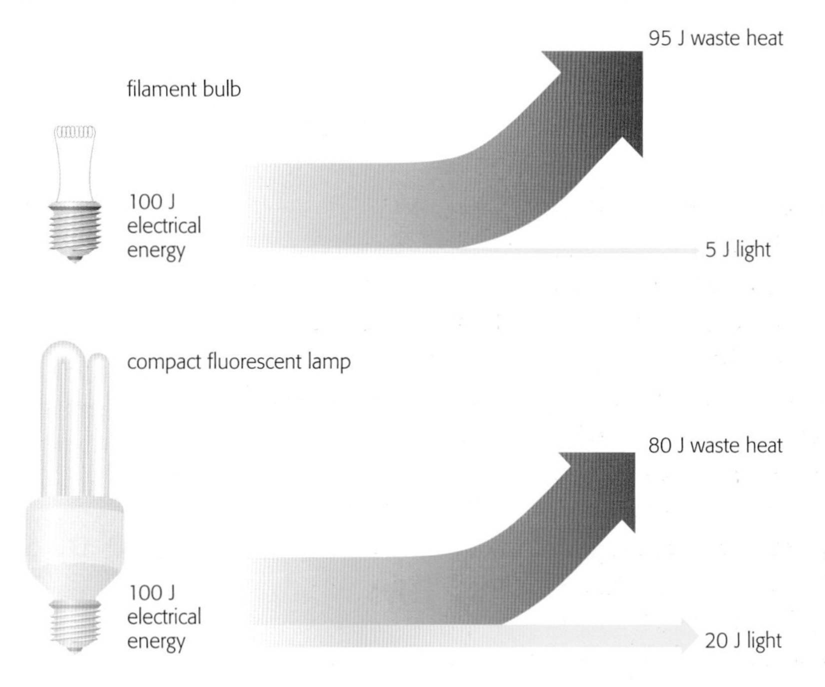

Calculating efficiency

We can express the **efficiency** of an energy change by thinking about what happens to each joule of energy input. Some becomes useful energy output, and some is inevitably wasted.

input energy → useful output + wasted output

Here is how to calculate efficiency, which is usually given as a percentage:

$$\text{efficiency} = \frac{\text{useful output}}{\text{total input}} \times 100\%$$

H

Worked example 1

A 60 W lamp uses 60 J of electrical energy every second. It gives out 3 J of light every second. What is its efficiency?

In each second the total input is 60 J, for which the useful output is only 3 J.

Efficiency = (useful output / total input) × 100% = 3 J / 60 J × 100% = 5%

So the lamp is only 5% efficient; 95% of the energy supplied to it is wasted as heat energy.

Efficient generation

We can apply the idea of efficiency to much larger systems. A coal-fired power station uses the energy stored in coal to generate electricity. Its efficiency might be 30%. Gas-fired power stations tend to be much more efficient – perhaps as high as 60%.

Worked example 2

A nuclear power station is 25% efficient. If its fuel provides 2000 MJ (megajoules) of energy every second, how much electrical energy does it supply per second? How much is wasted?

It helps to convert the percentage efficiency to a decimal:

efficiency = 25% = 0.25

Rearranging the equation for efficiency gives:

useful output = efficiency × total input

So electrical energy output = 0.25 × 2000 MJ = 500 MJ
and amount of energy wasted = 2000 MJ – 500 MJ = 1500 MJ

The more we can do to improve the efficiency of electricity generation, the better. Many new gas-fired power stations are of a type called Combined-Cycle Gas Turbine (CCGT). These use the technology of jet engines to get more of the energy out of the fuel.

Whenever we use electricity, we are being very wasteful. As much of 75% of the energy of the fuel may be wasted at the power station, and then we waste more when we use the electricity to power our appliances.

Making more of solar

How do solar cells measure up to the efficiency challenge?

1 The first solar cells in the 1970s had an efficiency of 1% or 2%. They wasted up to 99% of the energy of the sunlight falling on them! These were only used in specialist applications, such as spacecraft.
2 By the 1990s, the technology had progressed so that solar cells had efficiencies of between 10% and 15%. These have found much wider uses (see page 177). However, because they wasted so much energy, a large area was needed to generate significant amounts of electricity.
3 Since then, developments in the technology have led to improved efficiencies – up to 25%. This may sound like a small improvement, but it means that a solar cell can be half the size of its predecessor and produce the same amount of electricity.

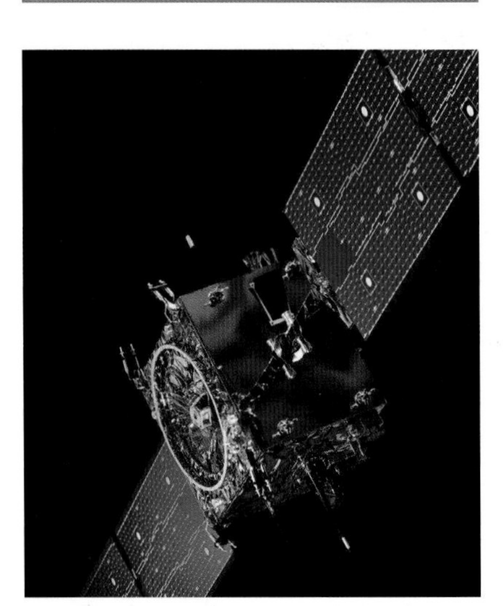

Figure 10.25 Spacecraft such as this, SMART-1, have large panels of solar cells

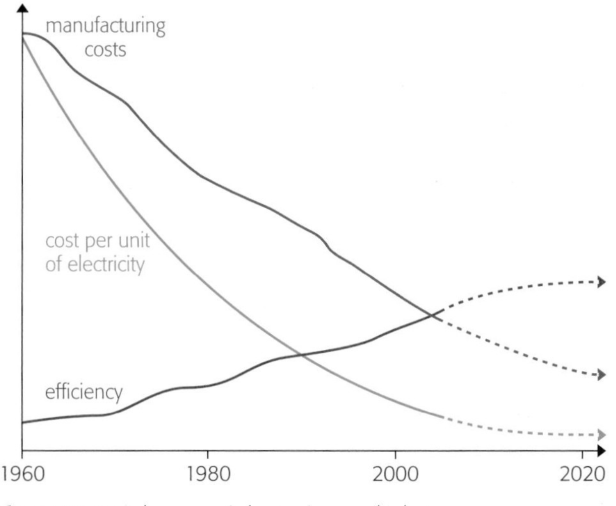

Figure 10.26 Solar power is becoming much cheaper

As well as worrying about energy efficiency, we need to think about the cost of actually making solar cells. Manufacturing methods have improved greatly so that solar cells are becoming cheaper. At the same time, their efficiency has doubled, so the cost of each unit of electricity produced has fallen sharply (see Figure 10.26). If this continues, solar cells will soon be able to compete with other means of generating electricity.

Energy-efficient appliances

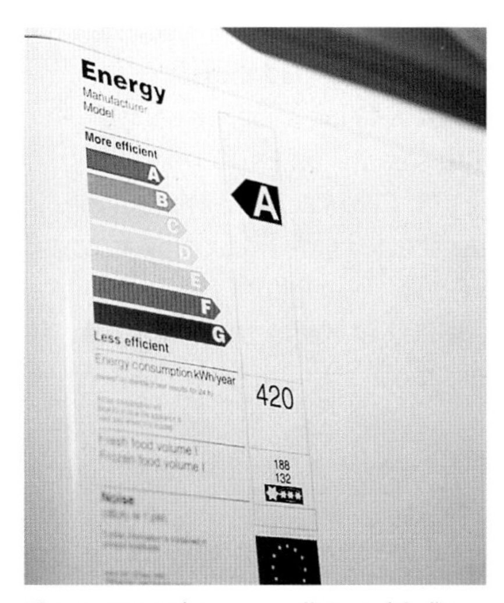

Figure 10.27 The energy-efficiency labelling scheme is compulsory in the European Union

The average consumer doesn't want to have to perform calculations of energy efficiency when deciding which washing machine or fridge to buy. To make life easier, appliances are labelled to indicate their efficiency, on a scale from A to G, where A is the most efficient. This means that an A-rated appliance uses least electrical energy for the same task. The ratings are based on standard tests. For example:

- for a washing machine – how much energy does it use to wash a standard load at a certain temperature?
- for a fridge – how much energy does it use to keep each kilogram of food at a standard temperature for a standard length of time?

Is it cost effective?

There are many campaigns that urge people to make better, more efficient use of energy. But a consumer may want to know whether it is worth, for example, switching to energy-efficient light bulbs, or buying a more efficient washing machine which may be more expensive. How can they decide?

They need to be able to weigh up all the costs.

Worked example 3

A householder is considering installing cavity-wall insulation. The cost is £400, and the salesman calculates that it will save 750 units of electricity each year. Is it worth it?

Assuming each unit of electricity costs 8p, the total saving each year will be $750 \times 8p = £60$.

Now we can calculate the **payback time** – this is the time taken to save the initial cost of the insulation.

Payback time = initial cost / annual saving = £400 / £60 = 6.7 years

So after nearly 7 years, the householder will have recovered the cost of the insulation and will be able to save £60 each year from then on.

Summary

- **The energy efficiency of a device can be calculated using:**
  efficiency = (useful output / total input) × 100%
- **As solar cells become more efficient and the cost of manufacture decreases, the cost of the electricity they produce decreases.**
- **An energy-efficiency measure is cost effective if it saves money in the course of its lifetime.**

Questions

10.15 Here are data for two light bulbs. Which is the more efficient?
Bulb A: uses 50 J of energy per second; produces 8 J of light per second.
Bulb B: uses 100 J of energy per second; produces 12 J of light per second.

10.16 A customer is considering two washing machines. One is A-rated for energy efficiency, and costs £400. The other is C-rated, and costs £350.
What other information would help the customer to decide whether it is cost effective to buy the A-rated machine?

10.17 A double-glazing salesman recommends fitting double-glazing throughout a customer's house. The cost will be £4000, and the annual saving on heating will be £125. Calculate the payback time.
How would you decide whether this form of insulation was cost effective? What other information would be helpful?

10.18 An electric oven is supplied with 4000 J of electrical energy each second. 1000 J of this escapes into the kitchen; the rest heats up the oven.

a How much energy heats up the oven each second?
b Calculate the efficiency of the oven.

10.19 A lamp consumes electrical energy at the rate of 150 W. If its efficiency is 8%, how much light energy does it produce each second? What happens to the rest of the energy supplied to it?

10.20 Twenty years ago, solar cells were both inefficient and expensive. Suggest why they were used in spacecraft.

E Electrical safety

Today, we are well aware of the hazards of electricity. Three hundred years ago, people were only just learning how to produce and use electricity, and they certainly didn't understand the nature of large currents and high voltage.

As with many new technologies, electricity found some surprising applications. In particular, many people claimed that electricity could effect wonderful cures of medical conditions. In the 1750s and 60s, people set up clinics where patients could come and be given electric shocks. It was claimed that this could cure all sorts of ailments, including rheumatism, paralysis, blindness and insanity. The photograph shows one of the instruments used to send electric currents through patients.

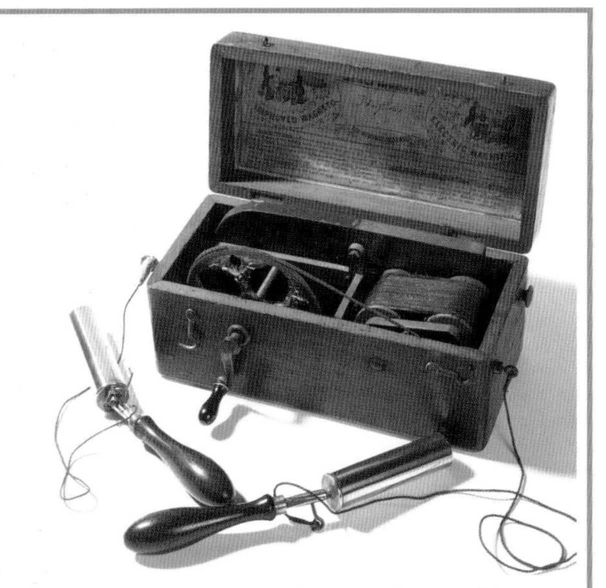

Figure 10.28 An electromagnetic induction machine, used for medical treatment in the 18th century. The doctor held the wooden handles (because they were insulated), while the patient gripped the brass rods

Reports of successful treatments were published in the scientific literature – at the time, there was little check on the truth of these reports, and no one carried out properly controlled tests. The shocks that the machines produced probably did no harm to fit people, but could have proved too much for those with a weak heart.

Since those early days, electricity has found many different, more controlled, uses in medicine. We have already seen how an electric current can sometimes jolt the heart into normal working after a heart attack (see page 189).

There is a more subtle use of electricity in medicine – to reduce chronic (long-term) pain. This is called electrotherapy, or TENS (transcutaneous electrical nerve stimulation) – see page 47. No one is quite sure how this works. It may be that it affects the way the electrical signals of pain travel along the nerves, or it may be that it releases natural painkilling substances in the brain.

Dangers at home

Home is a dangerous place. We use the 240 V mains electricity supply without thinking about it, but a single careless move could result in electrocution and even death.

The Royal Society for the Prevention of Accidents says that in the UK between 40 and 50 people a year die from accidents caused by faulty electrical installations at home, and almost 3000 are injured. Most of these are children. These accidents may involve electrocution or fire, and they arise when the domestic electricity supply has a fault because it has been badly installed. The fault may have occurred when the house was built, or when someone – perhaps a do-it-yourself enthusiast – made alterations.

In 2004, just 20% of electricians were registered with organisations which check their competence and make sure they are up-to-date with the latest regulations. That has now changed, with new rules introduced to make sure that fewer mistakes are made in domestic wiring. The Government now requires electrical safety to be checked as part of the Building Regulations; this means that, when a house is built or extended, the electrical work must be checked by an inspector in the same way that the structure, insulation and so on are inspected.

Think of the equation
power = current × voltage
The greater the voltage, the greater the power delivered by the current.

High voltage

The mains voltage in the UK is 240 V, quite high enough to electrocute you. We need a high voltage to deliver the power needed by many of the appliances we use.

Domestic electrical systems are designed to be safe; if you treat them with caution, you should get through life without getting a shock. However, carelessness or foolishness (or faulty installation) could be the end of you.

What happens if you get a shock? Suppose you touch the live wire of the mains supply with your hand. This wire is at a high voltage compared to the earth. This difference in voltage is enough to make a current flow through your body. The consequences can be:

- a nasty tingling feeling
- your muscles contract violently, throwing you across the room
- your skin is burned
- water in your cells boils
- you may die.

Figure 10.29 Birds can sit on high-voltage power lines because they do not form part of a complete circuit to earth, so no current will flow through them

The greater the current, the greater its effect. A few milliamps (thousandths of an amp) can give you a jolt. Just 50 milliamps may be enough to kill you. These currents are much smaller than the currents that flow when many domestic appliances are used.

Notice that it is the difference in voltage between the live wire and the earth that causes a current to flow. You may be lucky. If your resistance is high – for example, if your skin is dry and you are wearing rubber-soled shoes – the current flowing through you may be small, and its effect will be less.

Three wires

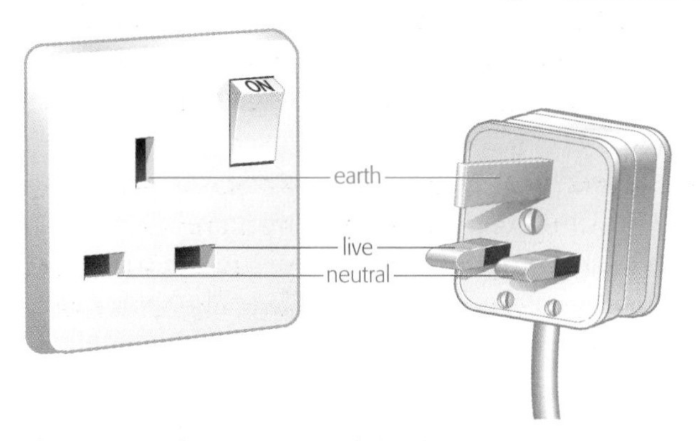

Figure 10.30 The UK mains supply has three connections

Think of a typical plug. It is known as a '13 amp plug' because it is designed to carry currents up to 13 A. It has three pins, which fit into the three holes in a socket. This means that an appliance has three connections to the mains supply. Why three? For a complete circuit, you would think only two should be necessary.

Here is what the three connections are for:

- The **live** and **neutral** connections are the two that make a complete circuit. The alternating current flows back and forth in these wires.
- The **earth** connection is for safety. If an appliance has a metal casing, this is connected to the earth wiring.

Figure 10.31 shows what happens to these wires outside the house.

Figure 10.31 The three wires outside the house

- The **live** wire is connected to the mains supply from the electricity substation. For a 240 V alternating supply, its voltage varies up and down between +340 V and −340 V. (Yes, its greatest value is 340 V, not 240 V.)
- The **neutral** wire is also connected to the mains supply, but it is earthed somewhere close to the house. This means that its voltage is close to 0 V.
- The **earth** wire is connected to a metal plate or rod which goes into the ground, securely connecting it to earth. Its voltage should always be 0 V.

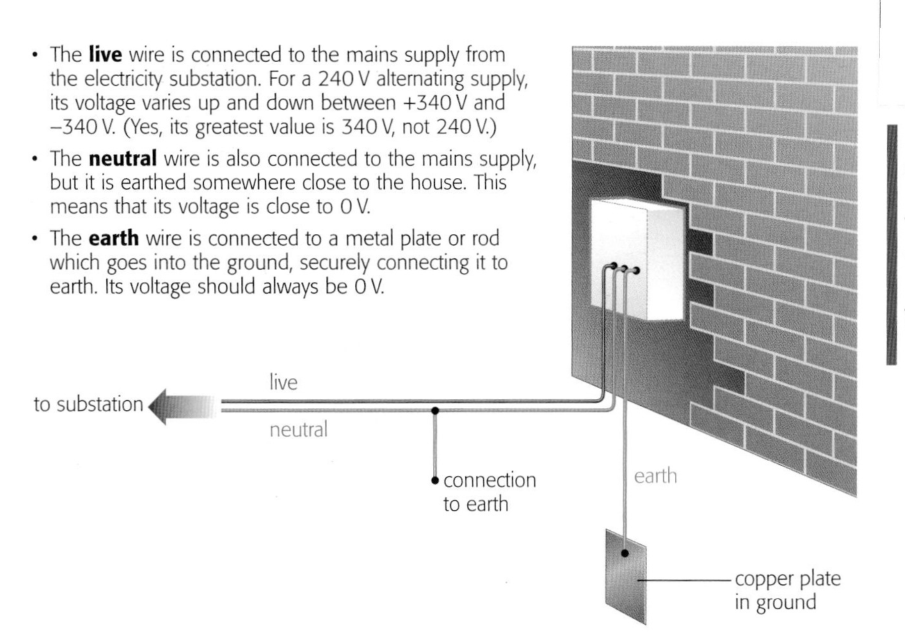

Fuses and trips

Every 13 A plug contains a **fuse**. This is in the live section of the circuit. The function of the fuse is to protect the appliance from currents that are too high.

The fuse is a small cartridge, and inside is a short length of thin wire (see Figure 10.32). If the current is too high, the wire heats up and becomes so hot that it melts. This breaks the circuit, and the dangerous current no longer flows.

The fuse wire gets hot because it has high electrical resistance.

Fuses have a range of different values, typically 3 A, 15 A and 30 A. The appropriate value must be chosen, so that it will allow the normal operating current to flow but it will 'blow' (burn out) if the current exceeds this.

Figure 10.32 Fuses – one has been opened to show the thin wire inside

Another type of fuse used to be fitted in domestic electrical systems. These were in the fuse box, close to the meter, where the mains supply enters the house. These were designed to blow if the current flowing through them was too high. They were there to protect the wiring in the house. It would be very inconvenient if a high current melted the wiring that travels around the house – it would be very awkward to replace wiring in the walls or under the floor, and it could even lead to a fire.

Nowadays, these fuses are likely to have been replaced by **trip switches** (see Figure 10.33); these use electromagnets and they switch off whenever the current gets too high. Once the problem has been sorted out, they can simply be switched back on again – much easier than replacing a fuse.

Laboratory power supplies often have a trip switch – look for the 'Reset' button.

Figure 10.33 Mains trip switches

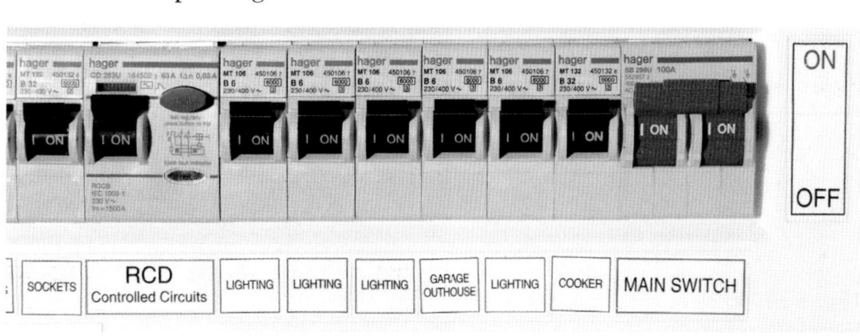

Earthing

Imagine that there was a fault in the wiring of a washing machine, so that the live wire became connected to the machine's metal case. If you touched it, you would be connected to the live mains supply, and you could get a severe shock. The fuse wouldn't protect you, because it could let through a fatal current of several amps without blowing.

For this reason, the metal case is connected to the earth wire. It is at 0 V, and would not push a current through you if you touched it. Current from the live wire would flow directly to earth.

Some appliances are designed so that it would be very difficult for the user to come into contact with the live wire. Radios, lamps, hairdryers and so on have plastic cases which act as very secure insulation, and so no earth connection is needed. These are described as **double insulated**, and they are marked with a 'double square' symbol.

Figure 10.34 The 'double square' symbol indicates that an appliance is double insulated, and so does not require an earth connection

H

Added safety – residual current devices

There is a second type of trip switch that can help to protect an electricity user. This is called a **residual current circuit breaker (RCCB)** or residual current device (RCD). It is advisable to use one of these if you are using tools such as an electric lawnmower or a drill. If you should happen to run the mower over its cable, for example, the blade might cut through to the live wire and you might get a serious shock. How can you avoid this?

Normally, the current flowing in the live wire is the same as the current in the neutral wire – current does not get used up as it flows through the mower. However, if even a small current flows through you (because of an accident), then one current will be bigger than the other. The RCCB detects this, and immediately (within milliseconds) switches off the supply.

An RCCB may be fitted where the mains supply enters the house; or you can plug one into a socket as shown in the illustration.

Figure 10.35 Use of an RCCB will prevent a fatal electric shock

Summary

- Our ideas about electricity have changed over time; this has influenced how electricity is used in medicine.
- Fuses and trip switches protect appliances and wiring from excessive currents.
- The earth wire or double insulation protects the user from accidentally touching a live wire.

- A residual current circuit breaker (RCCB) detects small currents flowing through the user and switches off the supply.

Questions

10.21 The earth wire is connected to the metal casing of an appliance such as a washing machine.

a What might happen to cause the casing of the appliance to become live?

b How does the earth wire protect the user?

10.22 Every 13 A plug contains a fuse.

a What happens when an excessive current flows through the fuse?

b How does this help to protect the appliance?

10.23 Explain how a residual current circuit breaker (RCCB) protects the user.

Now you see it, now you don't

A How safe is that phone?

PARENTS OPPOSE PHONE MAST

IN MUMBURY, protesting parents are down at the school gate. They don't like the idea that a mobile phone mast might be installed on the roof of the school. Rumours have been circulating for several weeks that new mobile phone network Gabble have approached the governors of Mumbury High School for permission to lease roof space for an aerial which will provide a service to subscribers in the Blackfields area of the town.

'There's radiation pouring out of those masts, 24 hours a day,' complained anxious parent Joyce Fillet. 'My Jo has enough trouble coping with her GCSEs without having her brain scrambled by microwave rays.'

The chair of governors, Mr John Wall, refused to confirm that the school would receive rental of £2000 each year from Gabble. He would only say that discussions had taken place with Gabble, and that the governors had been assured that no child would be put at risk.

Figure 11.1

How to decide?

Are mobile phone masts dangerous? Are phones themselves dangerous? How would you decide?

It's the **radiation** used by mobile phones that people are concerned about. This radiation is **microwaves**, similar to the microwaves used in a microwave oven. When you make a phone call, the aerial of your phone emits a string of pulses of microwaves. This radiation spreads out in all directions, and some is picked up by the nearest mobile phone mast (a type of aerial). The reply to your call is sent out from the mast, again in the form of microwaves.

Figure 11.2 Mobile phones send and receive signals using microwaves

The reason why people worry about mobile phones is that the microwaves can be **absorbed** by your body, just as microwaves in a microwave oven are absorbed by food. When you hold a phone to your ear, some of the microwaves radiate outwards into your head, where they are absorbed. Your head absorbs the **energy** of the microwaves, and becomes slightly hotter.

Figure 11.3 Some of the microwaves from your phone radiate into your head

The radiation from a mobile phone mast is more intense, because it is sending messages to many phones all at the same time. This means that masts could be more hazardous than phones. However, they are designed to ensure that no one can get closer to the aerial than 2 metres. At this distance, the radiation is weaker, and is believed to be at a safe level. So the radiation from a mast on a school roof should not be strong enough to affect the children working in the classrooms below.

Scientists who have investigated the safety of mobile phones believe that a user's brain may be slightly heated by the radiation it absorbs – perhaps by half a degree or so. But no one is sure whether or not such heating is harmful.

Ask the experts

The UK Government was concerned about possible hazards from mobile phones, and commissioned a report from a committee of experts. The committee took evidence from the phone industry, from independent scientists, and from other interested groups. In the end, they concluded that there was not, at present, any evidence that mobile phone use was causing any increase in diseases such as brain cancer. However, they did suggest that very young people (up to the age of 8), should be discouraged from using phones, as their brains are still developing rapidly.

Not everyone is happy with this. Some people think that, even if there were just a few deaths each year resulting from exposure to microwaves, this would be too many. Some suggest that phones should be redesigned with a separate aerial which would transmit the microwaves from waist level, rather than from next to the brain.

Mobile phones have a lot of uses. They have undoubtedly saved lives, for example, when passing motorists have called for help when they have seen an accident on the road. They also give many people a lot of pleasure. So it may be necessary to strike a balance – even if there is a small **risk** (or **cost** to health) in using a phone, this may be cancelled out by the **benefits** that come from having a phone. We must try to ensure that the benefits are worth more to people than the costs. And we can try to reduce the costs, by continued research into health risks and improved phone design.

Microwaves and radio waves

We make use of **radio waves** every day. They are used to transmit the radio and TV programmes that we watch. Radio waves spread out from a transmitter, and they are picked up by the radio or TV aerial at your home. You can be sitting indoors, and radio waves can still reach your radio, because they can pass through brick walls without being absorbed. The waves are **transmitted** (allowed to pass) through the walls.

Microwaves are very similar to radio waves, except that they have many more vibrations each second (see Figure 11.4).

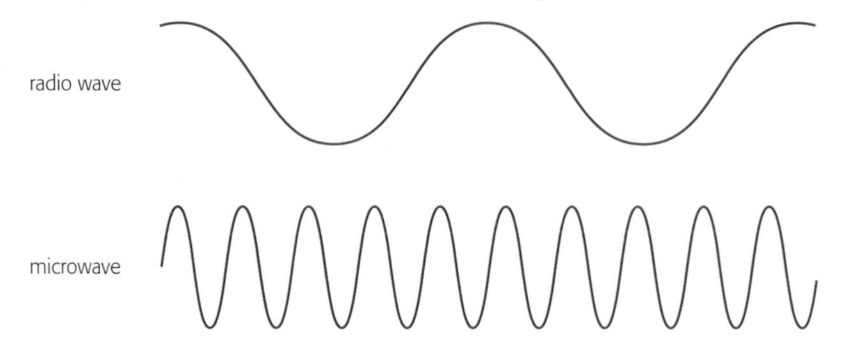

Figure 11.4 This is one way of picturing radio waves and microwaves. Radio waves have fewer vibrations per second

Another difference is that microwaves are more easily absorbed by some materials. When microwaves from a phone are absorbed by your head, the waves get smaller. We say that their **amplitude** decreases. Waves with a smaller amplitude carry less energy than waves with a large amplitude. Some of the energy carried by the microwaves has been absorbed by your head.

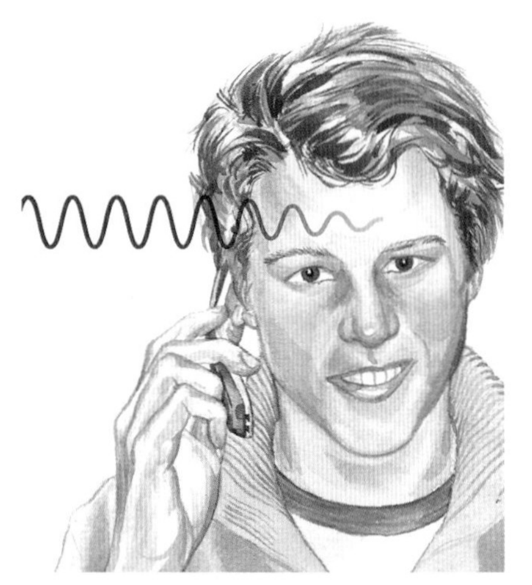

Figure 11.5 Microwaves are absorbed by living tissue, such as your brain. The amplitude of the waves decreases as they are absorbed

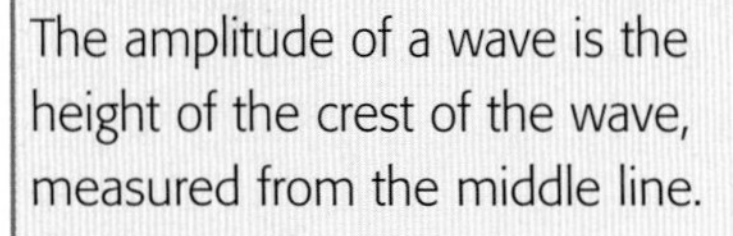

The amplitude of a wave is the height of the crest of the wave, measured from the middle line.

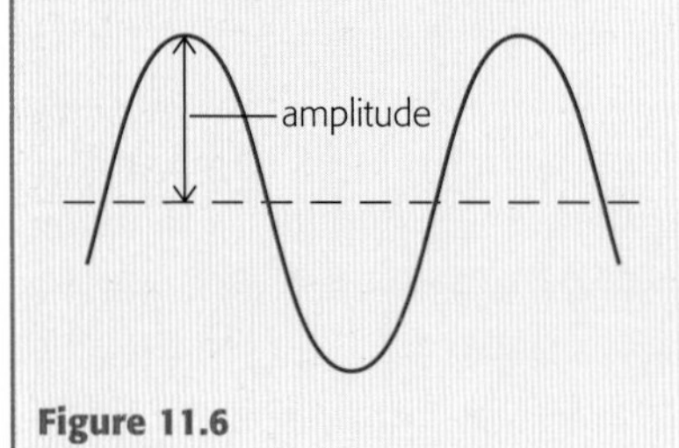

Figure 11.6

Summary

- Mobile phones use microwaves to send messages.
- Radio and TV programmes are carried by radio waves.
- Microwaves carry energy, which is absorbed by some materials, including body tissue.
- The more energy a wave carries, the greater its amplitude.
- When deciding whether to use a new technology, we have to ensure that the benefits outweigh the costs; we also try to reduce the costs.

Questions

11.1 Draw two waves, one with twice the amplitude of the other.

11.2 What is happening to the amount of energy carried by a wave if its amplitude is decreasing?

11.3 Why is it hazardous to get very close to a mobile phone mast?

11.4 How could you convince someone that the microwaves used by a mobile phone can pass through brick or glass?

11.5 List some benefits that come from using a mobile phone, and some costs.

B Ultrasound and seismic waves

Mr Williams was suffering from jaundice and nausea. His doctor diagnosed gallstones.

'We'll send you off for an ultrasound scan,' she said. 'That's the simplest way of detecting gallstones.'

Mr Williams, who thought that ultrasound scans were only for pregnant women, was not surprised to find himself in a waiting room with expectant mothers.

The technician smeared Mr Williams' stomach with jelly, and then moved the probe around until she could see a fuzzy image of his gall bladder on the screen. It showed up as a greyish bag, with a number of white dots inside – these were the gallstones that were causing Mr Williams' symptoms.

Mr Williams had another encounter with ultrasound. A beam of ultrasound waves was used to break up his gallstones, avoiding the need for an operation or drug treatment.

Figure 11.7

Baby's first photo

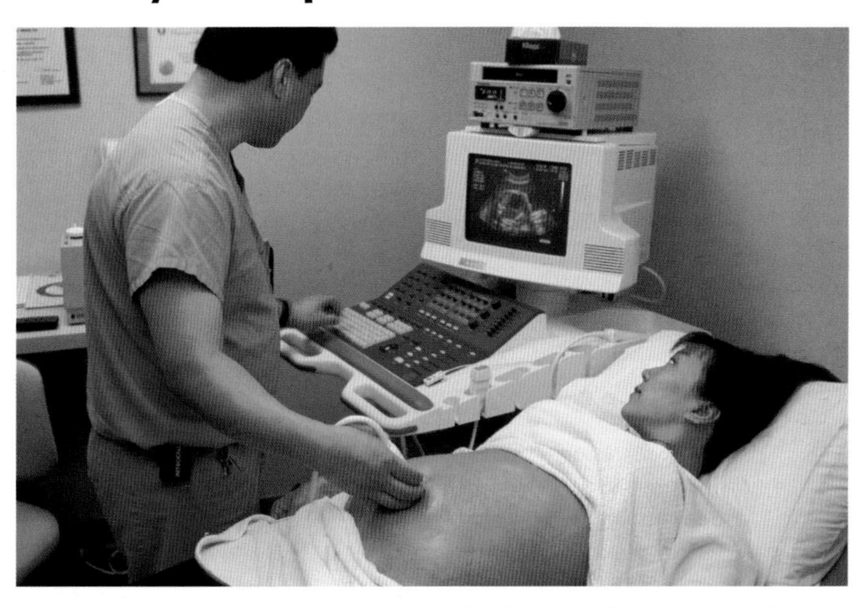

Figure 11.8 Scanning the unborn baby

When there's a new baby in the family, the first photo in the album is often an **ultrasound scan**, showing it as a fetus in the womb, months before birth. There have been many checks to find out if the use of ultrasound scanning can damage babies, but so far there is no evidence of any harmful effects.

An expectant mother has a similar experience to Mr Williams (see the box above); she lies on her back and has jelly smeared over her lower abdomen. The jelly makes it easier for the ultrasound waves to travel into her body as the probe is moved over her skin.

You should recall that sound travels as vibrations – 'sound waves' – and that we can hear sounds with **frequencies** from about 20 Hz (**hertz**) to 20 kHz. Ultrasound is simply sound waves that are too high-pitched for us to hear – their frequencies are above 20 kHz.

The frequency of a wave is the number of vibrations a second, measured in hertz (Hz).

Medical scanners use ultrasound waves with frequencies of about 10 MHz (10 megahertz); that's 10 million vibrations per second. So, every second, the probe sends millions of vibrations into the patient's body. What happens to these vibrations?

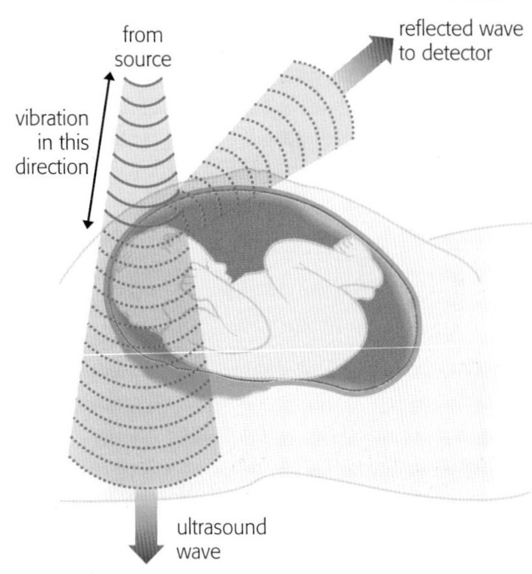

Figure 11.9 Ultrasound waves are partly reflected when the density of the material they are travelling through changes. A baby's head is denser than the fluid in the womb, so waves are reflected by the head

- They are partly **absorbed** by the patient's body.
- They are partly **transmitted** through the body.
- They are partly **reflected** back to the probe.

It's the reflected ultrasound waves that make scanning possible. They reflect from the surfaces of bones, muscles, the womb, the baby, and so on. A high-speed computer is needed to interpret the reflected waves and to draw an image on the screen.

Bones, muscles, blood and so on are materials with different densities. It is the difference in density that causes ultrasound waves to be reflected – a big change in density gives more reflection.

Ultrasound waves that are transmitted at a change in density may change direction. We say they are **refracted**, just as light is refracted when it travels from air into glass.

The wavelength of a wave is the distance from one wave crest to the next.

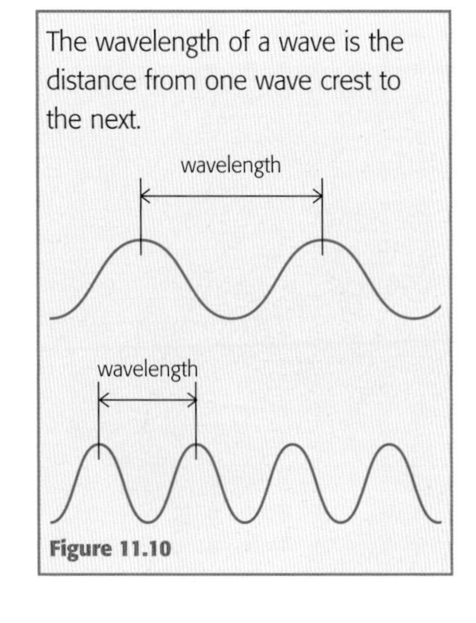

Figure 11.10

The wavelength of ultrasound

The features of babies in the womb are quite small, and gallstones are smaller still. This gives us a clue as to why we use ultrasound waves in medical scans. The important thing is their **wavelength**. The shorter the wavelength (Figure 11.10), the smaller the objects that can be seen on the scan.

The wavelength of the ultrasound waves used in scanning is about 1.5 mm. This means they are short enough to detect small objects such as gallstones, or a baby's fingers, a few millimetres across. But these ultrasound waves are not short enough to detect tiny blood vessels, less than a millimetre across.

Earthquakes

The tsunami of late 2004 wrought great havoc across the Indian Ocean. A tsunami is a giant wave; in this case, the cause was an undersea earthquake off the coast of Indonesia. The seabed ripped open and collapsed downwards, setting off a giant disturbance in the water. Within minutes, a towering wave had crashed down on nearby islands, causing many deaths and great destruction. The wave spread steadily outwards, reaching the coasts of Sri Lanka, India and eastern Africa several hours later.

Figure 11.11 A tsunami or tidal wave can result from an earthquake, a landslide of rocks into the sea, or a meteor impact

Earthquakes arise along the edges of the giant tectonic plates which make up the Earth's crust and upper mantle. As these vast masses of rock rub against each other, they produce shock waves in the ground known as **seismic waves**. Their effect is greatest close to the centre of the earthquake, but tremors of a large earthquake can be detected all around the world.

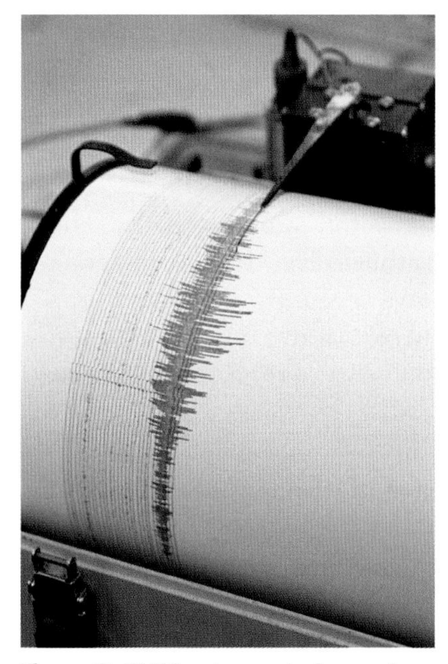

Figure 11.12 This seismometer is recording the activity of a volcano in Indonesia

Recording seismic waves

Scientists use **seismometers** to measure seismic waves. There is a network of labs around the world making these measurements; by comparing seismometer traces, the scientists can work out where the earthquake took place and how strong it was.

A seismometer trace is in two parts, showing that there are two types of seismic wave.

- Primary waves (P-waves) arrive first because they travel faster. These are longitudinal waves (see page 229), like sound and ultrasound waves.
- Secondary waves (S-waves) travel more slowly. These are transverse waves (see page 229).

While it is relatively easy to record earthquakes when they happen, they are much harder to predict. This is because the movement of tectonic plates is very unpredictable. The plates may pass each other in a series of small jerks, or they may stick, build up pressure, and suddenly move with a violent jolt. Similarly, volcanic eruptions are difficult to predict.

Earthscan

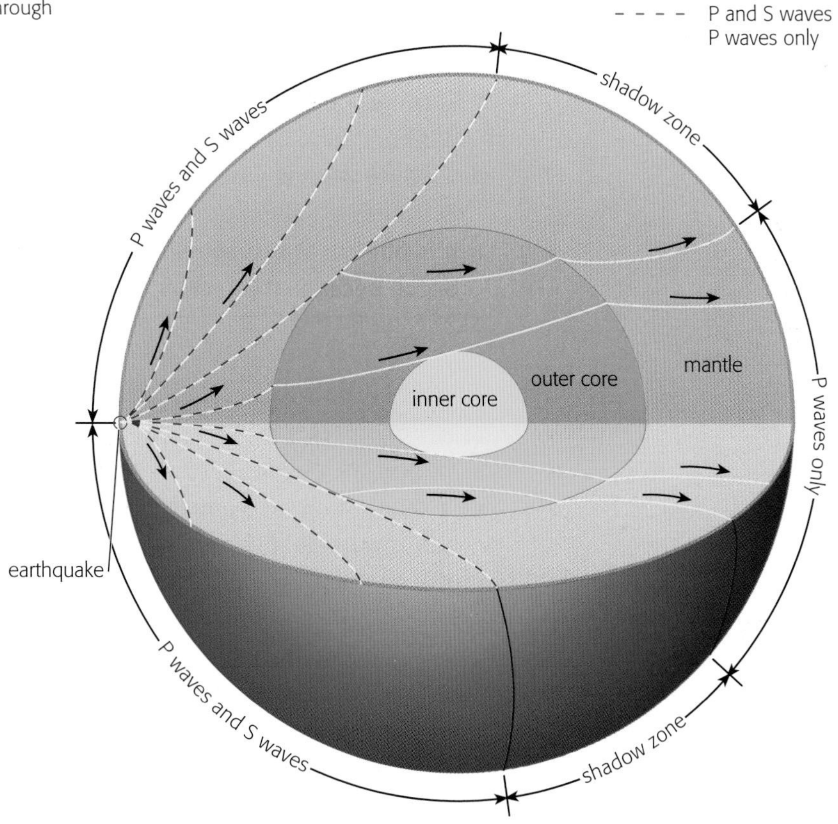

Figure 11.13 Paths of seismic waves through the Earth

The **mantle** is the part of the Earth just below the crust.

Figure 11.13 shows the structure of the inside of the Earth. We live on the outer crust – too thin to show on this scale – inside is the mantle (solid) and the core. The outer core is molten metal, and the inner core is solid metal. How do we know it's like that?

The information comes from seismic waves. As they spread out from the site of an earthquake, they travel through the interior of the Earth. P-waves can travel through both solid and liquid parts. However, S-waves can't pass through liquids. From the diagram, you can see that S-waves only reach about half of the Earth's surface; the rest of the surface is 'in shadow', because of the liquid core through which the S-waves cannot travel.

The diagram also shows that the waves follow curved paths. This is because they are refracted (bent) as they travel through materials of different densities, where they change speed. By comparing seismometer traces from around the world, earth scientists can work out the approximate dimensions of each part of the Earth's interior.

Summary

- The frequency of a wave is the number of vibrations per second, measured in hertz (Hz).
- Ultrasound is sound waves whose frequency is too high for us to hear.
- The wavelength of a wave is the distance from one wave crest to the next.
- When ultrasound waves pass into the body, they are partly transmitted, partly absorbed, and partly reflected.
- Reflected ultrasound waves can be detected and used to produce an image of the inside of someone's body.
- Ultrasound waves are reflected and refracted when they travel into a material of a different density.
- Earthquakes cause seismic waves, which are vibrations inside the Earth.
- S-waves cannot pass through liquids; this allows us to show that the Earth has a liquid core.
- It is difficult to predict when earthquakes, tsunamis or volcanoes will occur, because they result from sudden changes in the Earth's crust.

Questions

11.6 What three things can happen to ultrasound waves as they travel through a material?

11.7 Sketch two waves, both having the same amplitude, one with half the wavelength of the other.

11.8 An ultrasound scan can show up what's going on inside our bodies. In a similar way, fishing boats can use 'sonar' to detect shoals of fish and the depth of the sea, and bats can find insects to eat in the dark. Find out the similarities and the differences between these three ways of using sound waves.

11.9 Astronauts visiting the Moon have set off explosive charges to simulate a small earthquake. What sort of information about the Moon might this have revealed?

C X-rays and health

Alice Stewart was born in Sheffield in the early years of the 20th century and, unusually for a woman in those days, trained to be a doctor. In the 1950s, she became concerned that there was an increase in cancer among children. In particular, statistics showed that the number of children dying of leukaemia (cancer of the blood) had doubled since the 1930s (see Figure 11.15). Alice wanted to know why.

It's difficult to find the answer to a question like this. Alice tackled it by surveying the families of all the children who had died. She asked about their family medical history, diet, parents' occupations and so on. Eventually she spotted what she thought was the common factor: all of the mothers had been X-rayed while they were pregnant. This was fairly common at the time, if there was any concern about the health of the baby in the womb.

Figure 11.14 Alice Stewart (1906–2002)

Alice analysed the data. She was able to show that, even if an expectant mother had just one X-ray, the chance that her child would later suffer from cancer was doubled. This discovery wasn't popular with many in the medical profession, who found X-rays a useful way of investigating their patients. However, once Alice's findings were generally accepted, it was agreed that pregnant women should only be X-rayed if it was absolutely essential – if the possible benefit outweighed the risk.

Today, as we saw in section B, ultrasound scans are used to check on the baby's development. Ultrasound is much safer than X-rays.

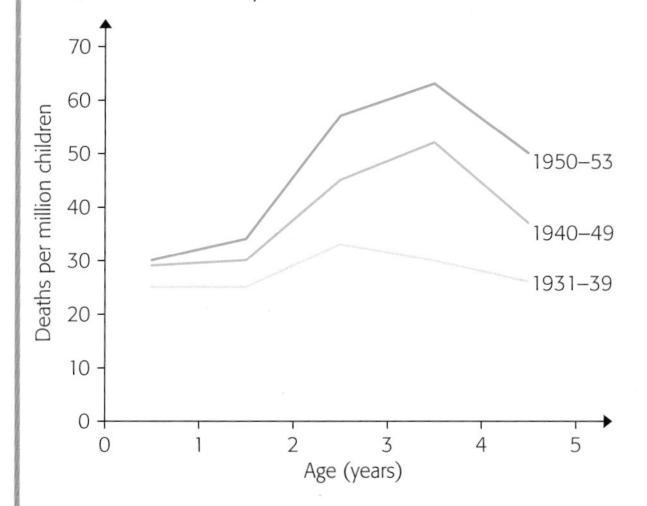

Figure 11.15 Deaths of young children from leukaemia

Cell damage

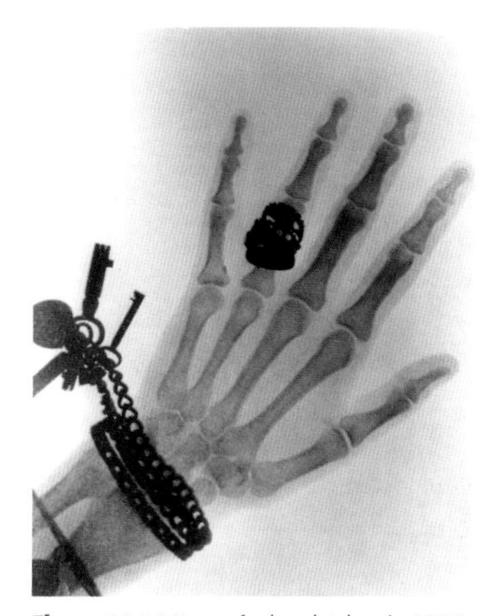

Figure 11.16 X-ray of a hand, taken in 1896

X-rays were discovered in 1895 by Wilhelm Roentgen. For scientists, this was a very exciting discovery. Roentgen had discovered how to produce invisible rays which could pass straight through solid matter, producing a 'shadow' image on a piece of photographic paper, or on a fluorescent screen. Everyone wanted to use these new rays, and it became fashionable among the rich to have an X-ray machine at home. At a dinner party, you could X-ray your guests; people brought their cats and dogs to join in the fun.

However, Roentgen soon discovered that there were problems with this. If he put his hand in the X-ray beam for too long, it went red and became painful. He had produced a radiation burn.

It's not surprising that X-rays are hazardous. They are beams of energy, and Roentgen's hand was absorbing the energy. The cells were being damaged. You could kill someone with a large dose.

Today, we understand X-rays a lot better. We can picture a beam of X-rays as a large number of tiny packets of energy, travelling at the speed of light. If the beam strikes your body, many of these packets of energy will pass straight through you. However, some

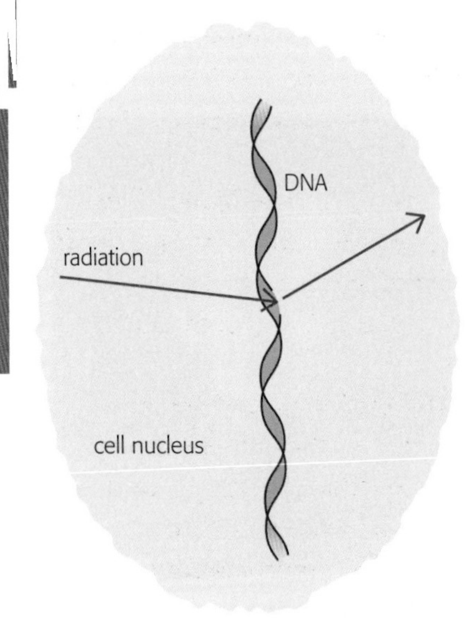

Figure 11.17 High-energy radiation can damage DNA

strike the insides of the cells of your body and are absorbed. In the process, they may damage the cells. Three things can result:

- Cells may be badly damaged, so that they die; that's what causes a **radiation burn**. New cells may grow to repair the damaged tissue.
- The genetic material of a cell may be damaged if a packet of energy strikes a DNA molecule in the nucleus of a cell. That could kill the cell, or it could make the cell grow and multiply in an uncontrolled way. That is how **cancer** starts. The multiplying cells form a tumour (see page 27).
- If the DNA in an egg or sperm cell is damaged, the damage can be passed on to future generations. Damaged DNA carries incorrect instructions for the working of cells; we say that a **mutation** has occurred. Most mutations result in cells that don't function correctly.

Shortly after X-rays were discovered, Henri Becquerel discovered another type of invisible radiation, produced by radioactive substances such as uranium. This radiation includes gamma rays, which are similar to X-rays except that the little packets of energy are even more energetic. Gamma rays can penetrate further than X-rays through solid materials – a metre or two of concrete, or 3 cm of lead. This means that engineers can use them to see inside solid objects. They can even 'X-ray' whole bridges to check for cracks.

Gamma rays can do great damage to living cells. That can be useful – many products are exposed to gamma radiation to kill off any harmful bacteria or other germs. For example, tampons and medical equipment such as syringes are sterilised in this way during manufacture.

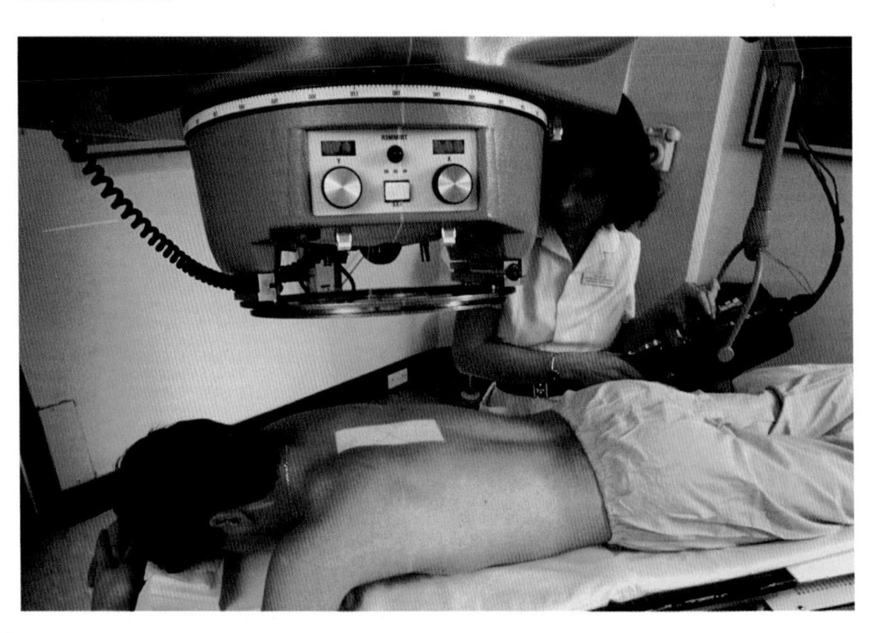

Figure 11.18 Gamma rays can cause cancer, but they can also be part of the treatment. This patient has a tumour; gamma rays are being directed at the tumour to kill its cells

How an X-ray works

Most people in the UK have an X-ray at some time in their lives. You don't feel anything (except perhaps a little nervous). The radiography technician directs the X-ray beam at the part of your body which is to be examined. The radiation passes through your body and is detected on the other side.

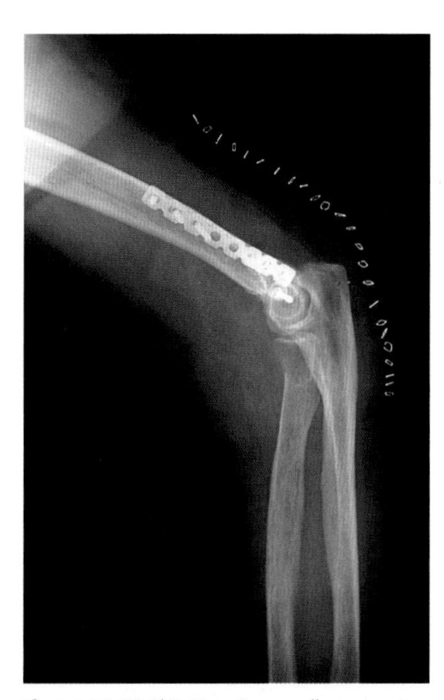

Figure 11.19 This X-ray image allows you to see inside a broken arm. Surgeons have fitted metal pins in the upper arm bone to hold it together while it heals. You can also see the metal staples which have been used to stitch up the wound where the surgeons cut into the patient's arm

Your bones are made of materials that absorb X-rays more strongly than flesh, so less radiation gets through where there is bone. Metals are even better at absorbing X-rays, as you can see in Figures 11.16 and 11.19.

Sometimes, photographic film is used to detect the X-rays. The result is a black-and-white photo. Electronic detectors are also used; these are similar to the detectors in a digital camera, and they produce an electronic image which can be stored in a computer and printed out later (or viewed on a monitor).

There are also X-ray **scanners**. Suppose that a patient has a tumour. Using a scanner, X-rays are directed at the tumour from different angles. This produces a number of different images of the tumour, and a computer can analyse these to produce a three-dimensional image of the tumour. That is much more useful for a doctor who wants to see its size and shape accurately. This technique is known as CAT scanning.

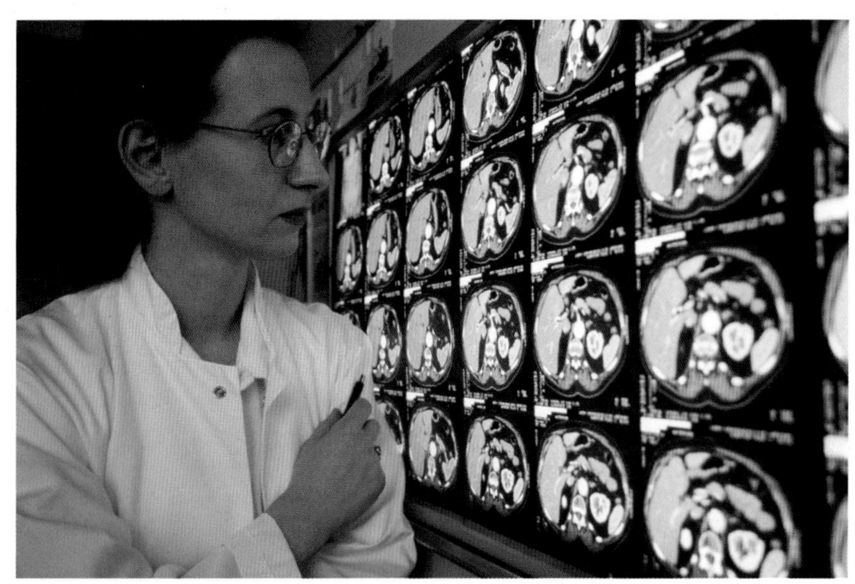

Figure 11.20 This doctor is analysing CAT scan images of slices through a patient's head. An ordinary X-ray would show none of this detail

Working safely

X-rays and gamma rays are hazardous, so the people who work with them have to observe various safety precautions. For example, when someone is about to have an X-ray, the technician retreats to a safe distance before switching on. If they stood next to the patient, they would receive a small dose of X-rays. The danger is that the technician works with X-radiation every day, and many small doses over a working lifetime would add up to a dangerously large dose.

The patient is at slight risk, too. Doctors and dentists try to avoid X-raying young people. They have to balance out the risk that the patient will get cancer against the benefit of finding out what's wrong with them. It's not such a worry with the elderly; even if a cancer starts as a result of an X-ray, the tumour will probably take years to develop.

To make the risk to the patient as small as possible, **sensitive** detectors are used, so that only a small dose of X-rays is needed. And only the affected part of the body is X-rayed, to protect other tissues.

At the dentist, for example, you may be asked to wear a lead apron to protect your body – X-rays don't pass through the lead.

X-rays, like gamma rays, can also be used to kill cancerous cells. The beam of X-rays is more energetic than one used for photography, and it is focused carefully on the affected part of the body.

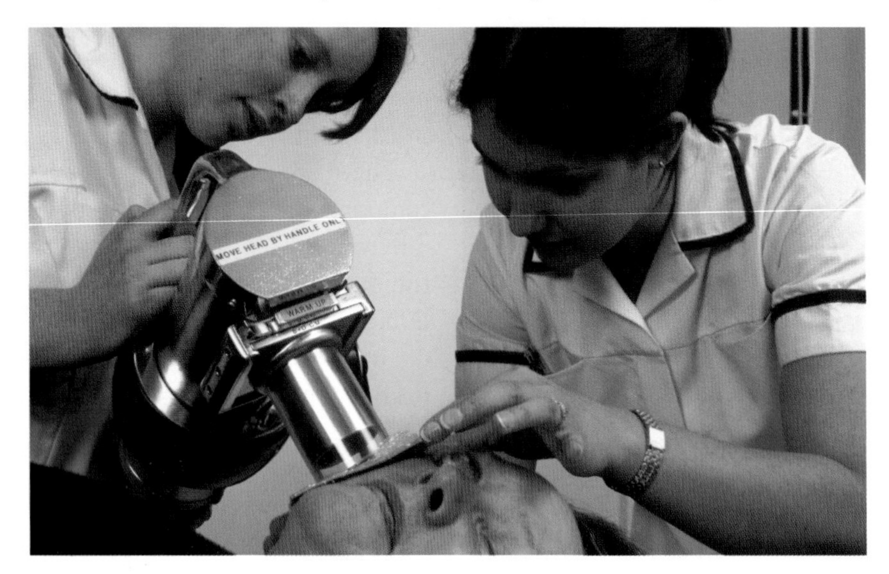

Figure 11.21 This patient has skin cancer. X-rays are being used to kill the cancer cells. To avoid exposing the rest of the patient's face to the radiation, a sheet of lead covers her cheek and a lead shield covers her eye. This treatment is usually 100% successful

Summary

- **X-rays can penetrate through the human body.**
- **X-rays are more strongly absorbed by some materials than others.**
- **X-ray images are useful in medical diagnosis.**
- **X-rays and gamma rays can cause damage to cells, and may cause mutation in the genetic material of cells.**
- **Gamma rays and X-rays can be used in the treatment of cancer.**

Questions

11.10 Make a list of ways in which X-rays and gamma rays are similar. Then give two ways in which they differ.

11.11 Many people are alarmed by the idea of radiation. A scientific understanding might help them to understand both the problems and the benefits of using X-rays and gamma rays.

- **a** Write a paragraph explaining how radiation can *cause* cancer.
- **b** Write a paragraph explaining how radiation can help to *detect* cancer.
- **c** Write a paragraph explaining how radiation can help to *cure* cancer.

11.12 Look back at the graph in Figure 11.15, showing how the number of children suffering from leukaemia changed in the middle of the 20th century.

- **a** Describe the patterns shown by the graph.
- **b** Explain whether you think the graph shows a large increase in the number of children dying from this disease, or a small increase.
- **c** Alice Stewart set out to find the cause of this increase. Make some suggestions of possible causes. Why do you think it was difficult to identify the actual cause?

11.13 Doctors are reluctant to X-ray a child; they are more likely to recommend an X-ray for an elderly person. Suggest reasons why this might be.

D The electromagnetic spectrum

Here are two things you should know about light:

- If you sit out in the sun, it warms your skin. Your skin absorbs the energy carried by sunlight.
- Sunlight looks white, but if you pass it through a prism, it is split up into a spectrum of different colours, from red to violet.

Figure 11.22 William Herschel (1738–1822) making his discovery of infra-red radiation

William Herschel knew these things as long ago as the year 1800. He wondered whether some colours of visible light carry more energy than others. To investigate this, he shone sunlight through a prism so that a spectrum fell on a piece of paper on the bench in his lab. Then he put thermometers at different points in the spectrum. As the thermometers absorbed the light, the mercury rose inside them.

What did William find? He discovered that red light made the temperature rise the most, and violet (at the other end of the spectrum) had the smallest effect. Then he discovered something completely unexpected. As the Sun moved across the sky, the spectrum moved across the bench. The thermometer which had been in the red light was now out of the spectrum – and its reading was even higher, showing that it was still absorbing energy!

Herschel realised that there was some kind of invisible radiation reaching the thermometer; it must be part of sunlight, but it was beyond the red end of the spectrum. He called it 'infra-red' radiation.

Infra-red and ultra-violet

Figure 11.23 Beyond the visible spectrum

William Herschel showed that heat can travel as radiation. We now know that all objects give out **infra-red radiation**, all of the time. The hotter they are, the faster they radiate. Think of the inside of an oven – it's very hot, and if you put your hand inside to remove some food, you can feel the heat. That's infra-red radiation.

People aren't special – we all radiate infra-red, too. If we are somewhere cold, we radiate faster than our surroundings, so we get cold. However, if we are somewhere hot, our surroundings radiate faster than us, and we absorb the radiation and get hot.

The police use special night-vision cameras to see in the dark. Cars and people are usually warmer than their surroundings, so they show up against the cold background. That can be useful if you are lost on a snowy hillside and need to be rescued. It's not so good if you are a crook escaping in a car with its headlights off. Police spotters flying overhead in a helicopter will find you.

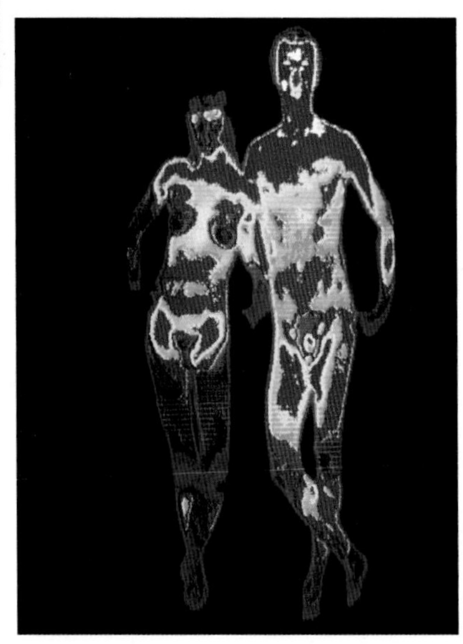

Figure 11.24 These people have been photographed using an infra-red camera. The colours tell you which are the warmest (white, yellow) and which are the coldest (blue, black) parts of their bodies

Figure 11.25 This euro banknote has special markings that only show up under UV radiation – they look bright red or blue. This is important for detecting forgeries

We all need *some* UV radiation on our skin, because it enables us to produce vitamin D.

There are types of radiation which are *not* electromagnetic waves – for example, alpha and beta radiation from radioactive materials. Sound and ultrasound are waves that radiate outwards from their source, but they are not electromagnetic.

Infra-red cameras have other uses. Doctors can measure the temperature of different parts of your body by taking an infra-red photo of you. If part of your body is infected, it will show up because it is warmer than the rest of you. It is even possible to take someone's temperature just by pointing an infra-red thermometer into their ear.

Ultra-violet is another type of invisible radiation. It is also part of sunlight, and was discovered beyond the violet end of the spectrum (see Figure 11.23). You may have seen ultra-violet used in clubs, discos or stage shows. Invisible radiation is shone on to the performers or the audience, and their white clothes glow purple. This is because the clothing absorbs the energy of the ultra-violet radiation, and then gives it out as visible purple light.

It is ultra-violet (also known as UV) that causes white skin to tan when exposed to sunlight. If your skin gets tanned, it shows that you have been exposed to hazardous radiation. UV is hazardous because it can cause skin cancer (in the same way that X-rays and gamma rays do). Skin cells produce a dark pigment called melanin in an effort to protect you from this danger. When your skin cells can't make enough of this to protect you, you get sunburnt. That's why it makes sense to use protective sun cream, which contains chemicals that are good at absorbing the harmful UV. It's no good thinking that you will be protected if your skin is wet after swimming – UV can penetrate several centimetres of water.

Out in space, the UV radiation from the Sun is very intense. Fortunately for us, the Earth's atmosphere absorbs most of it before it reaches us. The reason we worry about the hole in the atmosphere's ozone layer is that this is the part that absorbs most of the UV. Because man-made chemicals have damaged the ozone layer, we are exposed to higher levels of UV than in the past.

The complete spectrum

Today, we know that there are several types of radiation – including the ones you have learned about in this chapter – which make up the 'family' called the **electromagnetic spectrum**. Figure 11.26 shows the complete spectrum. It starts with gamma rays at one end and goes all the way to radio waves at the other end.

We can think of all these types of radiation as waves travelling along, carrying energy from a source.

- Gamma rays have the highest frequency (the greatest number of waves per second), and radio waves have the lowest frequency.
- Radio waves have the longest wavelength, and gamma rays have the shortest wavelength.

high frequency ← → low frequency

gamma radiation	X-rays	ultra-violet	visible light	infra-red	microwaves	radio waves

short wavelength → long wavelength

Figure 11.26 The electromagnetic spectrum

We can divide waves up into two types:

1 Waves that vibrate up and down as they travel along. These are called **transverse waves**. You can make a transverse wave travel along a stretched rope by waggling the end up and down. All electromagnetic waves are transverse.
2 Waves that vibrate back and forth as they travel along. These are called **longitudinal waves**. For example, a loudspeaker vibrates back and forth. It pushes on the molecules of the air, so that they vibrate back and forth, too, along the direction the wave is travelling in. So sound waves and ultrasound waves are longitudinal.

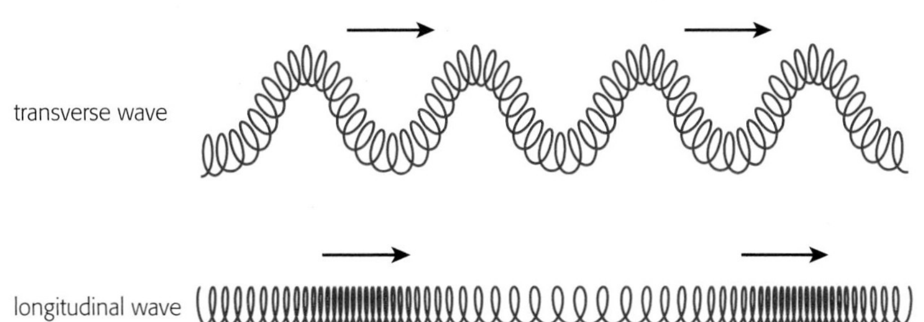

Figure 11.27 Transverse and longitudinal waves on a stretched spring

The speed of light

Electromagnetic waves all travel very fast. In fact, they all travel at the same speed in a vacuum, known as the **speed of light**.

Speed of light in a vacuum = 299 792 458 m/s
– that's about 300 000 000 m/s

Quantity	Symbol
speed	v
frequency	f
wavelength	λ (lambda)

Here is a way of working out the speed of a wave, if you know its frequency and its wavelength:

speed = frequency × wavelength $\quad v = f\lambda$

Just multiply frequency and wavelength together to find the speed.

Worked example 1

You sit on the end of a pier and watch the waves go past. Ten waves go past each minute and the waves are 20 m apart. How fast are they travelling?

Frequency = 10 waves per minute

Wavelength = 20 m

Now you can calculate the speed: speed = frequency × wavelength

Speed of waves = 10 × 20 = 200 metres per minute

We would usually give speed in m/s.

Worked example 2

A sound wave has a frequency of 500 Hz. As it travels through water, its wavelength is 3 m. What is the speed of these sound waves in water?

We know the wavelength and frequency of the waves.

So speed = frequency × wavelength
= 500 Hz × 3 m = 1500 m/s

The sound waves travel at 1500 m/s in water.

Summary

- The electromagnetic spectrum is a family of electromagnetic radiations which travel as waves, with a range of wavelengths. From short to long wavelengths the types of radiation are:

  gamma rays X-rays UV visible light
  infra-red microwaves radio waves

- Waves with short wavelength have high frequency; waves with longer wavelengths have lower frequency.
- Electromagnetic radiations of shorter wavelength (higher frequency) are more hazardous.
- All electromagnetic waves travel at the same speed in vacuum.
- Sound travels as a longitudinal wave; electromagnetic waves are transverse waves.
- For all waves:

  velocity = frequency × wavelength $v = f\lambda$

Questions

11.14 Copy the table on page 229 showing quantities and symbols relating to the speed of waves. Add a column showing the units of each of these quantities.

11.15 **a** What important property is the same for all types of electromagnetic waves?
b Which type has the highest frequency?
c Which type has the longest wavelength?
d Which type is in the middle of the electromagnetic spectrum?

11.16 **a** Name three types of electromagnetic radiation that can cause cancer. Where are these to be found in the electromagnetic spectrum?
b What does this suggest about the microwaves used by mobile phones – is it likely that they might cause cancer?

11.17 A piano plays a note whose frequency is 110 Hz. The sound waves produced have a wavelength of 3.0 m. What is the speed of these waves?

11.18 If you stretch a long spring along the bench, there are two ways to send waves along it.
a If you move the end of the spring from side to side, waves travel along the spring. What type of waves are these – transverse or longitudinal?
b If you push the end of the spring back and forth along its length, what type of waves are produced?

E On reflection

The Moon is our nearest neighbour in space. It travels around the Earth in an orbit which is almost circular, at a distance of roughly 400 000 km – that's about a quarter of a million miles.

Today we know a great deal about the Moon. Astronauts have even walked on its surface and brought back rocks for analysis. But visiting the Moon is no easy feat. You need to know precisely where it is – it would be a shame to travel all that way and find that you missed your target. So scientists have made careful measurements of the distance to the Moon and the shape of its orbit. These measurements have become more and more **precise** – today, we know the distance of the Moon from the Earth to within 3 cm!

Figure 11.28 People have had many different ideas about the Moon in the past. Was it the home of a goddess? Did it cause madness? Is it made of green cheese?

How is this done? It's like using radar to track aircraft, but much more precise. Astronomers send out radio waves from a transmitter towards the Moon. They reflect off the Moon's surface and return to Earth, where they are detected. The time it takes for the radio waves to make their round-trip to the Moon is measured, and from this the distance to the Moon can be calculated – because we know that radio waves travel through space at the speed of light.

This allows astronomers to give a very accurate picture of the Moon's orbit. They have even been able to show that the Moon is gradually drifting away from the Earth, at a rate of 4 cm every year.

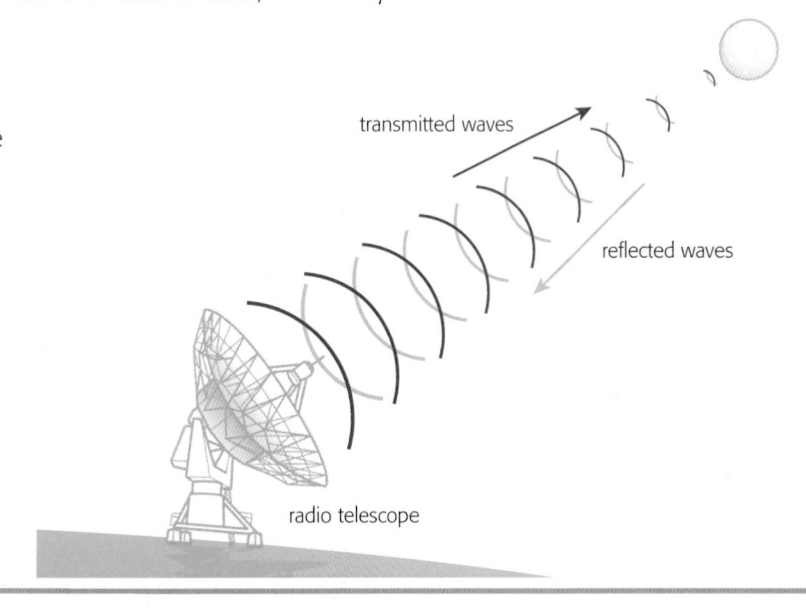

Figure 11.29 Reflecting radio waves from the Moon's surface

Time of flight

Spacecraft orbiting planets use radar to scan the surface beneath them. If the radio waves come back in a shorter time, the spacecraft must be passing over high mountains.

Radar was invented during the Second World War as a way of detecting enemy aircraft. Shortly after, 'sonar' was invented as a way of helping fishermen detect fish, and measure the depth of the sea. Fishing boats are fitted with a loudspeaker which sends pulses of sound waves down into the sea. They reflect off the seabed and are detected by a microphone.

An electronic system measures the time taken for the sound waves to travel to the seabed and back, and works out how far they have travelled. The results appear on a screen. If some of the sound waves reflect off a shoal of fish, the screen can show that, too.

You should recall the equation that relates speed, distance and time:

$$\text{speed} = \frac{\text{distance travelled}}{\text{time taken}} \qquad v = \frac{s}{t}$$

Take care with the symbols! v is for speed – think of *velocity* – and s is for distance.

We can use this equation in a different form:

$$\text{distance travelled} = \text{speed} \times \text{time taken}$$

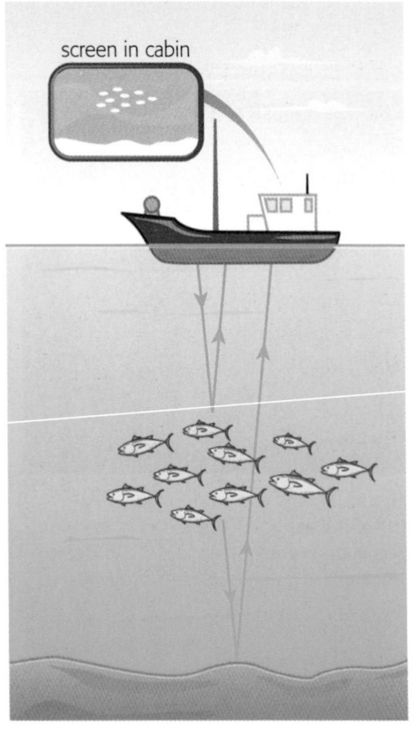

Figure 11.30 Using sonar

Worked example

Sound waves from a ship take 0.04 s to travel to the seabed and back. How deep is the sea at this point?

To solve this problem, we also need to know how fast the sound waves travel.

Speed of sound in water = 1500 m/s

$$\begin{aligned}\text{Distance travelled by sound waves} &= \text{speed} \times \text{time taken}\\ &= 1500\,\text{m/s} \times 0.04\,\text{s}\\ &= 60\,\text{m}\end{aligned}$$

This is the time taken for the sound waves to travel to the seabed *and back again*; that's twice the depth of the sea, so we have to divide by 2 to find the depth.

Depth of sea = 60 / 2 = 30 m

It's the same calculation when you measure the distance to the Moon using radio waves; you find the time take for the return journey, calculate the total distance travelled and divide by 2.

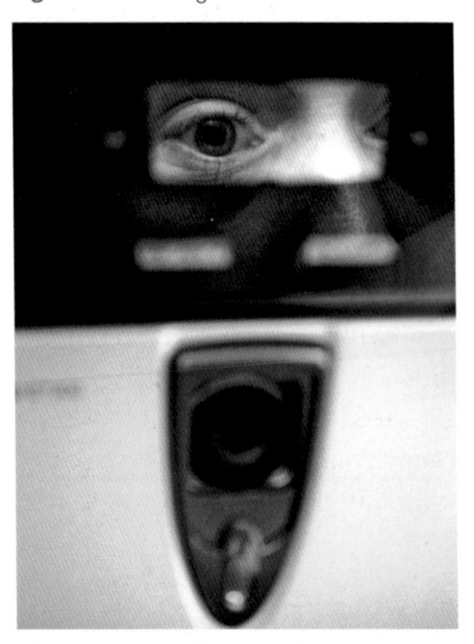
Figure 11.31 An iris scan being taken

Scanning with light

Radio waves are not the only electromagnetic radiation used for scanning. Light is used a lot, for example in bar code readers at checkouts. Red light from a laser scans across the label; the reflected light is detected and a computer looks for variations that represent the bar code.

That's quite a simple form of scanning with light, because a bar code is just a series of black and white lines. Today there are much more complex scanners which can scan things as varied as the iris of your eye – that's the coloured part of the eye. Because of slight variations in pattern and colour, each person's irises are unique. This means that iris scans can be used to check someone's identity, so in the future they may be included on identity cards.

Scanning with microwaves

Figure 11.32 Rainfall image from the TRMM (Tropical Rainfall Measurement Mission) satellite, showing the approach of Hurricane Rita to the southern United States in September 2005

Weather forecasters use scanning techniques, too. They have ground stations with large radar dishes which send out microwaves to scan clouds. Microwaves are reflected and absorbed differently by clouds, raindrops and ice crystals. Computers analyse the reflected microwaves, and work out how much rain is falling, and where. You have probably seen images produced in this way on TV weather forecasts.

Some weather satellites detect microwaves coming from clouds and rain, and this has provided vital information about how the Earth's climate is changing, particularly in tropical regions.

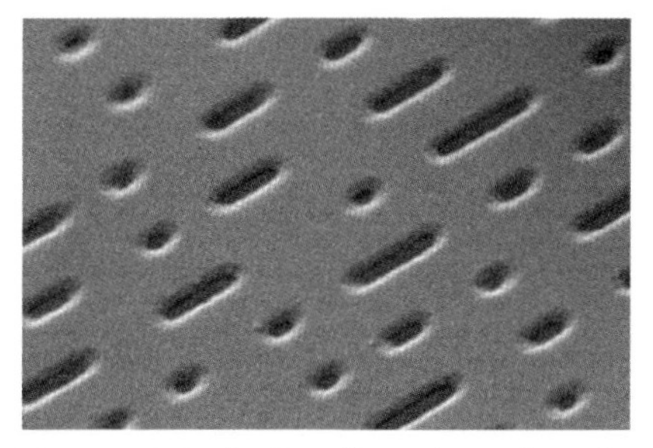

Figure 11.33 The tiny pits on the surface of a CD, magnified about 3000 times

Recall that loud sounds have a big amplitude; high-pitched sounds have a high frequency.

Figure 11.34 A fine beam of laser light reflects off the surface of the CD as it plays

Digital technology

The surface an audio compact disc (CD) is covered in tiny pits, arranged in a spiral; these carry the music in digital code. They are scanned with a laser to read them. The detector produces an electrical signal which is a series of on–off voltages (1s and 0s). An electronic circuit decodes this and drives the loudspeaker so that you can hear the music.

Music and speech travel through the air as sound waves. The waves vary in amplitude and frequency. The loudspeaker is driven by an electrical signal whose voltage varies up and down with the same pattern as the sound waves. This is called an **analogue signal**.

Nowadays, most music, speech and pictures are stored digitally. They are also transmitted digitally, as a series of 0s and 1s – that's a **digital signal**. Digital signals have several advantages over analogue signals:

1. If an analogue signal becomes degraded (distorted), it is difficult to restore it to its original form. Digital signals can be rescued if they are slightly degraded, because we know that the value of the signal is either 0 or 1. See Figure 11.35.
2. Digital storage takes up less space – a CD album is much smaller than an old-fashioned vinyl one. And thousands of tunes can be stored compactly in the memory of an MP3 player, so you can walk around listening to music without needing a bulky sound system.
3. Digital signals can be transmitted more compactly than analogue ones. So thousands of digital phone calls can be transmitted down a single optical fibre at the same time, compared with just one analogue phone call being sent as a varying voltage along a wire. Similarly, computer files (including music tracks) can be distributed at high speed – there's no need to buy a CD if you can download the contents straight to your computer or MP3 player.
4. Digital recordings can be easily manipulated in a studio, to alter their sound.
5. Digital technology can be used to create new sounds. For example, a keyboard can be programmed to work as a synthesiser, so that it makes the sound of any instrument.

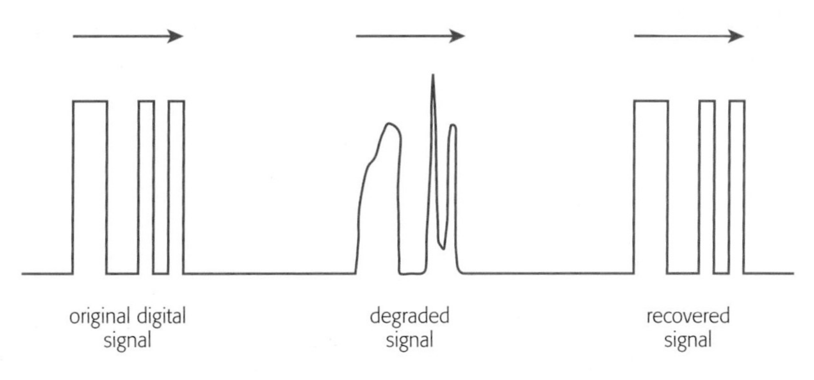

Figure 11.35 Electrical signals can become degraded as they are transmitted from place to place. Digital signals can be recovered if they are not too badly degraded

Figure 11.36 Bundles of optical fibres – you can see the light emerging from the ends of some of the fibres

Light signals

Light travels very fast – nothing travels faster through empty space. We can use light to send messages (transfer information) from place to place.

Imagine that it's a dark night. You are on a remote hillside, and you want to communicate with friends across the valley. You shine your torch in their direction; if you (and they) know Morse code, you can flash them a message.

That's a very simple way of transferring information using light. Nowadays, we use a much more sophisticated version to carry phone messages over long distances at high speed, using optical fibres.

An optical fibre is a very long fibre of high-purity glass, thinner than a human hair. Light or infra-red radiation can travel from one end to the other; by sending the light as pulses, we can send messages in digital code. This is similar to Morse code, but many millions of on–off flashes are sent every second.

Light round the bend

To understand how an optical fibre works, you need to know what happens when light travels through glass and reaches the edge of the glass. Usually, we expect the light to escape from the glass; however, it depends on the angle at which the ray of light hits the surface of the glass.

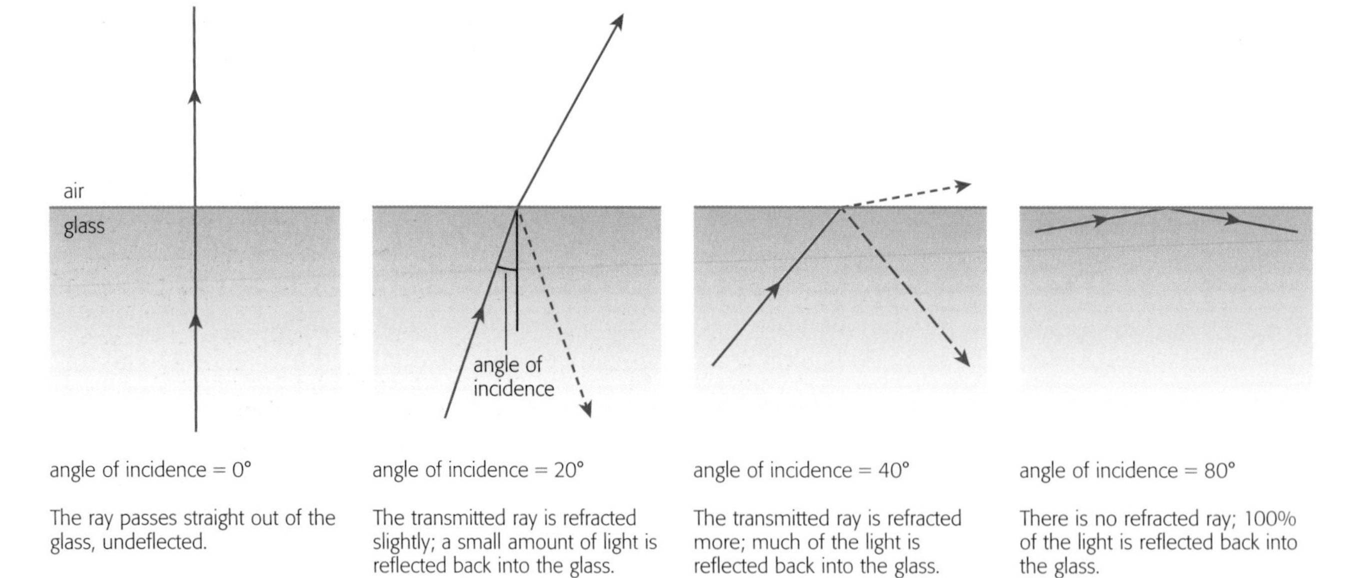

Figure 11.37 What happens when light strikes a glass–air interface

You can see that when the ray of light strikes the surface of the glass at a steep angle, it is entirely reflected back into the glass. This is known as **total internal reflection (TIR)**:

- *total*, because *all* of the light is reflected
- *internal*, because the ray is reflected back *inside* the glass.

To achieve total internal reflection, the ray must strike the inner surface of the glass at an angle of incidence greater than the **critical** angle. At a smaller angle, some of the light will emerge from the glass and be lost.

This means that a ray of light can travel along inside an optical fibre, reflecting from side to side, so that it follows the curve of the fibre. The ray loses no energy as it reflects; in fact, in the purest glass fibres, a ray can travel over 100 km and still be detected when it emerges at the far end.

Today, optical fibres are used to carry phone messages around the country as a series of high-speed digital light signals. They are even used in undersea cables, linking one continent to another.

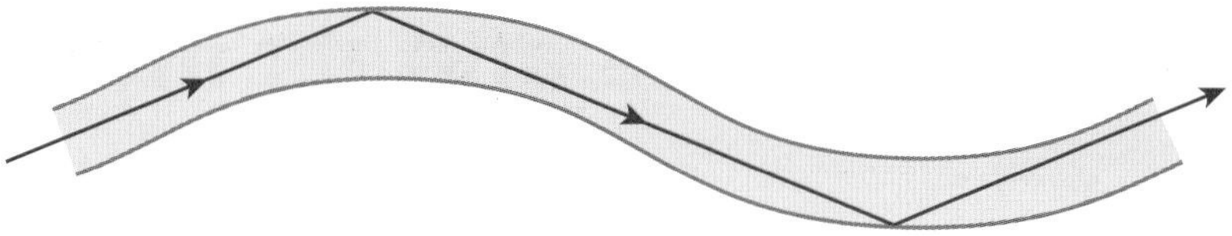

Figure 11.38 A ray of light can travel along a curved optical fibre. Each time it strikes the inside of the fibre, it undergoes total internal reflection (TIR)

Summary

- **Speed, distance and time are related by the equation**

  speed = distance / time $v = s / t$

- **We can use this equation to calculate the distance to a reflecting surface, if we know the time taken by the radiation and its speed.**
- **CDs store music and other information in digital form; they are played by scanning their surface using a laser.**
- **Music, photographs and other forms of information are now transmitted as well as stored in digital form. This means they can be transmitted quickly and stored very compactly.**
- **Total internal reflection of light allows a ray of light to travel along an optical fibre, transferring information over long distances.**

Questions

11.19 Scientists have measured the width of the Atlantic Ocean by finding how long it takes for radio waves to travel from one side to the other. Each year, the radio waves take a little longer. What does this tell you?

11.20 We cannot see the surface of the planet Venus because it is covered in dense clouds. Light cannot penetrate the clouds, but radio waves can. Draw a diagram to show how an orbiting spacecraft can measure the heights of mountains on Venus. Explain how to calculate the distance from the spacecraft to the planet's surface.

11.21 Surveyors use ultrasound waves to measure the sizes of buildings. A surveyor stands facing a distant wall; it takes 0.2 s for a pulse of ultrasound to travel to the wall and back to the surveyor. How far away is the wall? [Speed of ultrasound in air = 330 m/s]

11.22 When we build up a picture of something by looking at the radiation coming from it, we often call this 'scanning'. Make a list of the different examples of scanning from this chapter. Research at least two more examples.

Divide your list into two types of scanning:

- scanning by looking at radiation that an object gives out
- scanning by looking at radiation reflected by the object.

11.23 Music is often transmitted electronically in the form of a digital signal.

a Describe a digital signal.

b In what form is digital information stored on the surface of a CD?

c Describe how the digital information is read from a CD.

12 Space and its mysteries

A The longest journey

The first astronauts stepped onto the Moon in July 1969. In all, six NASA Apollo missions landed on the Moon, the last in December 1972. Since then, people have been into Earth orbit, but no one has ventured to the Moon or to another planet.

However, the USA has now pledged its commitment to future manned exploration of the Solar System, starting with a manned mission to the Moon scheduled for 2018.

From the Office of the President's Press Secretary

January 14, 2004

President Bush Announces New Vision for Space Exploration Program
A Renewed Spirit of Discovery

Today's Presidential Action

* Today, President Bush announced a new vision for the Nation's space exploration program. The President committed the United States to a long-term human and robotic program to explore the Solar System, starting with a return to the Moon that will ultimately enable future exploration of Mars and other destinations.

* The President's vision affirms our Nation's commitment to manned space exploration. It gives NASA a new focus and clear objectives. It will be affordable and sustainable while maintaining the highest levels of safety.

* The benefits of space technology are far-reaching and affect the lives of every American. Space exploration has yielded advances in communications, weather forecasting, electronics, and countless other fields. For example, image-processing technologies used in life-saving CAT scanners and MRIs trace their origins to technologies engineered for use in space.

www.whitehouse.gov/news/releases/2004/01/20040114-3.html

Figure 12.1

Three years in space

President Bush has set in motion the planning of a future manned mission to Mars. What are the challenges of such a daring journey?

- A crew of six would spend three years in space, far longer than any astronauts have ever achieved before.
- The journey each way would take about 8 months.
- Four astronauts would spend 18 months on the surface of Mars, constructing a home for themselves.
- The other two would remain in the orbiting mother craft.

You can probably guess many of the things the astronauts would have to take. They would need to take three years' worth of supplies – 6 tonnes of oxygen (because space is a **vacuum**), 15 tonnes of food and 150 tonnes of water. They would also need fuel for the outward and return trips, plus a landing craft, materials for building their base, an explorer vehicle, and so on – perhaps 500 tonnes of stuff.

The current space shuttle can carry a maximum of 30 tonnes, so you can imagine that a spacecraft capable of completing a return trip to Mars would have to be considerably bigger.

There's another problem – it's cold in space. While the side of a spacecraft facing the Sun gets hot (perhaps too hot), the side that is in shadow cools down. The spacecraft must be fitted with air conditioning so that the astronauts are neither fried nor frozen. Air conditioning puts quite a demand on the craft's electricity supply, which would come from its solar panels.

Keeping fit in space

Take care! When a spacecraft is in orbit around the Earth, it is *not* beyond the pull of gravity. In fact, the Earth's gravity holds the spacecraft in its orbit. The astronauts experience weightlessness because the force of gravity is 'used up' in holding them in their curved orbit. If there were no gravity, the astronauts and spacecraft would travel off in a straight line.

Astronauts experience weightlessness in space. There are two situations where weightlessness occurs:

- in orbit around the Earth
- on long flights, beyond the pull of the Earth's **gravity**.

Weightlessness sounds like fun. You can walk upside down, twist and turn in mid-air, hang from the ceiling. But weightlessness is also a problem, because your muscles and bones may start to waste away.

In everyday life (unless you are a complete couch potato), you will walk and run around, lift and pull, and so on. These activities put a demand on your body, to which it responds by building up your bones and muscles so that they can carry out your normal daily routine.

In space, if you are weightless, your muscles no longer need to hold you upright or help you run around. Your bones no longer experience the regular stresses of moving which cause them to become strong. Astronauts in orbit find that they can lose up to 2% of their bone mass each month. Weakened bones are vulnerable – a 15% loss will leave you with bones that are much more likely to fracture. Muscles waste away even more quickly – as much as 4% in a week! Your heart has to do less work in pumping blood around your body, so it also deteriorates. Astronauts can soon become dangerously weakened.

Figure 12.2 Astronauts need to keep fit. This crew member of the Space Shuttle is exercising his arm muscles while his feet are strapped because of weightless conditions

There are two ways round this. One is to use exercise machines, so that you increase the demands on your body. The International Space Station, where astronauts may spend months, is equipped with the sort of equipment you would expect to find in a gym.

The second way, which has yet to be tried out, is to create 'artificial gravity'. A spinning spacecraft (see Figure 12.3) could act a bit like a 'wall-of-death' at the fairground. The astronauts would find that they would 'stick' to the outer wall of the craft, so this would become the 'floor' of the craft. It would push upwards on their feet, just as the ground pushes on our feet when we stand on the Earth.

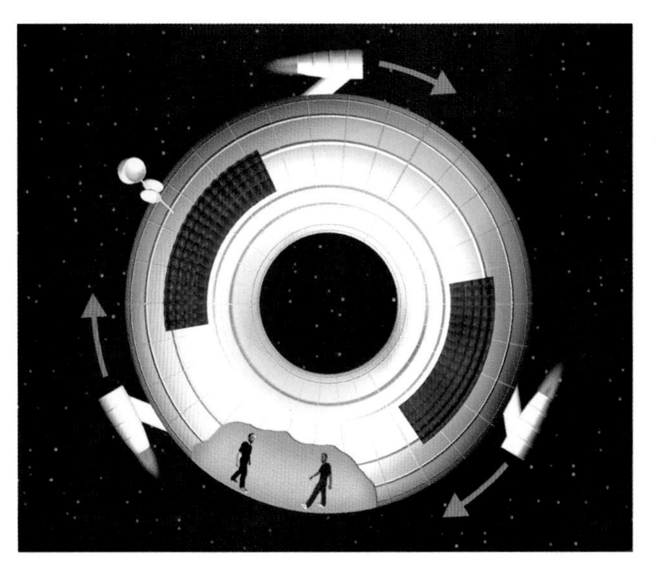

Figure 12.3 Creating artificial gravity in a spinning spacecraft

Summary

- **A manned space mission to another planet would need to take account of the fact that there is no atmosphere in space, that space is cold, and that the lack of gravity leads to weightlessness.**
- **Weightlessness leads to a deterioration in muscles and bones. This can be overcome using exercise machines or (possibly) artificial gravity.**

Questions

12.1 **a** What problems can weightlessness cause for the human body?
b What equipment can astronauts use to reduce these effects?

12.2 Astronauts visiting Mars might set up a greenhouse to grow some fresh food. What materials would they need to take with them? What would be available locally? Make two lists.

H

12.3 A spacecraft in interplanetary space will have one side facing the Sun; this side will tend to heat up. The other side, facing the cold darkness of space, will tend to cool down. Draw a diagram to show this. Indicate how the spacecraft should be made to rotate to reduce the temperature difference between its two sides.

12.4 A spacecraft could be designed to spin in order to create artificial gravity. Why is this important? What would be the effect of spinning faster?

B Astronauts on the move

It might be very exciting to be an astronaut on the Moon or on another planet, zooming around, thanks to a jet pack strapped to your back. Unfortunately, the cartoon misrepresents how this would work.

To fly horizontally, you would need to have two jets. One would point downwards, to counteract the pull of the Moon's gravity. The other would point backwards, to push you forwards.

Figure 12.4

Figure 12.5 The Mars Polar Lander firing its retro-rockets to leave the surface

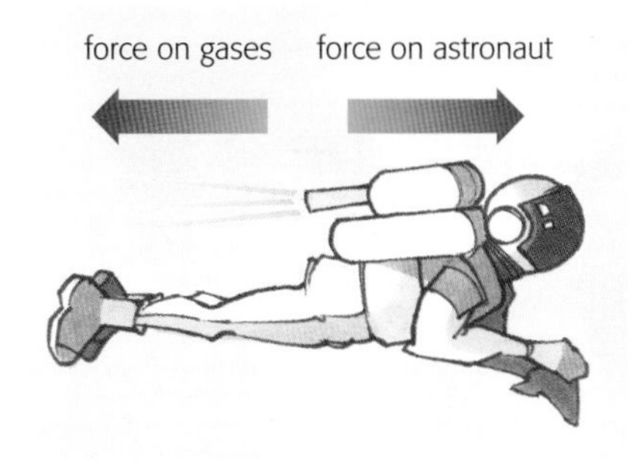

Figure 12.6 Jet forces are in pairs

Jet forces

A jet pack works just like a miniature version of the rockets that launch a spacecraft. There are two possibilities:

- It can burn fuel with oxygen, and the hot exhaust gases provide the force to push the astronaut around.
- It can contain a compressed gas, such as air, which pushes out backwards, resulting in a forward force on the astronaut.

In either case, there are *two* forces at work.

1 One force pushes the astronaut forwards.
2 The other force pushes the gases out backwards.

These two forces are equal in size and opposite in direction (Figure 12.6). We call them an action–reaction pair. If we think of the force pushing the gases out backwards as the **action**, then the forward force on the astronaut is the **reaction**.

There are many other examples of objects that move because of a pair of forces like this. If you blow up a toy balloon and let it go, it flies through the air. There is a force pushing the air out of the balloon; the balloon moves because of the opposite reaction force. Similarly, a jet engine works by blowing hot gases in one direction so that there is a force in the opposite direction.

Figure 12.7 Acceleration depends on mass

Force, mass, acceleration

The force produced by a jet pack will make an astronaut accelerate – his or her speed will change. The acceleration will depend on the astronaut's mass. A large (high mass) astronaut will accelerate less than a small (low mass) astronaut.

The greater the mass of an object, the harder it is to change its speed. If the captain of an oil tanker wants to slow down or stop, the engines have to provide the necessary force. The tanker has a huge mass – hundreds of thousands of tonnes – so it takes a very long time for it to slow down or stop.

We can show how force, mass and acceleration are related in an equation:

force = mass × acceleration $\quad F = ma$

Table 12.1 shows the units of these quantities.

Quantity	Unit
force F	newton N
mass m	kilogram kg
acceleration a	metres per second squared m/s^2

Table 12.1

Worked example 1

An astronaut of mass 80 kg has a jet pack, which gives him an acceleration of 0.5 m/s^2. What force does the jet pack provide?

Write down the equation:

force = mass × acceleration

Substitute values and calculate the answer:

force = 80 kg × 0.5 m/s^2 = 40 N

So the jet pack provides a force of 40 N.

H

Worked example 2

A car of mass 500 kg has an engine that can give a maximum force of 3000 N. What acceleration will this give the car?

Write down the equation:

force = mass × acceleration

Rearrange so that the quantity we want to calculate is on the left:

acceleration = force / mass

Substitute values and calculate the answer:

acceleration = 3000 N / 500 kg = 6 m/s^2

So the car's maximum acceleration is 6 m/s^2.

Changing speed, changing direction

Small rockets can allow satellites to adjust their speed and direction. To make things simple, imagine an astronaut with a jet pack in deep space, well away from the gravity of any planet.

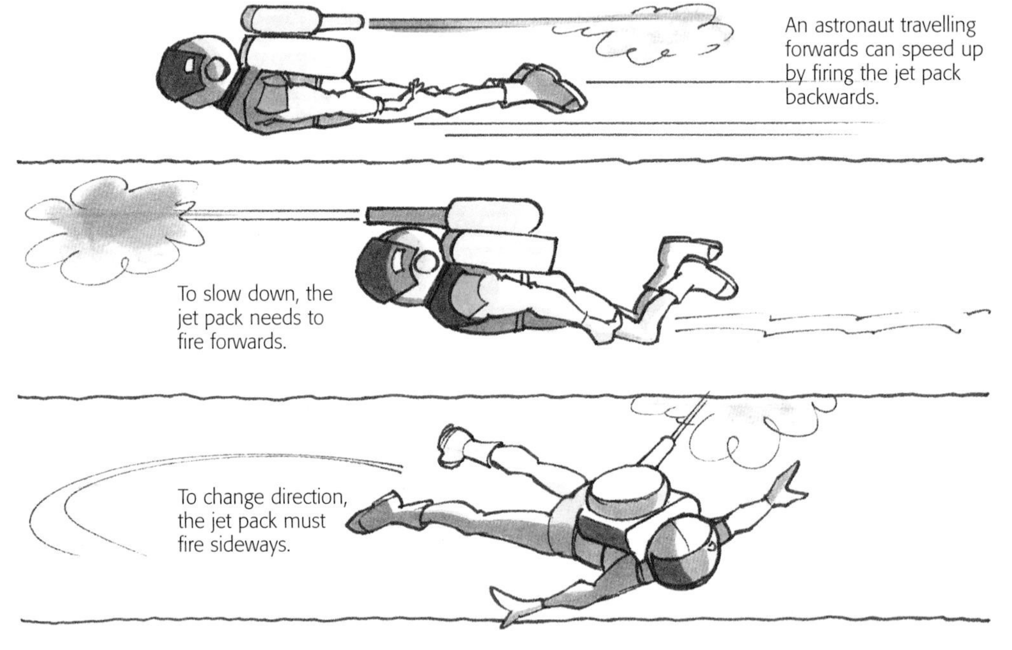

Figure 12.8 Controlling speed and direction with jet forces

Losing weight on Mars

The first astronauts on Mars will find that they feel quite a bit lighter. This doesn't mean that they have slimmed down a lot on the long journey. It is a sign that Mars' gravity is a lot weaker than Earth's.

If the astronauts measure their masses, they will find that they haven't changed. You should recall that mass and weight are not the same thing, although they are related.

- The **mass** of an object tells us how much matter it is made of; it's measured in kilograms (kg).
- The **weight** of an object is a force, the pull of gravity on the object; it's measured in newtons (N).

The greater your mass, the greater your weight, because gravity pulls on your mass to create your weight. But your weight also depends on the strength of gravity (the gravitational field) – so it can vary from place to place. You don't notice this on Earth, but you certainly would if you visited the Moon or another planet.

The Moon is much smaller than the Earth, so its gravity is much less. Mars is in-between, so its gravity is less than Earth's but greater than the Moon's.

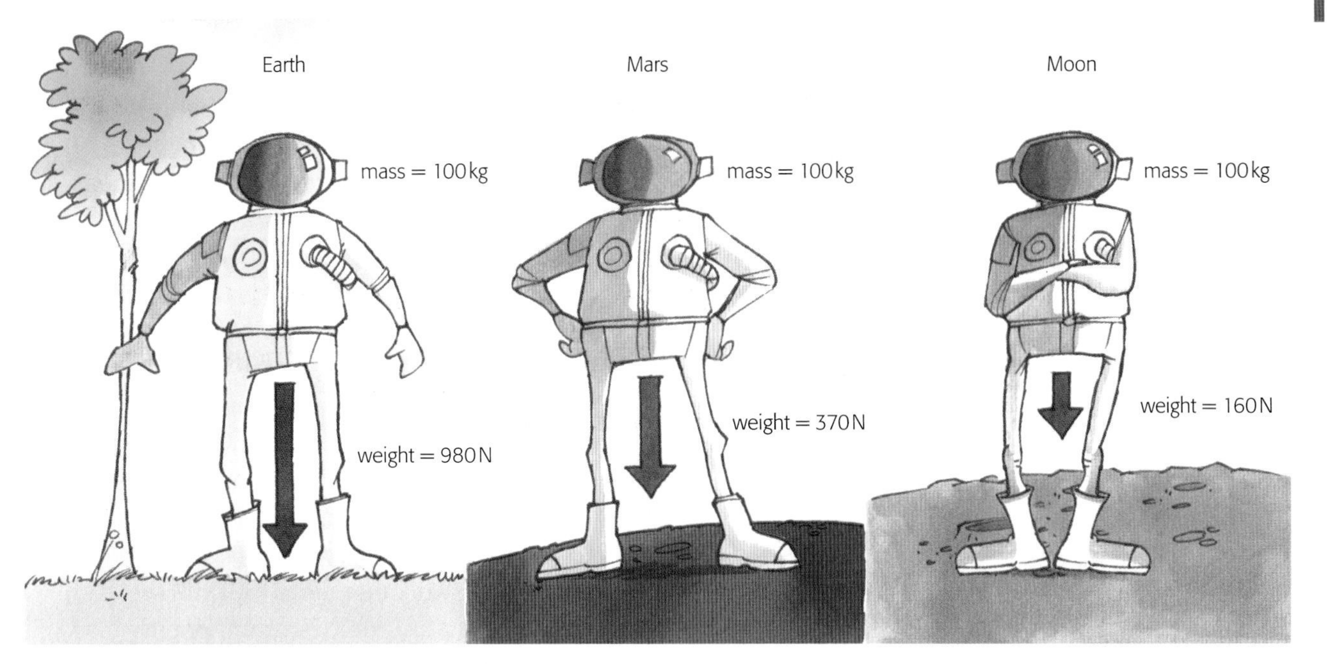

Figure 12.9 You would weigh less on Mars, and even less on the Moon

Planet/ Moon	Gravitational field strength (N/kg)
Mercury	3.7
Venus	8.9
Earth	9.8
Moon	1.6
Mars	3.7
Jupiter	23.1
Saturn	9.0
Uranus	8.7
Neptune	11.0
Pluto	0.6

Table 12.2 How gravity varies on different planets

Note that another name for *gravitational field strength* is *acceleration due to gravity*.

Table 12.2 shows the gravitational field strength of the planets and of the Moon. You can see that Jupiter, the giant planet, has the strongest gravity of all.

Calculating weight

Gravitational field strength (symbol g) is measured in newtons per kilogram (N/kg). This means it tells us the weight (in N) of each kilogram of mass. We can write this as an equation:

weight = mass × gravitational field strength $\quad W = mg$

The table shows that Earth's gravitational field strength is 9.8 N/kg, or almost 10 N/kg. This means that each kilogram of your mass is pulled on by about 10 N of force.

Worked example 3

Calculate the weight on Earth of a student whose mass is 50 kg.

Weight = mass × gravitational field strength
= 50 kg × 9.8 N/kg
= 490 N

So the student's weight is 490 N.

The importance of gravity

For astronomers, gravity is the most important force.

- Gravity holds the planets in their orbits around the Sun, and moons in their orbits around the planets.
- Gravity is the force that caused the Solar System to form in the first place, from a cloud of dust and gas in space (see page 251).
- Gravity keeps the stars of our galaxy clustered together.

Figure 12.10 This artist's impression shows what happens when a black hole passes close to a star. The black hole itself is hidden in the centre of the spinning white-hot disc on the right. It is pulling matter from the star on the left. Two jets of high-energy particles are fired out from the centre of the disc

Black holes

H

It was over two centuries ago that astronomers first suggested that **black holes** might exist. A black hole is a small, massive object in space. It is very dense so that its gravitational field strength is exceedingly high, so high that nothing can escape from it, not even light – hence the term black hole. Light is the fastest thing in the Universe, so the gravitational pull of a black hole must be enormous to prevent light from escaping.

Nowadays, we know that black holes exist even though they can't be seen. Some form when massive stars explode at the ends of their lives; these are quite small black holes. If they pass close to another star, their gravity may start pulling material from that star (see Figure 12.10), and this is what astronomers look for.

Supermassive black holes contain the mass of perhaps millions of stars. There may be one lurking at the centre of our galaxy, but no one has a clear idea of how it might have formed.

Summary

- A spacecraft's rocket engine works by pushing out hot gases in one direction. This force (the action) results in an equal, opposite force (the reaction) on the spacecraft.
- An unbalanced force on an object with mass causes an acceleration. These are linked by:

  force = mass × acceleration $F = ma$

- Mass is a measure of the amount of matter of which an object consists.
- Weight is the force of gravity, acting on an object's mass.

  weight = mass × gravitational field strength $W = mg$

- Gravitational field strength g (in N/kg) is also known as acceleration due to gravity.
- The force of gravity acts on all objects with mass. Because astronomical objects have large masses, gravity is an important force in astronomy.
- H If the mass of an object is compressed into a small enough volume, it may become a black hole, an object from which even light cannot escape.

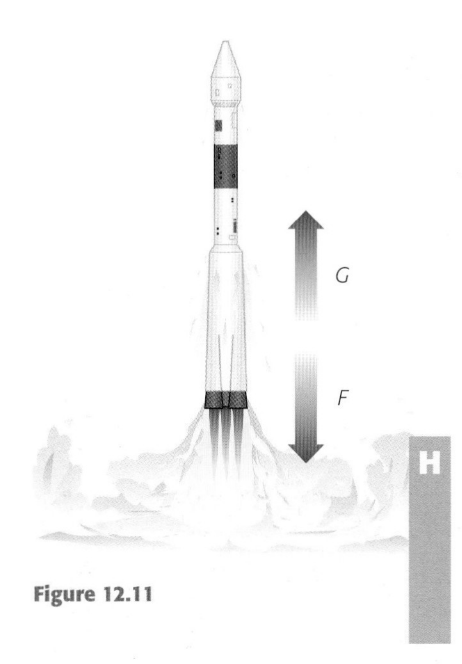

Figure 12.11

Questions

12.5 Calculate the weight of a rock of mass 5 kg on the Moon, where the gravitational field strength is 1.6 N/kg.

12.6 What force is needed to give a spacecraft of mass 600 kg an acceleration of 2.5 m/s^2?

12.7 Figure 12.11 shows a space rocket. Hot gases are pushed out of the rear end by the force *F*; this results in a force *G* acting on the rocket. What can you say about:

a the sizes of these forces?

b the directions of these forces?

H

12.8 A rock is weighed on the Moon; its weight is 3.2 N. What will its weight be when it is brought back to Earth? Use data from Table 12.2 (page 241) to help you work this out.

12.9 An astronaut of mass 100 kg uses a jet pack to move around in weightless conditions. It pushes him with a force of 40 N. What will his acceleration be?

C Remote sensing in space

In January 2005, an extraordinary space journey reached its climax. The Cassini spacecraft, in orbit around Saturn, released the Huygens probe. This small spacecraft parachuted down to the surface of Titan, the largest of Saturn's moons.

As it descended, the probe sent back images of Titan's surface. These hint at why Titan is of interest. The landscape is somewhat Earth-like. Perhaps one day Titan might become an outpost of human life in the Solar System, maybe even a staging post on the way to colonising a planet orbiting a distant star.

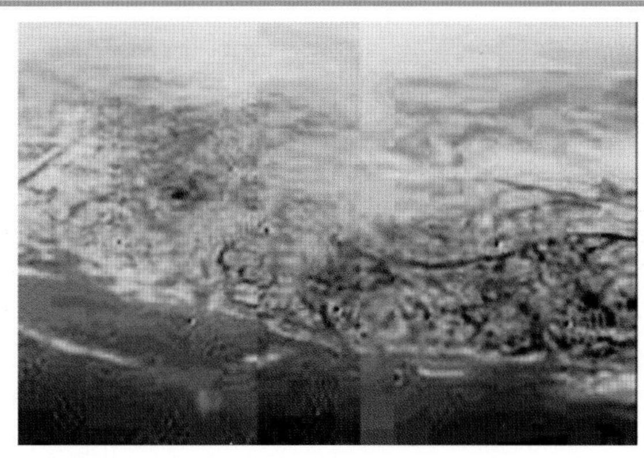

Figure 12.12 This composite photo was taken by the Huygens probe when it was 8 km above the surface of Titan. There seems to be higher ground towards the back, with a 'sea' in the foreground

The view from Earth

You don't have to be an astronaut to learn about space. You can find out quite a lot by observation from Earth. **Telescopes** have been used for four centuries, and they have been greatly improved in that time.

Telescopes capture more light than the human eye, so they can reveal many more stars than we normally see. They also allow us to see some of the more distant planets – Uranus, Neptune and Pluto, if you know where to look. With a camera attached, a telescope can gather light over several hours, so that even fainter objects can be spotted.

The Earth's atmosphere absorbs much of the radiation reaching us from space; it lets through mainly visible light, radio waves and microwaves. So as well as telescopes that make use of light, we have radio telescopes ('dishes') which gather radio waves (and microwaves).

Figure 12.13 Radio telescope at Jodrell Bank in Cheshire

Because the atmosphere causes some distortion of images, and prevents us from 'seeing' many types of radiation (infra-red, for example), telescopes have been sent up into space, above the atmosphere. The Hubble Space Telescope is the most famous of these, and it has produced photographs of some of the most distant, faint objects in the Universe, never seen before. Other space telescopes gather infra-red, X-rays and even gamma rays to give images of some of the most energetic objects in space.

Figure 12.14 Each speck of light in this Hubble Space Telescope photograph is a distant galaxy

Measurements in space

All of the other eight planets in the Solar System have been visited by spacecraft. Sometimes the spacecraft have landed; sometimes they have gone into orbit, or simply flown close by. These craft (or 'probes') carry instruments which can make useful measurements of various aspects of the planets, and radio their data back to Earth. Here are some of the measurements they can make:

- Cameras can photograph the planet's surface. By taking measurements of infra-red and ultra-violet radiations, they can find out about the types of rocks, whether water is present, and so on.
- Temperature sensors measure the temperature at different points above the planet's surface.
- The planet's magnetic field can be measured, as well as its gravity.
- The planet's atmosphere can be analysed, including its pressure.
- By reflecting radio waves from the planet's surface, maps can be made of mountains, valleys, craters and so on.

Spacecraft like these are very complex. They carry on-board datalogging computers which process and store the data from the different sensors, and send it back to Earth at a suitable opportunity.

Being there

The Apollo programme landed six spacecraft on the Moon. The astronauts returned to Earth bringing a total of 382 kg of rock and soil samples. These have been analysed to determine their composition; scientists have also looked at the effects of radiation on them. Mostly they proved to be basalt, a volcanic rock formed three billion years ago when the Moon was young. One type of moon rock, not found on Earth, has been named 'kreep' because it consists of potassium (symbol K), rare earth elements and phosphorus (P).

No one yet has brought back rocks from another planet. However, spacecraft have landed on the surface of Mars and sent out small rover vehicles to examine the rocks. These rovers carry cameras, which enable operators on Earth to control where they should go. They also carry instruments for analysing samples of rocks and soil. They can look for tell-tale signs of life – for example, carbonate rocks might indicate that, in the past, Mars was home to living organisms which produced carbon dioxide.

Figure 12.15 (right) A Mars Exploration Rover examining a rock

Figure 12.16 Astronauts were sent into space to repair the Hubble Space Telescope, something which a robot would find hard to do. Here one of them is training underwater, an easy way of simulating the weightless conditions they had to work in

There is a hot debate among astronomers and space scientists – is it worth the effort to send astronauts to other planets, or should we rely on unmanned spacecraft and robotic landers? Perhaps a human astronaut might notice interesting features that a robot lander might miss. But is it worth risking human lives?

H

What's the point?

People debate the question of space exploration, whether this includes manned space flight or simply unmanned craft. What do we gain from it? Is it worth the expense, or should we spend the money on other things – better health services, or the alleviation of poverty? Here are some areas in which space exploration is said to benefit us – what do you think?

1 **Economic/social:** Spacecraft orbiting the Earth can tell us where new natural resources might be found. Some guide aircraft, ships and even cars and lorries. Others transmit TV and telephone and other information signals.
2 **Technological:** Space is a hostile environment, and space exploration puts great demands on our engineering capabilities. Many of the technologies developed for space exploration have found other uses back here on Earth.
3 **Environmental:** Monitoring spacecraft can tell us how human activities are altering the Earth – changing the climate and destroying natural environments. They can act as a warning system, telling us when we are endangering our own future.
4 **Philosophical:** Our developing picture of space gives us a new view of our place, as humans, in the Universe. Where once we seemed to be at the centre of things, we now realise that we inhabit a small planet orbiting an average star in just one of many billions of galaxies.

Summary

- We can learn about the Solar System and the rest of the Universe using telescopes, unmanned spacecraft, robotic landers, and manned spacecraft.
- Spacecraft carrying sensors and datalogging computers can collect information about other planets and their moons.
- H Space exploration and our increasing knowledge of the Universe can bring economic, social, technological, environmental and philosophical benefits.

Questions

12.10 a Unmanned spacecraft can orbit other planets. Give some examples of the sorts of information they can send back to us.

b How is this information gathered?

c What additional information might be gathered by robotic landers, sent down onto the surface of the planet?

H **12.11 a** List some uses of satellites in orbit around the Earth.

b Why might similar satellites in orbit around other planets be useful?

D The scale of things

An international group of astronomers has formed a body called the Spaceguard Foundation. They are worried – very worried – that one day we may all die in an appalling impact, when a giant rock from space crashes into the Earth. Even if we did not all die, a lesser impact would have the most serious effects on our way of life.

What sort of 'giant rock' do these astronomers imagine might strike the Earth? There are two possibilities: comets and asteroids.

Could this possibly happen? The Earth seems a pretty safe sort of place, with little evidence of earlier impacts from space. However, look a little more closely and the evidence begins to emerge.

Figure 12.17 Meteor Crater, Arizona, which is 1200 m across

- In the US state of Arizona, the huge Meteor Crater is where a space rock landed, roughly 49 000 years ago. The rock's size is estimated at just 40 m across; there are many rocks much bigger than that out in space.
- Other craters have been identified, although most of them have been eroded over time so that they are hard to spot.
- Mystery still surrounds an event at Tunguska in Siberia, in 1908. Local reindeer herders saw a brilliant fireball in the sky. It seems that a comet may have exploded shortly before it hit the ground. The blast destroyed houses and whole forests. People and animals were swept away.

The Spaceguard astronomers use telescopes to watch out for Near-Earth Objects (NEOs). These are comets or asteroids whose orbits might bring them close to Earth. Newspapers and TV usually report any sightings. Then the astronomers make further observations until they are sure of the object's orbit. If it will miss Earth, they sound the 'all clear'. If not, brace yourselves! We have no way of stopping a space rock heading our way.

Comets and asteroids

What are comets and asteroids? They are very different objects, made of different materials, although both can be thought of as 'debris' left over from when the Solar System formed. Both orbit the Sun, but their orbits are rather different.

A **comet** is a frozen ball of ice and dust, usually a few kilometres across. The orbit of a comet is an elongated ellipse around the Sun. Comets spend most of their time far out in the coldest reaches of the Solar System, beyond Pluto. However, from time to time a comet will travel in towards the Sun. As it does so, it warms up and starts to evaporate. The material that streams off it forms its tail. It passes around the Sun and then speeds back out into the depths of space.

An **asteroid** is a rock from the asteroid belt, between Mars and Jupiter. The largest asteroids are hundreds of kilometres across, but some are less than 10 m across. They are thought to be planets that never formed; the gravitational pull of Jupiter prevented them from sticking together. Occasionally, two asteroids may collide with each other. This may send one tumbling in towards the Sun, and that could bring it close to Earth.

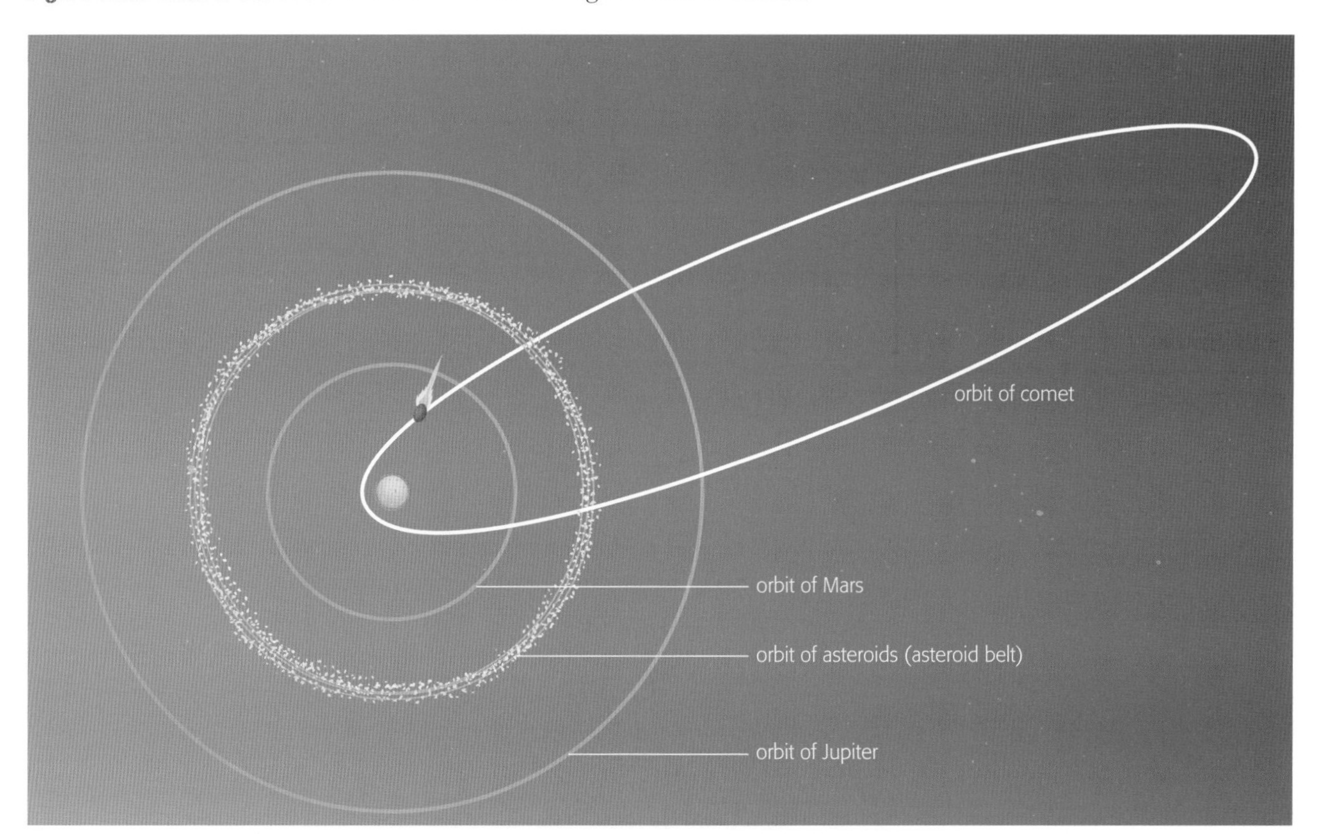

Figure 12.18 Orbits of asteroids and comets

Target Earth?

Fortunately for us, the Earth is a very small target for a Near-Earth Object to hit. The usual pictures we see of the planets orbiting the Sun can give a false impression of the size of the planets – they cannot be shown to the same scale as their orbits, or they would be far too small to see. The Earth is really a tiny speck moving in the emptiness of space. And, although the Solar System seems vast to us humans,

- the Sun is just one of billions of stars in our galaxy, the Milky Way
- the Milky Way is just one of billions of galaxies in the Universe.

Table 12.3 gives an idea of the great distances involved when we think of the Solar System and its place in space.

Table 12.3 Distances in space

Object	Diameter (km)	Distance from Earth (km)
Earth	12 760	
Moon	3475	384 000
Sun	1 490 000	149 600 000
Pluto	2390	5870 000 000
Solar System	20 000 000 000 000	
Next nearest star		40 000 000 000 000
Milky Way	1000 000 000 000 000 000	
Next nearest galaxy	300 000 000 000 000 000	1600 000 000 000 000 000

The Universe itself, which consists of all the matter and energy we can ever know of, has an estimated diameter of 100 000 000 000 000 000 000 000 kilometres!

Light-years

Here's another way in which astronomers think of these enormous distances.

1 **light-year** is the distance travelled by light in 1 year.
1 **light-minute** is the distance travelled by light in 1 minute.

- It takes light from the Sun about 8 minutes to reach us. We say that the Sun is 8 light-minutes away.
- It takes light from the next nearest star (Proxima Centauri) about 4 years to reach us. We say that Proxima Centauri is 4 light-years away.
- The next nearest galaxy (Andromeda) is about 170 000 light-years away.
- And the Universe is about 14 billion light-years across!

Light travels faster than anything else (as far as we know). That is why astronomers choose the time taken for light to travel as a way of measuring distances. Spacecraft travel much slower than light – at about one-thousandth of the speed of light. (That seems slow, but it's about one thousand times the motorway speed limit.) So a spacecraft would take about 4000 years to reach the next nearest star, and 170 million years to reach the next galaxy.

Taking a chance

An asteroid or comet would have to be on a very precise course to strike the Earth. To get an idea of this, imagine looking up into the night sky. You see the planet Venus, shining brightly. Now imagine aiming a rocket straight at it. The chances are you would miss.

So the chance that a particular space rock would hit us is small. The problem is, if we were hit, the consequences would be appalling. The relatively small rock that crashed into Arizona 49 000 years ago produced a crater over a kilometre wide and hundreds of metres deep.

An enormous plume of dust and fire would have been thrown up. What are the consequences of an impact like this?

- Forest fires are ignited and spread wildly.
- Water evaporates, causing dense clouds and heavy rainfall.
- The material thrown up into the air darkens the skies all round the world, as dust spreads through the upper atmosphere.
- Sunlight is reflected back into space, so that the Earth cools down.
- An impact in the sea would cause a giant tsunami, far bigger than the tsunami of 2004.

Many animals and humans far from the actual impact might survive in the short term. But it is likely that many plant species would be wiped out, which could destroy the basis of whole ecosystems. So a major impact would eventually result in many thousands or even millions of human deaths – perhaps a large fraction of the human population. An asteroid impact 65 million years ago is thought to have been responsible for the extinction of most dinosaur species. The global cooling caused by the impact was too much for them to adapt to.

Table 12.4 compares the probabilities that an individual (it could be you) will die from various causes. You can see that you are more likely to be killed by the effects of an asteroid or comet impact than in a plane crash. This seems surprising. Here is how to think of it:

- A plane crash is a frequent event that kills a small number of people.
- An asteroid or comet impact is a rare event that kills a very large number of people.

Cause of death	Probability of death
Road accident	1 in 100
Murder/ manslaughter	1 in 300
Fire	1 in 800
Electrocution	1 in 5000
Asteroid or comet impact	1 in 10 000
Airliner crash	1 in 30 000
Tornado	1 in 60 000
Fireworks accident	1 in 1 000 000

Table 12.4 Some causes of death and their probabilities

So although there is only a small chance that a space rock will hit Earth while you are alive, if one does, the effects will be devastating. The astronomers of the Spaceguard Foundation think we should be prepared. They would like to see telescopes dedicated to keeping a constant look-out for Near-Earth Objects, and they would like to develop methods of diverting any NEO that is heading our way – perhaps by blasting it with nuclear weapons. All this would cost a lot of money, but it might save us from a global disaster.

Figure 12.19 One way of diverting an asteroid as it heads for Earth could be to deflect it with a rocket

Summary

- An asteroid or comet impact could cause a global catastrophe.
- Planets and most asteroids orbit the Sun in roughly circular orbits; a comet's orbit is a long ellipse.
- The Solar System is mostly empty space, with vast distances separating the orbits of the planets around the Sun.
- The Sun is just one of billions of stars in our galaxy, the Milky Way.
- There are billions of widely separated galaxies in the Universe.
- 1 light-year is the distance travelled by light in 1 year.

Questions

12.12 Put the following objects in order of size, starting with the smallest:
Earth Universe Sun Solar System Moon
Milky Way Halley's comet

12.13 Draw a diagram to show:
- the orbits of the Earth, Mars and Jupiter around the Sun
- the asteroid belt
- the orbit of a comet.

12.14 **a** What is meant by the term 'Near-Earth Object'?
b What might be the consequences of an impact of a Near-Earth Object with the Earth?

E Is there life out there?

The SETI Institute is based in California. It employs over 100 people in the Search for Extra-Terrestrial Intelligence. ('Extra-terrestrial' means 'beyond Earth'.) These astronomers, engineers and scientists are hoping to be the first to detect signals coming from space that would show there is intelligent life out there, somewhere.

It seems unlikely that there is life of any kind on any of the planets of the Solar System. But in the last few years astronomers have discovered planets orbiting around other stars, and it is not impossible that some of these are home to intelligent life.

There are two ways in which the SETI people think they may be able to detect intelligent life on a distant planet:

Figure 12.20 The Parkes radio telescope in Australia, often used by the SETI Institute (www.seti.org) to gather radio signals from space

- They use radio telescopes to pick up radio waves arriving from space. Then they analyse them to see if there are any patterns, which might suggest that they are signals from an alien civilisation.
- They use optical (light) telescopes to look for brief flashes of light from other stars. An alien intelligence might know that we exist here on Earth, and could be flashing powerful lasers in our direction.

Space is full of radio waves. Some come from Earth, because some of the energy of our radio and TV broadcasts passes out through the atmosphere. Other, more random, bursts of radio waves come from stars or from energetic clouds of dust and gas. The trick is to spot a signal that seems like a deliberate attempt to communicate.

The SETI Institute has some radio telescopes of its own, and it shares others which are also being used to study distant stars and galaxies. Because they gather masses of data, the SETI scientists need help in analysing it all. Anyone with a computer can download some SETI data and the software to analyse it, thus helping in the search.

H

Planet Goldilocks

We know that there are planets which orbit other stars. Could any of them be home to intelligent life? Or to any form of life, however simple? To start with, we should think about what makes Earth suitable for life.

No one can be certain, but it seems likely that a fundamental requirement for life is liquid water. This means that a planet must not be too hot or too cold. Its average temperature must be similar to Earth's (which is around 15 °C). The planet must therefore not be too close to its star, or too far away. For an average star like the Sun, this means that there is a zone, between 100 million km and 200 million km from the star, in which a planet might exist with liquid water on its surface. This zone is sometimes called the Goldilocks Zone, because a planet in the zone is neither too hot nor too cold; its temperature is just right.

The steady Sun

Other requirements are that the star around which a planet orbits must be stable, and it must shine for many millions (or even billions) of years.

Some stars vary in brightness quite regularly, and this means that the temperature of any planet would vary up and down rather dramatically, perhaps between boiling hot and freezing cold – not ideal for life.

A star needs to shine for a long time, because it probably takes a long time for life to evolve. Certainly, life on Earth seems to have taken 3 billion years to evolve to the stage of intelligent (human) life. The Sun has been around for 4.5 billion years, and is estimated that it will last for at least the same amount of time again.

The life of a star

How does a star form? Space is not empty but contains gas and dust. This material is thinly spread out, so its gravitational pull is very weak. In a place where the material is denser than average, however, it may start to collect together, pulled in by its own gravity. The result is a swirling, collapsing cloud called a **nebula**.

As this cloud collapses further under its own gravity, it becomes denser and hotter. Eventually, the densest material at the centre starts to glow and then burn brightly – a star has formed. Around it, lumps of colder matter may join together to form planets.

Figure 12.21 This nebula is a cloud of dust and gas in which about ten new stars are forming

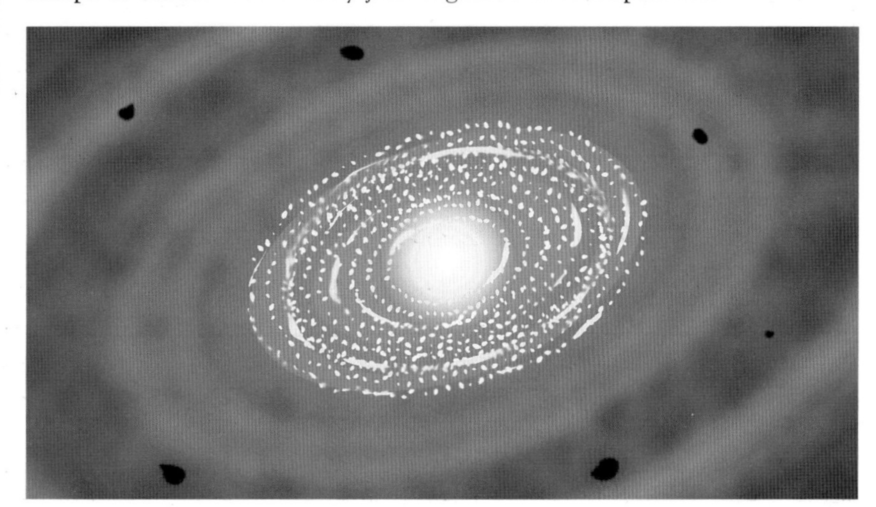

Figure 12.22 A star and planetary system being born

Eventually – after millions of years – the star burns steadily. Surprisingly, the biggest stars burn out most quickly – perhaps in just 20 million years. That doesn't leave time for intelligent life to evolve on any of their planets. Smaller stars, similar to the Sun, burn for billions of years.

The death of a star

What will eventually happen to the Sun? By studying many other stars, astronomers have a good idea. The Sun will eventually expand into a cooler **red giant** star, big enough to engulf the inner planets, perhaps including Earth. The outer layers will then blow off into space, leaving the hot, white core – a **white dwarf**. This will gradually fade away.

Stars much bigger than the Sun behave differently. At the end of the red giant stage, the core of the red giant contracts, leading to an explosion called a **supernova**, which throws vast amounts of material out into space (see page 87). What's left behind is a tiny, very dense **neutron star**, or even (for the biggest stars) a **black hole** (see page 242).

This pattern of the formation, life and death of a star is known as **stellar evolution** (although it is very different from the evolution of living organisms).

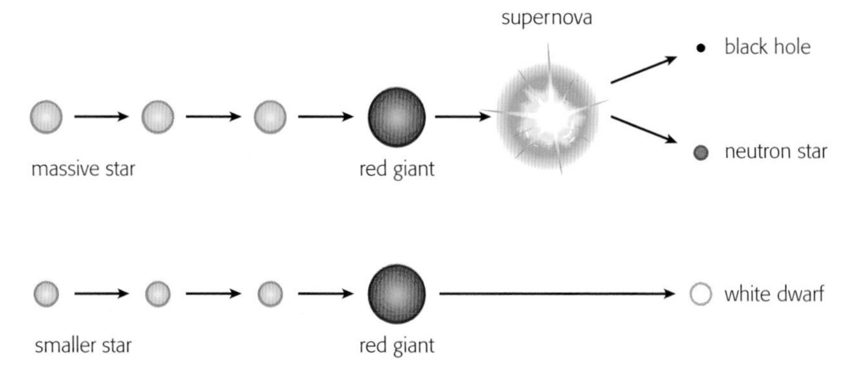

Figure 12.23 The possible fates of a star after it has finished its main stage of burning. A supernova may result in a neutron star or a black hole, depending on the size of the original star

Intelligent life

People argue heatedly about whether or not there might be intelligent life elsewhere in the Universe. Here are some of the points people make:

- There must be billions of stars with planets orbiting around them – some of these must have life on them.
- If there really is intelligent life out there, it would have visited us by now.
- Our human civilisation has only recently reached the point where we can travel through space, after billions of years of evolution. Our civilisation may end soon, probably because of climate change. Perhaps all alien civilisations end in this way, before they can make contact.
- If there is intelligent life elsewhere, it may have the sense to avoid contacting us.
- Our planet is a tiny speck in space. It would be difficult for any alien life to find us, even if they wanted to.

Even if we do believe that there is intelligent life elsewhere in our galaxy, we are still left with the question: Would it be sensible to try to make contact, or should we try to keep ourselves to ourselves?

Summary

- Stars and their planets form from a nebula – a cloud of dust and gas.
- Many stars shine steadily for millions or even billions of years.
- An old star expands to become a red giant. This may collapse to become a white dwarf, or it may explode as a supernova, leaving behind a neutron star or a black hole.
- H For life to evolve on a planet, the star must shine steadily for a long time. The planet must orbit at a suitable distance, so that liquid water can exist.
- We do not know whether intelligent life exists elsewhere in the galaxy. Attempts have been made to detect signals (radio waves and light) coming from such life.

Questions

12.15 a Which three of the following are possible final stages in the life of a star?

white dwarf   supernova   nebula
black hole   red giant   neutron star

b Draw a diagram to show how all of the stages listed above are linked in the evolution of a star.

H **12.16** Why are we unlikely to find living organisms on Neptune, or on Mercury?

12.17 A massive star shines for only a few tens of millions of years. Why is life unlikely to be found on a planet orbiting such a star?

F The life of the Universe

Edwin Hubble was an American astronomer. In the 1920s he worked at the Mount Wilson Observatory in California, where he could use a much larger telescope than he had used previously. It had a concave mirror 2.5 m in diameter to collect the light of distant stars and galaxies. And it was galaxies that interested Hubble. Because he had such a powerful telescope, he was able to make clear photographs of many distant galaxies, far beyond the Solar System. Before this, these other galaxies had appeared as dim smudges in space; now Hubble could see their shapes in detail.

By studying the light from these galaxies, Hubble could work out two things about them:

- how far away they were
- how fast they were moving.

He discovered an extraordinary thing: most of the galaxies he looked at were moving away from us, and the farther away they were, the faster they were moving. He had discovered that the Universe seems to be expanding.

Figure 12.24 Edwin Hubble (1889–1953) looking through the telescope's eyepiece at the Mount Wilson Observatory

Figure 12.25 A computer graphic of the expanding Universe. Space itself is expanding like a growing bubble, carrying the galaxies farther and farther apart from each other

Thinking backwards

We know from Edwin Hubble's observations that the Universe is expanding. It doesn't take much thinking to realise that, in the past, the galaxies were much closer together. In fact, at some time in the distant past, all of the matter and energy in the Universe must have been packed together into a tiny space. This is one of the reasons why most astronomers now believe that the Universe began with a Big Bang, roughly 14 billion years ago.

It's hard to picture the Big Bang. All of the matter and energy in the Universe was compressed together in an infinitely small space. It burst outwards, an enormously hot soup of sub-atomic particles. As it got bigger, it cooled. The particles slowed down and clustered together, forming atoms and molecules, and eventually stars and galaxies – which all moved apart with great amounts of kinetic energy.

Not everyone agrees with this **Big Bang theory** of the origin of the Universe. Some astronomers are uncomfortable with the idea that the Universe had a beginning – they want to know what happened before that. Here are a couple of alternative theories:

- The **Steady State** theory says that the Universe has always been like this – there was no beginning. Instead, as the Universe expands, new matter appears to fill in the empty space created by the expansion.
- The **Oscillating Universe** theory agrees that the Universe is expanding now, but that it might have been contracting in the past, and that it will contract again in the future. So the Universe expands, contracts, expands again – there was no Big Bang, and no beginning to the Universe. It will oscillate forever.

Thinking ahead

If the Big Bang theory is correct, how will the Universe change in the future? One idea is that the expansion of the Universe will gradually slow down; then it will start to contract, until all of the galaxies come together in a very hot Big Crunch. We can show this idea as a graph.

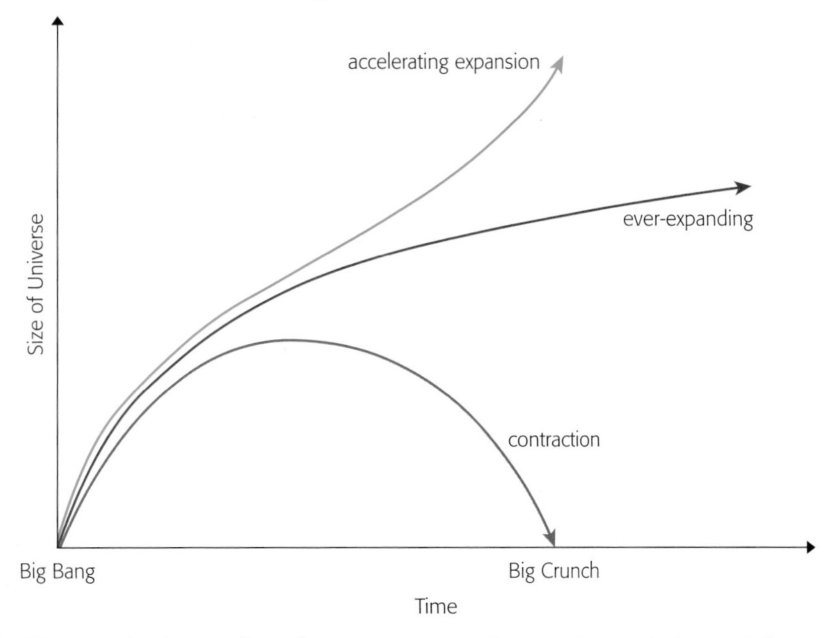

Figure 12.26 Possible fates of our expanding Universe

The graph shows that this is just one of several possibilities. The Universe might expand forever, getting bigger and bigger and cooler and cooler. The latest measurements suggest something different. It seems that the Universe's expansion is speeding up, so that we live in an accelerating Universe. There will be no Big Crunch.

What we don't know

This book is full of scientific ideas which we think are well established. Most scientists agree with most of these ideas, and they are unlikely to be overthrown by new evidence. However, we could fill another book with all of the things that scientists don't yet know. This chapter has hinted at some areas of uncertainty. These are some questions that remain for the next generation of scientists to tackle:

- How will the Universe end?
- What existed before the Big Bang?
- Is there life on other planets?
- Is there intelligent life elsewhere in the Universe?

H

Measuring galactic speeds

Edwin Hubble's observations led to the Big Bang theory. How could he tell that distant galaxies were moving away from us? How could he measure their speed?

He looked at the light coming from the galaxies. He noticed that it appeared redder than the light from our own galaxy; that is, the light waves had longer wavelengths – they were shifted towards the red end of the spectrum. This effect is called the **red shift**, and Figure 12.27 shows why it happens.

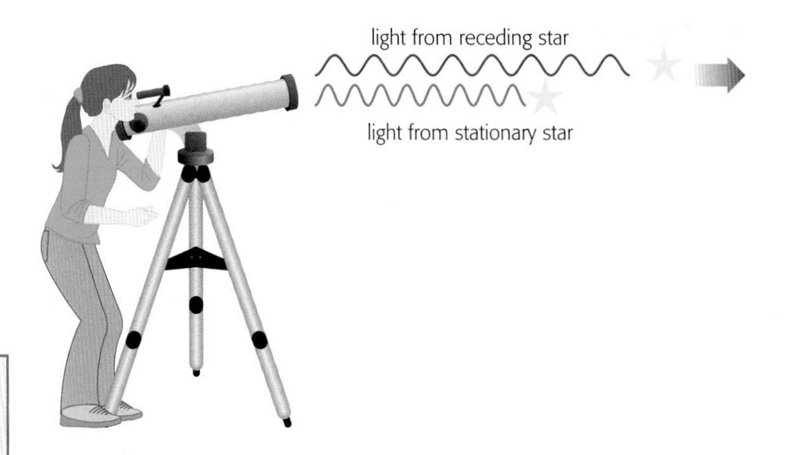

Figure 12.27 The red shift effect

This is similar to the Doppler effect for sound: if an ambulance is sounding its siren, it sounds more high-pitched as it approaches – the sound waves are squashed together – and more low-pitched as it moves away – the sound waves are stretched out.

Light detected from a stationary star or galaxy has a particular wavelength. If the star is moving away from the detector, its light waves are stretched out. Their wavelength increases, which means the light looks redder.

Hubble measured the red shifts for many galaxies and used this data to calculate the galaxies' speeds.

Support for the Big Bang theory

The Big Bang was very hot – billions of degrees. There was a lot of electromagnetic radiation throughout space – light, ultra-violet, even X-rays and gamma rays. As the Universe expanded, this radiation cooled. Today, the Universe is really cold and dark. The radiation has become invisible – now it has become microwaves.

The Big Bang theory predicts that the temperature of the Universe is now about –270 °C. That's just 3 degrees above absolute zero, the coldest temperature of all. Astronomical satellites have been used to study this 'background microwave radiation' and to measure its temperature. Measurements show that the prediction is correct – the temperature of the microwaves is just 2.73 degrees above absolute zero.

As the Universe expanded and cooled, atoms formed. The Big Bang theory predicted that the most common atoms would be hydrogen, followed by helium. When the Hubble Space Telescope looked at distant galaxies, formed when the Universe was very young, this was just what it observed – more support for the Big Bang theory.

The mystery of dark matter

A lot of science involves calculations. When scientists want to test a theory, they use the theory to make predictions. Making predictions often involves calculating what we would expect to observe – such as the amounts of hydrogen and helium in the Universe – and then making measurements to test the predictions.

When we look at the Universe, we see the light from stars. By estimating the total number of stars in the Universe we can estimate its total mass. When astronomers wanted to test the Big Bang theory, they realised that the theory predicted a lot more matter in the Universe than was calculated from these observations.

We now know:

- The matter that we can see is only a small fraction of the matter that makes up the Universe.
- There is also a large amount of invisible energy out there.

Astronomers conclude that there is some kind of **dark matter** which makes up most of the mass of the Universe, but they can't be sure what it is. It must be cold, so that it doesn't glow like a star. And it must be something that doesn't interact much with ordinary matter, otherwise we would have detected it long ago. There are many theories as to what this dark matter might be, and experiments are under way to try to identify it.

Figure 12.28 A researcher in a laboratory in North Yorkshire that is more than 1 km underground. The equipment is designed to detect any dark matter that penetrates the Earth

Summary

- **Observations tell us that galaxies are moving apart, showing that the Universe is expanding.**
- **The Big Bang theory explains this by saying that all matter and energy emerged from a giant explosion at the start of time.**
- **We cannot be sure how the Universe will end; it may continue to expand, or it may contract to a Big Crunch.**
- **The red shift of light from distant galaxies provides evidence that they are moving away from us.**
- **The Big Bang theory is supported by the existence of microwave radiation throughout space, left over from the early days of the Universe.**
- **Much of the mass of the Universe is made up of 'dark matter', but no one yet knows what this is.**

Questions

12.18 **a** What *two* measurements did Edwin Hubble make when he looked at galaxies beyond our own?

b Why was he able to make these measurements when earlier astronomers were not?

12.19 Figure 12.26 shows possible ways in which the size of the Universe may change, according to different theories. Another theory suggests that the Universe will expand, contract, expand again, and so on. Draw a graph to represent this possibility.

12.20 Explain how Edwin Hubble was able to show that distant galaxies are moving away from us.

12.21 Space is filled with cold microwave radiation. Explain why this provides support for the Big Bang theory of the origin of the Universe.

Acknowledgements

The following have supplied photographs or have given permission for photographs to be reproduced.

p.2 Scripps Inst. Oceanography/Oxford Scientific; **p.4** *both* Mary Jones; **p.6** *tl* © Niall Benvie/Corbis, *tr* Frank Lane Picture Agency/Corbis, *b* Peter David/Natural Visions; **p.7** *t* Manor Photography/Alamy, *b* Alan Carey/Science Photo Library; **p.8** Paul Hobson/Alamy; **p.10** AP Photo/Conservation International; **p.11** © Dung Vo Trung/Corbis Sygma; **p.13** *t* Agus Suparto/AFP/Getty Images, *b* Tom McHugh/Science Photo Library; **p.15** © Galen Rowell/Corbis; **p.16** *t* © Archivo Iconografico, S.A./Corbis, *bl*, *br* John Doebley; **p.19** *both* Torsten Blackwood/AFP/Getty Images; **p.21** Profimedia.CZ s.r.o./Alamy; **p.22** *tl* © Bill Stormont/Corbis, *tr* Mauro Fermariello/Science Photo Library, *b* The Photolibrary Wales/Alamy; **p.24** *t* Michael Donne/Science Photo Library, *b* Andrew Syred/Science Photo Library; **p.25** *t* Dept. of Clinical Cytogenetics, Addenbrookes Hospital/Science Photo Library, *b* Peter Menzel/Science Photo Library; **p.26** Simon Fraser/RVI, Newcastle-Upon-Tyne/Science Photo Library; **p.28** *t* Doug Steley/Alamy, *b* Holt Studios International Ltd/Alamy; **p.34** © Max Rossi/Reuters/Corbis; **p.35** Victor Habbick Visions/Science Photo Library; **p.36** Pascal Goetgheluck/Science Photo Library; **p.37** AP Photo/Nancy Palmieri; **p.39** ILPO MUSTO/Rex Features; **p.40** Jim Dowdalls/Science Photo Library; **p.44** *t* BSIP VEM/Science Photo Library, *b* Hattie Young/Science Photo Library; **p.45** *t* Simon Fraser/Newcastle Hospitals NHS Trust/Science Photo Library, *b* AP Photo/John Cetrino; **p.47** *t* © Mike Hutchings/Reuters/Corbis, *b* CC Studio/Science Photo Library; **p.52** Ian Shaw/Alamy; **p.54** Ian Hooton/Science Photo Library; **p.55** Scott Camazine/Science Photo Library; **p.57** Mark Thomas/Science Photo Library; **p.60** Mauro Fermariello/Science Photo Library; **p.61** Andrew McClenaghan/Science Photo Library; **p.62** © Stephen Roberts/Alamy; **p.63** Mark Thomas/Science Photo Library; **p.64** SHOUT/Alamy; **p.65** Matt Meadows, Peter Arnold Inc./Science Photo Library; **p.66** Martin Rickett/PA/Topham; **p.67** *l* Science Museum, *r* Philippe Psaila/Science Photo Library; **p.68** Dave Reede/Agstock/Science Photo Library; **p.71** AP Photo/Nick Ut; **p.72** *t to b* Dr Klaus Boller/Science Photo Library, Steve Gschmeissner/Science Photo Library, Dr M.A. Ansary/Science Photo Library, Steve Gschmeissner/Science Photo Library; **p.73** *clockwise from tl* Faye Norman/Science Photo Library, Dennis Galante/Corbis, BSIP, Keene/Science Photo Library, Paul Whitehill/Science Photo Library, PHOTOTAKE Inc./Alamy, Mitch Diamond/Alamy, Bob Pardue/Alamy; **p.79** Peter Treanor/Alamy; **p.80** © Karen Kasmauski/Corbis; **p.82** AP Photo/Sava Radovanovic/Empics; **p.83** Andrew Lambert Photography/Science Photo Library *except 1st from l* Jerry Mason/Science Photo Library, *3rd from l* David Taylor/Science Photo Library, *b* Last Resort Picture Library; **p.85** *tl* National Museums & Galleries of Wales, *r* Ed Bailey/AP/Empics, *bl* © Galen Rowell/Corbis; **p.86** Alex Bartel/Science Photo Library; **p.87** NASA; **p.88** www.TopFoto.co.uk; **p.91** smart-elements.com - Juergen Bauer; **p.92** *all* Andrew Lambert Photography/Science Photo Library; **p.96** *tl to r* Andrew Lambert Photography/Science Photo Library, © sciencephotos/Alamy, Claude Nuridsany & Marie Perennous/Science Photo Library, *b* Andrew Lambert Photography/Science Photo Library; **p.97** © Corbis **p.99** Courtesy of IBM Researach, Almaden Research Centre; **p.101** Vaughan Fleming/Science Photo Library; **p.102** Dave Reede/Agstock/Science Photo Library; **p.103** David Aubrey/Science Photo Library; **p.106** *both* Last Resort Picture Library; **p.107** *both* Last Resort Picture Library; **p.108** *both* © West Semitic Research/Dead Sea Scrolls Foundation/Corbis; **p.110** *t* Andrew Lambert Photography/Science Photo Library, *bl to r* © allOver photography/Alamy, Carlos Dominguez/Science Photo Library, Bill Longcore/Science Photo Library; **p.113** Andrew McClenaghan/Science Photo Library; **p.114** © The Anthony Blake Photo Library/Alamy; **p.115** © *t* Leslie Garland Picture Library/Alamy, *b* © Stephen Olner/Photographers Direct; **p.117** Mary Evans Picture Library; **p.118** *t* Mary Evans Picture Library, *b* Last Resort Picture Library; **p.119** *both* Last Resort Picture Library; **p.121** *l* Garry Watson/Science Photo Library, *r* © Photofusion Picture Library/Alamy; **p.122** Louise Forrester/Photographers Direct; **p.125** 20th Century Fox/The Kobal Collection; **p.126** Mary Evans Picture Library; **p.128** © Jeff Morgan/Alamy; **p.129** Simon Fraser/Mauna Loa Observatory/Science Photo Library; **p.130** Pasquale Sorrentino/Science Photo Library; **p.133** Valerie Taylor/Ardea; **p.134** *t* Martin Bond/Science Photo Library, *b* Thomas Hollyman/Science Photo Library; **p.135** © 2005 TopFoto; **p.136** *t* Cordelia Molloy/Science Photo Library, *b* David Parker/Science Photo Library; **p.138** © INSADCO Photography/Alamy; **p.139** *l* © Mireille Vautier/Alamy, *r* © Network Photographers/Alamy; **p.140** © Jim West/Alamy; **p.143** *t to b* David R. Frazier Photolibrary, Inc./Alamy, Paul Doyle/Alamy, Carlos Dominguez/Science Photo Library, Photodisc, Photodisc, Photodisc, Photodisc; **p.145** *l* © Andre Seale/Alamy, *r* Andrew Lambert Photography/Science Photo Library; **p.146** *t* © Photofusion Picture Library/Alamy, *b* Paul Whitehill/Science Photo Library; **p.148** Courtesy of BOC; **p.149** Martin Rickett/PA/empics; **p.152** *t* © Larry Kasperek/NewSport/Corbis, *b* Jon Buckle/Empics; **p.153** *t* V & A Images, *b* © Tim Wimborne/Reuters/Corbis; **p.154** *l* Copyright of 3M™ Company, *r* © Nick Hanna/Alamy; **p.155** *t* Copyright © 2005 W.L. Gore & Associates, *b* Max Stuart/Alamy; **p.156** Rob Griffith/AP Photo/Empics; **p.157** © Topham/AP; **p.158** © Bill Varie/Corbis; **p.159** *t* Photolibrary.com, *b* Cordelia Molloy/Science Photo Library; **p.160** *t* John Olive/Chemistry Review, *b* Jennie Hart Alamy; **p.161** *t* Pilkington Building Products – UK, *b* Eye of Science/Science Photo Library; **p.165** © Robert Harding Picture Library Ltd/Alamy; **p.166** *l* © Charles O'Rear/Corbis, *r* © Cephas Picture Library/Alamy; **p.170** Ripesense Limited; **p.171** © Robert Battersby/BDI Images; **p.172** Martyn F. Chillmaid; **p.175** Bernie Mordorski/Photographers Direct; **p.176** Cordelia Molloy/Science Photo Library; **p.177** Peter Bowater/Science Photo Library; **p.178** *t* Ashley Cooper/Alamy, *b* Volker Steger/Science Photo Library; **p.179** *t* Stockfolio/Alamy, *b* Lino Wchima/Photographers Direct; **p.181** Martyn F. Chillmaid/Science Photo Library; **p.182** © Bettmann/Corbis; **p.185** Andrew Lambert Photography/Science Photo Library; **p.186** © Crown Copyright 2000. Reproduced with the permission of the Controller of HMSO and the Queen's Printer for Scotland/National Physical Laboratory; **p.188** *both* Science Photo Library; **p.189** *l* © Lester Lefkowitz/Corbis, *r* Steve Allen/Science Photo Library; **p.190** DPA/Empics; **p.192** Science Photo Library; **p.193** David Cheskin/PA/Empics; **p.194** Hank Morgan/Science Photo Library; **p.195** *l* Michael Donne/Science Photo Library, *r* Image from Aquatera.co.uk; **p.197** *l* © Photography by Chris Davies, *r* © Judy Griesedieck/Corbis; **p.198** Image supplied courtesy of National Grid **p.199** Martin Bond/Science Photo Library; **p.200** Andrew Lambert Photography/Science Photo Library; **p.203** Sheila Terry/Science Photo Library; **p.204** Robert Llewellyn/Alamy; **p.206** Rob Bowden/EASI-Images/Chris Fairclough Worldwide; **p.207** © 2005 TopFoto; **p.208** European Space Agency/Science Photo Library; **p.209** Energy Saving Trust; **p.211** CC Studio/Science Photo Library; **p.212** © Eric and David Hosking/Corbis; **p.213** *l* Martyn F. Chillmaid, *b* Sheila Terry/ Science Photo Library; **p.219** © Chad Ehlers/Alamy; **p.220** Eric Skitzi/AP Photo/Empics; **p.221** Zephyr/Science Photo Library; **p.223** *t* © 2005 TopFoto/AP, *b* Science Photo Library; **p.224** Martin Dohrn/Science Photo Library; **p.225** *tl* Mauro Fermariello/Science Photo Library, *r* © Jens Haas/ Corbis; **p.226** Simon Fraser/NCCT, Freeman Trust, Newcastle-Upon-Tyne/ Science Photo Library; **p.227** Seymour/Science Photo Library; **p.228** *t* Dr Ray Clark (FRPS) & Mervyn de Calcina-Goff (FRPS)/Science Photo Library, *b* Mauro Fermariello/Science Photo Library; **p.231** Frank Zullo/Science Photo Library; **p.232** *t* © Photonews Service Ltd/Topfoto, *b* NASA/Science Photo Library; **p.233** *t* Andrew Syred/Science Photo Library, b Steve Horrell/ Science Photo Library; **p.234** Lawrence Lawry/Science Photo Library; **p.237** NASA/Science Photo Library; **p.239** © Reuters/Corbis; **p.242** Science Photo Library; **p.243** European Space Agency/Science Photo Library; **p.244** *t* Martin Bond/Science Photo Library, *b* NASA/ESA/STScI/Science Photo Library; **p.245** *both* NASA/Science Photo Library; **p.246** Francois Gohier/Science Photo Library; **p.249** Chris Butler/Science Photo Library; **p.250** Dr Seth Shostak/Science Photo Library; **p.251** NASA/Science Photo Library; **p.253** Emilio Segre Visual Archives/American Institute of Physics/Science Photo Library; **p.254** Mark Garlick/Science Photo Library; **p.257** James King-Holmes/Science Photo Library.

t = top, *b* = bottom, *r* = right, *c* = centre

Every effort has been made to contact copyright holders but if any have been inadvertently overlooked the Publishers will be pleased to make the

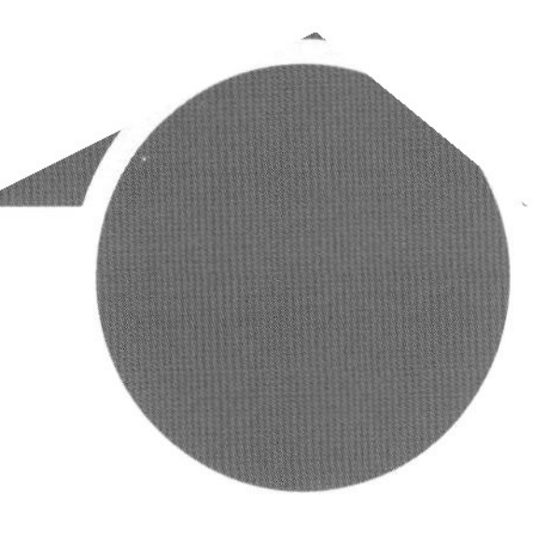

Index